Springer Series in Reliability Engineering

Series editor

Hoang Pham, Piscataway, USA

More information about this series at http://www.springer.com/series/6917

Mauricio Sánchez-Silva · Georgia-Ann Klutke

Reliability and Life-Cycle Analysis of Deteriorating Systems

Springer

Mauricio Sánchez-Silva
Department of Civil and Environmental Engineering
Universidad de Los Andes
Bogotá
Colombia

Georgia-Ann Klutke
Department of Industrial and Systems Engineering
Texas A&M University
College Station, TX
USA

ISSN 1614-7839 ISSN 2196-999X (electronic)
Springer Series in Reliability Engineering
ISBN 978-3-319-20945-6 ISBN 978-3-319-20946-3 (eBook)
DOI 10.1007/978-3-319-20946-3

Library of Congress Control Number: 2015950899

Springer Cham Heidelberg New York Dordrecht London

Printed on acid-free paper

Springer International Publishing AG Switzerland is part of Springer Science+Business Media (www.springer.com)

To
Silvia, Cami and Ale
Mauricio

To
John and Alan ... my lights
Georgia-Ann

Preface

The concepts behind the design and operation of engineered systems have evolved significantly over the last decades. Engineering design has historically been conceived as an optimization problem consisting of selecting the physical characteristics of a system[1] that satisfy predefined functional requirements at minimum cost. The cost-based optimization approach, fundamentally deterministic in nature, has at the same time recognized that the performance of the system is uncertain and potentially hazardous. During the nineteenth century and the beginning of the twentieth century, safety factors where used implicitly or explicitly to cover design, construction, and operational uncertainties. For example, [1] reports that in the nineteenth century in the UK the average ultimate tensile strength for cast iron beam designs was computed using safety factors between 4 and 5 [1]; similar safety factors were typically used for other type of structures as well. These large safety factors became smaller with time as there were better knowledge of the materials and the mechanical performance of engineering devices; and also as the need to reduce costs became more important. By the mid twentieth century, probability theory began to play an important role in the characterization and management of uncertainties and probabilistic techniques began to augment safety factors in the assessment of engineering safety. The concept of component and system reliability was introduced in industrial manufacturing and later in buildings and civil infrastructure in the form of distributional estimates and risk assessment (e.g., load and resistance partial factors).

As the balance between cost and safety has become more important, industry recognizes that design and construction, based on a deterministic cost minimization objective under certain reliability constraints, lead to suboptimal solutions and higher capital expenditure in the long run. This realization creates an increasing awareness of the importance of future investments (i.e., inspection, maintenance, and repair) for project cost evaluation and brings attention to the assessment of all the uncertainties associated with the lifetime operation; specially, in the case of

[1]The term *system* is used generically to describe any engineered artifact or device.

long-lasting projects. This also reinforces the significance of using stochastic processes in engineering design and life-cycle analysis. This new understanding of design and operation of large infrastructure projects opens many new research questions and challenges. This book is intended as a contribution to this important discussion.

A new engineering project management paradigm, where projects are evaluated throughout their lifetime, requires, in addition to the mechanical models, the integration of complex probabilistic tools and operational decisions (e.g., policy to carry out preventive maintenance). Under the assumption that people act rationally, the objective of this book is to present and examine the tools of modern stochastic processes to provide appropriate models to characterize the system's performance over time so that engineers and planners have better evidence to inform their decisions. It should be clear to engineers that mathematical models are only tools that provide input to decision-making. Model-based evidence is not necessarily the most valuable or the most relevant for the overall decision, but we contend that it is essential when it comes to characterizing the system's performance measures in an uncertain operating environment.

This book compiles and critically examines modern degradation models for engineered systems and their use in supporting life-cycle engineering decisions. In particular, we focus on modeling the uncertain nature of degradation, considering both conceptual discussions and formal mathematical formulations. The book also presents the basic concepts and modeling aspects of life-cycle analysis (LCA). Special attention is given to the role of degradation in LCA and in optimal design and operational analysis. Given the relationship between operating decisions and the performance of the systems condition over time, part of the book is also concerned with maintenance models.

The book is organized into ten chapters and one appendix. Chapters have been arranged to take the reader from the basic concepts up through more complex and multidisciplinary aspects. The book is intended for readers with basic knowledge of the fundamentals of probability. However, we have included a brief introduction to the concepts and terminology of probability theory in the appendix and some details on various stochastic process models in the chapters themselves. We do not intend this book to be a monograph on applied probability or stochastic processes, but rather a book on modeling degradation to support decision-making in engineering. The book chapters are organized in four main parts; (see Fig. 1):

1. Conceptual and theoretical basis (Chaps. 1–3).
2. Degradation models (Chaps. 4–7).
3. Life-cycle analysis and optimization (Chaps. 8–9).
4. Maintenance models (Chap. 10).

In the first part of the book, we discuss conceptual aspects that are essential for making predictions and to provide information to decision makers (Chap. 1). Furthermore, we provide an overview of the concepts of risk and reliability and present various approaches used in engineering practice to estimate reliability (Chap. 2). In Chap. 3 we describe, both conceptually and in formal mathematical

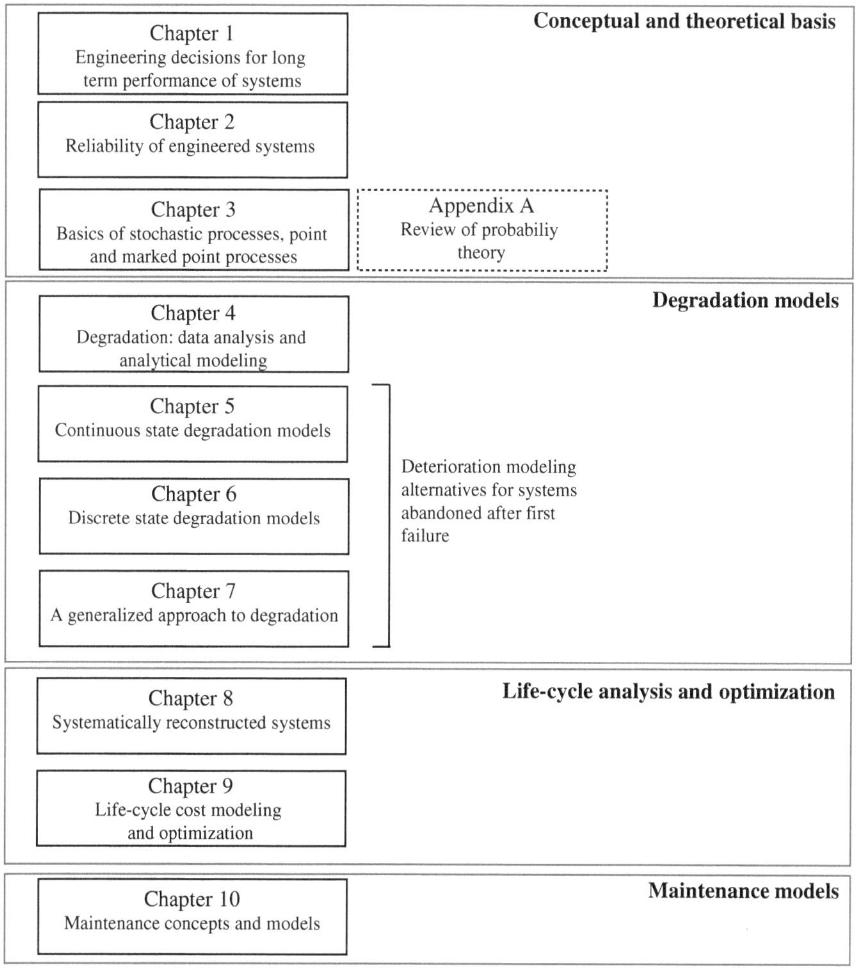

Fig. 1 Book organization

terms, important aspects of selected stochastic process as a tool for prediction; and emphasize the underlying assumptions to provide some context as to when these particular models are relevant or useful. These results will be used in the models developed for degradation in subsequent chapters.

Predicting the performance of engineered systems involves characterizing changes in the system state as it evolves over time; in particular, this includes how system performance degrades over time, which is the main topic of this book. Then, the second part of the book, Chaps. 4–7, deals with degradation models. Chapter 4 discusses the foundations of degradation from a conceptual and theoretical point of view. In this chapter we also review briefly the problem of obtaining and analyzing degradation data, while in Chaps. 5–7 we are concerned with modeling degradation mechanisms for systems that are not maintained and are abandoned after failure. In particular we distinguish between continuous and discrete space state degradation models. In Chap. 7, we present a general approach to degradation based on the Lévy process, which is a flexible approach to accommodate most models presented in previous chapters. The models presented in these chapters are illustrated with cases that are of interest in engineering applications.

With the background on degradation models presented in Chaps. 2 through 7, in the third part of the book, i.e., Chaps. 8 and 9, we present the conceptual and theoretical bases behind life-cycle analysis (LCA). First, as a preamble, in Chap. 8 we describe the performance of systems that are successively intervened or reconstructed. By doing this we include in the analysis the concept of system interventions (e.g., maintenance and repair), which clearly modify both the system's performance and the future investments. Afterwards, in Chap. 9, both LCA and life-cycle cost analysis (LCCA) are introduced. In particular we focus on LCCA as a project evaluation techniques conceived to study the performance (and the associated costs) of an engineered system within a given time-window. They are used to estimate system availability and maintenance needs in order to make better investment and operational decisions. Life-cycle analyses can also be used as a stochastic optimization technique to determine the design parameters and maintenance strategy that maximize the benefit derived from the existence of the system. The value of LCCA is that they are able to integrate the mechanical performance with the financial and economic considerations within a framework of uncertainty.

Finally, in the last part of the book, Chap. 10, we address the task of defining optimum intervention strategies; in other words, defining maintenance programs that maximize the profit derived from the existence of the project while ensuring its safety and availability. Maintenance activities are understood to include all physical activities intended to increase the useful life of the system. These activities may be initiated because the system is observed to be in a particular system state, e.g., failure state (e.g., corrective maintenance), or they may be initiated before such a fault is observed (e.g., preventive maintenance). After a conceptual discussion about some key aspects of maintenance, we address traditional maintenance models. Finally, towards the end of the chapter, we study the case of maintenance of systems that exhibit nonself announcing failures, as well as systems that are continuously monitored.

The book is intended to be used by educators, researchers, and practitioners interested in topics related to risk and reliability, infrastructure performance modeling, and life-cycle assessment. The concepts and models presented have applications in a large variety of engineering fields such as civil, environmental, industrial, electrical, and mechanical engineering. However, special emphasis is given to problems related to managing large infrastructure systems.

More specifically, this book is aimed at two main audiences. First, it can be used as reference for research in topics involving degradation of a variety of large, complex engineered systems. Some examples include civil infrastructure, such as bridges, buildings, water distribution systems, sewage systems, pipelines, ports and offshore structures, and so forth. Other examples include complex consumer products, such as automobiles, and large-scale commercial undertakings, such as aircraft, ships, and power generation and distribution systems.

The second use of the book is as a guide for a graduate course on infrastructure modeling and management. In this regard, the book compiles and explains, both conceptually and formally, key aspects for modeling the stochastic nature of degradation. In this regard, we view the book as a major contribution to the field, since many courses in design and operation of civil infrastructure focus exclusively on management aspects and do not do justice to the performance modeling and analysis of the problem.

July 2015

Mauricio Sánchez-Silva
Georgia-Ann Klutke

Reference

1. A.N. Beal, T. Leeds. A history of the safety factors. Struct. Eng. **89**(20), 1–14 (2011)

Acknowledgments

The authors would like to acknowledge the constructive comments and suggestions made by the many colleagues who reviewed several drafts of the book. In particular, we wish to thank Javier Riascos-Ochoa, whose Ph.D. thesis provided the basis for Chap. 7, and Professor Mauricio Junca (Mathematics Department at Los Andes University), for his invaluable research insights, shared through many constructive discussions on these topics. We would also like to recognize the help of Edgar Andrés Virguez, and the comments and suggestions made by many graduate and undergraduate students over the years that have contributed in different ways to make this book possible.

Finally, we would like to acknowledge the Department of Civil and Environmental Engineering at Los Andes University (Bogotá, Colombia), and the Department of Industrial and Systems Engineering at Texas A&M University (College Station, USA) for their support of this project.

Mauricio Sánchez-Silva
Georgia-Ann Klutke

Contents

Abbreviations

AFOSM	Advanced First-Order Second-Moment
ALARP	As Low As Reasonable Practical
COV	Coefficient of Variation
CPP	Compound Poisson Process
CTMC	Continuous-Time Markov Chains
DFR	Decreasing Failure Rate
DTMC	Discrete-Time Markov Chains
FHWA	Federal Highway Administration
FMECA	Failure Mode, Effects and Criticality Analysis
FORM	First-Order Reliability Method
FOSM	First-Order Second-Moment
FTA	Fault Tree Analysis
GP	Gamma Process
IFR	Increasing Failure Rate
KRT	Key Renewal Theorem
LCA	Life-Cycle Analysis
LCCA	Life-Cycle Cost Analysis
LD	Linear Deterministic drift
LQI	Life Quality Index
ML	Maximum Likelihood
MM	Moment Matching method
MTBF	Mean Time to Repair
MTTF	Mean Time to Failure
NBU	New Better than Used
NBUE	New Better than Used in Expectation
NIST	U.S. National Institute of Standards and Technology
PCI	Pavement Condition Index
PH	Phase-Type
PRA	Probabilistic Risk Analysis
PSI	Present Serviceability Index
QBI	Quantile Based inspection

SDR	Social Discount Rate
SMP	Semi-Markov process
SOC	Social Opportunity Cost
SORM	Second Order Reliability Method
SRI	Sufficiency Rating Index
SRTP	Social Rate of Time Preferences
SVLY	Societal Value of Statistical Life-Year
SVSL	Societal Value of Statistical Life
SWTP	Societal Willingness to Pay
UBDI	Utah Bridge Deck Index
WTP	Willingness to Pay

Chapter 1
Engineering Decisions for Long-Term Performance of Systems

1.1 Introduction

The objective of engineering practice is to provide solutions to human needs by developing and deploying technologies that make life better. Engineering is part of almost everything we do—from the water we drink and the food we eat, to the buildings we live in and the devices we use in our daily lives [1]. It has been an essential part of human history and plays a central role in building our future.

In essence, engineers use ingenuity to make things work more efficiently and less expensively by converting scientific knowledge into actual objects. For that purpose, they need to make decisions. This means that decision making and engineering are strongly interconnected. Although this book emphasizes that models provide valuable and relevant evidence to develop engineering products, we also recognize that its value strongly depends on the characteristics of the decision process. This chapter outlines some basic concepts related to the decision-making process in long-lasting engineeried systems so that the theory presented in subsequent chapters can be understood in context.

1.2 Engineering: A Decision-Making Discipline

Traditionally engineering has been regarded as a problem-solving discipline. However, although problem solving capabilities are important, the concept that is really central to modern engineering is that of *decision making*; i.e.,

> the process of choosing between alternative courses of action; this means, selecting between available options defined according to a set of restrictions (e.g., technical, economic, social) to optimally assign the resources available.

Decision making is what distinguishes engineers from scientists. While engineering focuses on technological development, the purpose of science is to understand

M. Sánchez-Silva and G.-A. Klutke, *Reliability and Life-Cycle Analysis of Deteriorating Systems*, Springer Series in Reliability Engineering,
DOI 10.1007/978-3-319-20946-3_1

and provide explanations for how the world works—science is a search for truth. Blockley [1] put it as follows:

> "The purpose of science is to know by producing "objects" of theory or "knowledge". The purpose of mathematics is clear, unambiguous and precise reasoning. The purpose of engineering and technology is to produce useful physical tools with other qualities such as being safe, affordable and sustainable."

According to Hazelrigg [2], engineering is built on three important principles. The first is the idea that problems can be described through a set of laws and boundary conditions. Secondly, engineering techniques (e.g., mechanics) lead to products (usually physical devices) with a purpose. And thirdly, the final product is the best option (e.g., mechanical and operational) among a set of feasible solutions, within a set of external constraints (e.g., constructive, social and economic). This last principle implies that engineering projects are, to a large extent, an exercise of optimal resource allocation. Therefore, well-engineered products result not only from effective modeling and understanding of the problem, but also from good engineering decisions. Engineering decisions appear at different levels within the design and manufacturing process; from initial conceptual design to the details of the construction. A structured and rational engineering decision-making process is the way in which the hierarchy of manufacturing (constructing) becomes efficient and leads to satisfactory products.

Contrary to what is frequently taught in engineering schools, design goes beyond selecting a set of parameters with the purpose of fulfilling certain mechanical laws. It requires the understanding of the context; the knowledge of construction materials, processes and requirements; the recognition of the uncertainty in both the models and the variables; and the economic and financial restrictions, among others. More often than not, design requirements are dominated by restrictions that cannot be mathematically formulated. For instance, there may not be a dependable formulation to define an acceptable level of risk or the life of an artifact. These are central design requirements that most engineers take as *given* just because, for example, they have been established in codes of practice or in device specifications. Thus, making engineered artifacts that fulfill mechanic laws and functional requirements is a mathematical problem but engineering conception and design is a decision problem.

As an example, let us consider the construction of a highway. The selection of its location and capacity is based on estimates of demand and plans for the highway network expansion. However, this selection is also the result of regional political interests and socioeconomic needs and restrictions that must be reconciled with the technical requirements and budget constraints. Once the the road layout has been determined, the initial design phase commences. Here, decisions are made about the geometry, the materials, and the various geotechnical aspects, as well as legal aspects regarding land acquisition and management. The next step includes detailed designs of the pavement structure, blueprints of bridge structures, construction planning, building contract requirements (e.g., warranties, length of construction), and so forth. Finally, after construction, a plan for inspection and maintenance is laid out. It should be clear from this brief description that in this entire process the quality of decision making is what finally leads to a good product. Note that not only the planning but also the technical engineering aspects of this process require making decisions. For example, the lifetime of the highway is a fundamental design parameter. However, it cannot be defined precisely since variables such as traffic frequency and loading, material properties, and soil characteristics cannot be determined with certainty; and mechanical models, while helpful, are not precise enough. Thus, engineering solutions require making decisions whose consequences may be significant in terms of the highway's ability to fulfill its function within given safety and socioeconomic restrictions.

Engineering decisions are accompanied by substantial responsibilities; they generally have consequences to both the enterprise (e.g., affecting the income and opportunity for growth) as well as to society at large [3] (e.g., impact on the environment and sustainability). Thus it is of great importance for engineering practitioners to understand both the physical laws that characterize artifact performance as well as the tremendous responsibility their decisions entail. Because of the many details that influence our decisions in engineering, we heartily endorse the notion that the study of the framework and mathematics of decision making is vital to becoming a better engineer [4].

1.3 Decision Making

1.3.1 The Nature of Engineering Decisions

As mentioned in Sect. 1.2, the term *decision making* is concerned with the process of selecting[1] the best choice from a set of available (feasible) options to meet one or multiple objective criteria. This definition highlights the need for determining what the particular decision criteria are, as well as deciding what constitutes the set of feasible options. From an engineering perspective, decisions should be the result of a well-structured train of thoughts (e.g., inductive/deductive reasoning) that justifies the selection of the final solution. Decisions made as a result of a logical, scientifically structured process will be referred to as *rational* decisions in this book. It is important to stress that we do not want to imply that other ways of making decisions (i.e., nonscientific approaches) are not rational in the broader sense of the word, nor do we want to imply that other decision processes cannot lead to good decisions.

There are actually many structured, mathematically rigorous (i.e., rational) approaches to decision making. A common approach employed in engineering is known as Decision Analysis (DA), a term coined by Howard in 1966 [5] to describe a framework for applied decision making, which has its foundations in the work of mathematical economists Von Neumann and Morgenstern [6] . Decision Analysis is

---

[1] The selection should be made according to the values and preferences of the decision maker.

now viewed as a discipline on its own, encompassing a scientific and philosophical approach that serve as the basis for much of modern decision making in engineering design [2, 4, 7].

Because engineering decisions are almost always made in the presence of uncertainty, it is important to distinguish between the *decision* and its *outcome*. A decision is simply a choice among alternatives. An outcome is what happens as a result of that choice. Note that even when decisions are made rationally, i.e., within a structured and rigorous framework, they may result in outcomes that are undesired. Thus, decisions are good if they represent the best choice among a set of risky alternatives, while outcomes are good if the decision maker is satisfied with the consequences of the decision; i.e., the bet pays off. Careful modeling and analysis may help us make good decisions, but because outcomes are rarely certain, any decision may still result in an undesirable outcome.

The fundamential notion underlying Von Neumann and Morgenstern's axiomatic treatment of decision making under uncertainty [6] is that an individual can express a preference (i.e., a preferred choice) between all possible pairs of outcomes, and that these preferences are *transitive*. This means that if there are three possible outcomes A, B, and C such that option A is preferred over option B and option B is preferred over C; then, option A will be preferred over option C. Furthermore, a decision made *under uncertainty* means that the decision maker selecting a particular (risky) alternative will achieve a given outcome only with a certain (known) probability. Thus, given certain technical details (the so-called Von Neumann-Morgenstern axioms that define rational behavior), they prove that there exists a real-valued function, known as the *utility* or *cardinal utility*, that allows a decision maker to choose the "best" alternative in the following sense: If the decision maker selects the risky alternative with the highest expected utility (in the sense of mathematical expectation), he is always acting consistently with his own true tastes, as determined by his pairwise preferences. The precise result, known as the *expected utility* theorem, is a powerful theoretical result that provides a basis for a unified understanding of the role of uncertainty, prediction, information gathering, and modeling to support engineering decisions. Since most decisions are not a logical derivation of the assessment of possible various outcomes, expected utility theory has become a very good descriptor of the choices that people, and animals [8], make in a variety of simple situations [7, 9, 10].

1.3.2 The Decision-Making Process

The overall decision-making process in engineering includes the following basic steps:

- define the problem and the objective(s) of the decision;
- identify the set of feasible alternative actions;

- predict and evaluate the performance of the selected alternatives; and
- select the best alternative from the set of feasible alternatives.

In what follows we will mention some key aspects of each step that are important conceptually and within the context of this book. A detailed discussion on decision theory in engineering and other fields is beyond the scope of this book and can be found in, for example, [6, 11–14].

Definining the Decision Objective

The first step of the decision-making process requires the complete and clear definition of the objective (purpose) of the decision, which includes not only defining the target but also the scope and the boundaries (constraints). Clearly, in most "hard" engineering[2] (e.g., mechanics) this may be easier than in cases where the problem includes many components or involves a stronger interaction with personal aspects or social organizations [14].

Defining the objective of the decision involves establishing the decision criteria. We should note first that decision criteria may not always be easily quantifiable (e.g., in monetary terms), so that choosing the objective should include a thorough understanding of the decision problem and the sensitivity of outcomes to the decision variables [12]. Another important aspect to be considered is that although most engineering decision problems are described in textbooks as having a unique solution, in practice, decisions involve multiple, possibly conflicting criteria, which may vary in their scope and explicitness [16]. Furthermore, in some cases the objective of the decision might not result in a single decision but lead to a set of sequential decisions; this cases are usually referred to as *dynamic* decision situations [11].

Identifying Feasible Alternatives

An *alternative* is a possible choice that meets the purpose of the decision. In practice, there is not a standard procedure to identify the set of alternatives; on the contrary, the choice of alternatives is generally the result of intuition, experience, or a brainstorming process [17]. Furthermore, more often than not, in selecting the alternatives creativity plays an important role. It is important to stress that although many possible solutions can be identified, the set of *all possible* alternatives will always be incomplete.

In decision analysis we need to distinguish, among all possible alternatives, the subset of those that are *feasible*; i.e., those that satisfy the constraints of the problem. Some authors argue that the clear definition of the objective and the selection of the feasible alternatives are the result of an iterative process, which guarantees the consistent formulation of the problem [11].

[2] *Hard* systems refer to structured physical systems whose performance can be described by well established mechanical laws [14, 15].

Predicting the Performance of Feasible Alternatives

Among the set of feasible alternatives we are further interested in those that have the highest probability of success or effectiveness; thus, the dependability of the decision is conditioned on our ability to predict the system's future performance of each of the feasible alternatives. This step generally requires the construction of predictive models that help us to understand the performance of the system for every option in the set of feasible alternatives. Contrary to decisions where the result leads invariably to a specific outcome, in decisions under uncertainty[3] the actual result is unknown [17, 19]. In order to take into account uncertainty, we use probability theory. Thus, we are interested in building models that consider the uncertainty in the system performance and the external conditions so that we can predict its behavior over a given time span (i.e., finite/infinite). For instance, we will discuss later in this book the problem of degradation; in this case, we want to predict the expected time at which the system will reach a given state (e.g., failure). A more detailed discussion of the nature of prediction will be presented in Sect. 1.5.

Selecting the Best Alternative

Selecting the "best" feasible alternative is not an easy task and requires considering many different aspects. In this section we briefly discuss the nature of the decision maker and the context within which the decision is taken, the representation of the decision process, the criteria for evaluating the feasible alternatives, and the importance of decisions that involve future actions. However, if the reader is interested, there is a vast literature on decision making in engineering that deals with these and many other relevant aspects in more detail; see for instance [4, 11, 13, 14, 16, 17, 20].

There are two aspects that are particularly relevant regarding who makes the decision: the nature of the decision maker and his/her relationship with the context. First, it is necessary to differentiate between decisions made by an individual or by a group. The former describes a single person or an organization acting as a single unit (e.g., a corporate position); in contrast, a group is defined as a collection of individuals, who may have conflicting interests. This distinction is important since the *utility* criterion used to make the decision changes among different types of individuals, or changes in the eventuality in which an agreement among different parties is required; further details can be found in [14]. In addition to the nature of the decision maker, the context within which the decision is made is also important. For instance, in engineering problems, decisions might depend on the physical environment (e.g., climatic conditions, the topography, or the geology), the technology available (or accessible), the availability of resources, the social implications of the solution, etc. The relationship between the decision maker and the context

[3]Uncertainty is "a state of not knowing whether a proposition is true or false" [18]. Uncertainty may result from a lack of knowledge or from *randomness*—i.e., lack of a pattern in the system behavior [1].

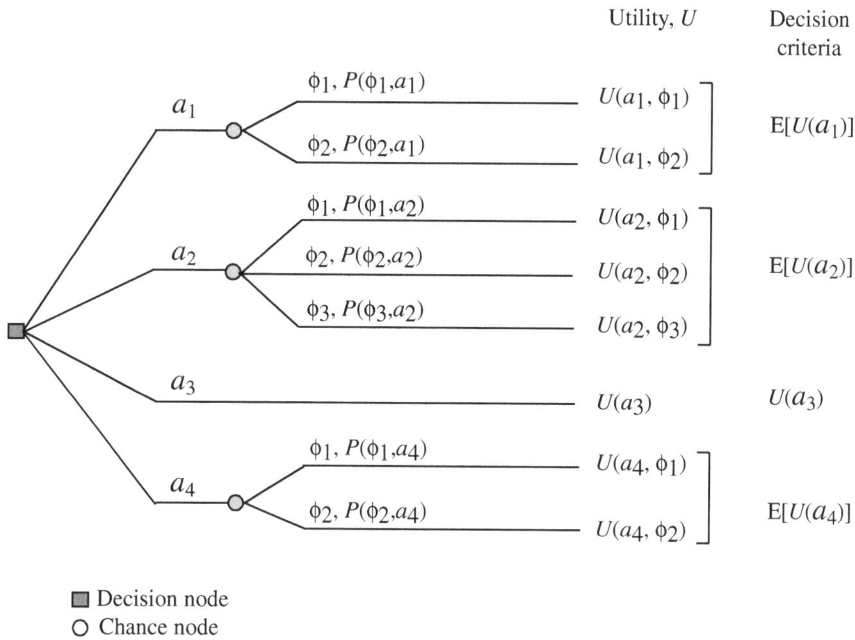

Fig. 1.1 Example of a decision tree

(i.e., restrictions or criteria under which the decision is made) defines, to a large extent, the characteristics of the decision. A detailed discussion about these and many other aspects that influence our decisions can be found in, for example, [11, 13].

In classic decision theory, when there is a set of distinct *feasible* alternatives, the decision problem is often structured as a *decision tree*; see Fig. 1.1. In a decision tree, there are decision nodes (denoted by squares in Fig. 1.1) where the decision maker must choose from a set of alternatives $\{a_1, a_2, ...\}$. The set of alternatives, also called the *option space*, may be finite or infinite; and once it is defined the problem is bounded [2]. Note that when decisions are made at different points in time, the set of possible alternatives may change also with time. For instance, for systems that deteriorate, the set of possible intervention measures depends on its condition at the time of evaluation. For every feasible alternative a_i (Fig. 1.1), there may be several possible outcomes $\{\phi_1, \phi_2, ...\}$ (derived from the chance nodes) defined in terms of some probability function. For completeness, the outcomes from a chance node must be mutually exclusive and collectively exhaustive; this means that the sum of the conditional probabilities must add to one. Finally, the outcome at the end of every branch of the tree is measured in decisions units; e.g., economic value or utility, which are organized according to a decision criteria to choose the best option [21].

Over the years, economists have worked on developing models to describe what *rational* agents, as defined at the beginning of this section, should do when confronted with a choice between two or more options. A widely used approach for selecting the best option is the relative comparison of the *expected value* with respect to some evaluation criteria. Typical criteria include costs (i.e., value of gains or losses) and, in the case where human preferences are involved, an *utility* measure [22]. Note that these two measures (i.e., costs and utility), or any other criteria for that matter, do not lead necessarily to the same output.

For the particular case of decisions that involve actions in the future, the metrics used to compare alternatives should take into account the fact that decisions affect the system at different points in time. Regardless of the evaluation metrics (e.g., costs or utility), these type of problems should take into account the concept of *discounting*. This is a way of weighting the importance of decisions in the future. This can be interpreted as a way to value current decisions within the context of possible future scenarios. Discounting is also an essential element to define risk-acceptability criteria of engineering decisions that evolve with time. There has been a debate as to how to discount the many factors involved in decision making. For example, some ethical and economical arguments regarding discounting from the public interest perspective can be found in [3, 23, 24]; a discussion on interest rates for life-saving investments in [25]; a discussion on the ethical problems associated to inter-generationally discounting are discussed in [26]; and additional discussion on discounting can be found in [27–29]. A more detailed discussion on this topic will be presented in Chap. 9.

Finally, it is important to stress that an essential element of the decision-making process is the uncertainty as to whether the final decision will actually lead to the best outcome. This uncertainty comes from the fact that we cannot predict (model) accurately the scenarios that will be derived from our decisions. Therefore, engineering is mostly about *good enough* (satisfactory [30]) decisions[4]—i.e., grounded on a dependable evidence and on a scientifically justifiable derivation, and not concerned with correct decisions, since this concept is impossible to assess.

1.4 Decisions in the Public Interest

The term *public interest* refers to all aspects that may affect a community (i.e., public) grouped under a certain political structure under which they share common resources [31]. For example, countries are societies that gather around basic socioeconomic principles (e.g., constitution) and normative (e.g., law). Then, *decisions in the public interest* are those concerned with the welfare or well-being of the general public.

[4]Note that *satisfactory* decisions are somehow sub-optimal.

Within the context of decisions in the public interest, Natwani et al. [3] state that "the basic principles and requirements [for making decisions] that serve the public interest are:

- comprehensive evaluation of options and alternatives;
- transparent and open process(es), iterative as necessary; and
- defensible outcome(s), defined as positive net benefit to society."

Because not all societies are organized along the same principles, we must realize that decisions in the public interest cannot be formulated under a unique framework.

With regard to public investment in engineering infrastructure projects, two aspects are particularly important [23]: the resources committed to make this developments, and its sustainability. The first aspect is related to the fact that the resources used to develop this project come from what the entire society has agreed upon to contribute for their overall well-being and development, usually via taxes [3]. Therefore, their use should be based on constitutional and ethical considerations [23] and the profit should be reinvested in society.

The second aspect is concerned with the fact that by building large engineering projects we are using mostly limited and nonrenewable natural resources. Due to their expected long operation times, the damage to the environment that they may cause and the impact on future generations become relevant. Therefore, "our generation must not leave the burden of maintenance or replacement [*of engineering devices*] to future generations. In addition, we must not use more of the financial resources than are really available. We can use only those which are available and affordable in a sustainable manner and discounting with its many myopic aspects must be done with utmost care." [20, 23]. This statement clearly emphasizes the basic sustainability principle expressed by the Brundland Commission [32]; i.e., a sustainable development is a development "that meets the needs of the present without compromising the ability of future generations to meet their own needs." Therefore, according to Rackwitz et al. [23] "intergenerational equity is the core of the new ethical standard the Brundland Commission [32] has set."

In summary, it is important to stress that when dealing with decisions in the public interest, and especially when these decisions involve long-term projects, engineering decisions should be optimal from both a technological and a sustainability point of view [23, 33].

1.5 Prediction

A decision is made based on the analysis of our *predictions*. Thus, the decision of a rational agent depends to a large extent on its ability to collect information about the behavior of the system (e.g., possible failures and investments) and to make relevant inferences.

There are three important aspects that influence our predictions:

- time horizon;
- ability to make inferences; and
- evolution of knowledge.

First, the accuracy of our predictions depends on how far into the future we want to go. Clearly, our ability to predict diminishes as the time horizon increases. For example, under normal conditions, it may be possible to make a reasonable estimative of tomorrow's variations in the stock market, but very difficult to predict what would be its state in 5 years' time. Secondly, our ability to make predictions is generally based on past experiences and observations; our predictive models rely to a large extent on observed data. We may be unable to envisage events that have not been previously observed, which does not mean that such events will not occur. For example, recently, there has been much interest in so-called "black swan" events [34] and the limitations on decision making imposed by classical notions of probability. Our predictions often rely on the notion of causality; however, inferences about causality that are not properly grounded scientifically should be carefully analyzed. Hume in the *Treatise of Human Nature* [35] criticizes the existence of causality and argues that it cannot be proven by either logic or experience. Finally, making predictions is a dynamic process. It changes permanently as new information and new technological developments become available. Furthermore, predictions may possibly change as our understanding of the system performance evolves.

Despite the practical and conceptual difficulties in making predictions, they are unavoidable in decision making. Good predictions require the appropriate understanding and management of uncertainty. Thus, in most engineering problems, the *stochastic* nature of the "laws" that describe the system performance (e.g., *stochastic mechanics*) plays a major role. Most of this book is about making predictions of the performance of systems that deteriorate over long periods of time.

1.6 Choosing Preferred Alternatives

1.6.1 The Role of Optimization in Engineering Decisions

Decisions involved in managing large engineering projects are often associated with selecting effective *operating* strategies during what is often referred to as the *gate-to-grave* phase of an engineering project, as opposed to the *cradle-to-gate* phase (i.e., conception, design, and construction) [36]. Operational decisions include, for instance, intervention measures through activities such as maintenance (retrofitting), repair (after failure), and decommissioning or replacement (at the end of the system's life cycle). Future investments in any of these activities not only carry economic costs but may also have an impact on other aspects of project life, such as sustainability and climate change, whose effects can be estimated through indicators such as CO_2

emissions and embodied energy [36–38]. Then, deciding on the best design alternative or operation strategy depends on our ability to model the system performance over time, which is uncertain by nature. The models and analytical procedures that form the basis of this book are primarily focused on predicting the performance of various design alternatives (e.g., selection of design parameters, operating and maintenance strategies, and infrastructure replacement). It is then argued that the results of these models provide the rational bases over which better decisions can be made.

The economic framework for rational decision-making asserts that the best alternative is the one that maximizes expected utility; thus, in the engineering framework, selecting the best design or operating alternative involves optimization. In the sections that follow, we briefly investigate the mathematical formulation of an optimization problem and provide a framework for optimization under uncertainty in the context of making engineering decisions.

1.6.2 The Constrained Optimization Problem

The mathematical optimization problems associate with engineering design decisions are inherently constrained by available resources . In this setting, decision variables are determined so that they maximize or minimize a predefined decision criterion, or in mathematical terms, an *objective function*, subject to constraints of the design space that characterize a feasible region. Let $\mathscr{X}$ denote the set of feasible decision variables (generally an n-dimensional space), and let $f : \mathscr{X} \rightarrow \mathbb{R}$ denote a scalar objective function. The constrained optimization problem can be expressed mathematically as [39],

$$\begin{aligned} &\min_{\mathbf{x}\in\mathscr{X}} f(\mathbf{x}) \\ &\text{subject to:} \\ &h_i(\mathbf{x}) \geq b_i, \quad i = 1, \ldots, n \\ &g_j(\mathbf{x}) = c_j, \quad j = 1, \ldots, m \end{aligned} \tag{1.1}$$

where the functions h_i and g_j determine constraints ("subject to") that must be satisfied. Discrete optimization problems deal with the case in which the optimization function is defined on a discrete variable space, while in the continuous case decision variables are allowed to take any value within a finite/infinite range. In the engineering decision framework, the objective function represents the *utility*, which is typically formulated as the value of the return/cost of the alternative $\mathbf{x} \in \mathscr{X}$.

Depending on the mathematical form of the objective function and the constraints, there are many techniques leading to determining optimal solutions. Constrained optimization can be solved by linear programming in the special case that the objective and constraints are linear functions, and more generally, by branch and bound, penalty methods, and Lagrange multipliers, among many other techniques; see [40, 41].

1.6.3 Multi-Criteria Optimization

Most complex engineering decisions, including those that are the subject of this book, involve complex trade-offs between a number of conflicting objectives, such as cost, performance, societal benefit, safety, etc. Often these problems can be formulated as so-called *multi-criteria (or multi-objective) optimization* problems.

Again, let $\mathscr{X}$ denote the set of feasible decision alternatives (a subset of the decision space), and let the set of decision objectives be defined by the functions $f_i : \mathscr{X} \rightarrow \mathbb{R}$, $i = 1, 2, \ldots$ (e.g., functionality, cost, CO_2 emissions). Then, the multi-criteria optimization problem can be expressed mathematically as [39],

$$\begin{aligned} &\min_{\mathbf{x}\in\mathscr{X}}\{f_1(\mathbf{x}), f_2(\mathbf{x}), \ldots, f_n(\mathbf{x})\} \\ &\text{subject to :} \\ &h_i(\mathbf{x}) \geq b_i, \quad i = 1, \ldots, n \\ &g_j(\mathbf{x}) = c_j, \quad j = 1, \ldots, m \end{aligned} \tag{1.2}$$

where the functions h_i and g_j describe the constraints of the problem. Although these problems may be formulated in a straightforward way, their solution involves quite different techniques than those described in the single objective case. These techniques revolve around determination of efficient (or *Pareto optimal*) solutions that explicitly take the conflicting nature of the objectives into account. The set of non-dominated solutions define the Pareto *frontier* along which all solutions are feasible and additional decision criteria are needed to select the best alternative. Because of the conceptual and mathematical complexity of these models, most tractable engineering problems are limited to a single or very few objectives, often through the imposition of a weighting scheme that determines the relative importance of each objective. Additional literature on this subject can be found in [40, 42, 43]. In addition, the basis and some advanced multi-criteria optimization models can be found in, for instance, [39, 44].

1.6.4 Incorporating Randomness into the Optimization

As we have emphasized, while in classical (deterministic) optimization it is assumed that the system performance is fully known and that there is perfect information, the nature of engineered systems is that they are subject to randomness, and hence reward/risk are described by random variables. In this case, the performance measure is formulated in terms of both decision variables and random quantities; i.e., $f(\mathbf{x}, \mathbf{w})$, where $\mathbf{w}$ is a vector of random variables with given joint probability distribution F. In this case, the objective is to miminize the expected value of the objective function [45, 46], which can be written as

$$\min_{\mathbf{x} \in \mathscr{X}} \mathbb{E}[f(\mathbf{x}, \mathbf{w})] \tag{1.3}$$

where $\mathbb{E}$ is the expectation operator; i.e., $\mathbb{E}[f(\mathbf{x}, \mathbf{w})] = \int_0^\infty f(\mathbf{x}, \mathbf{w}) dF(\mathbf{w})$. In Chaps. 8 and 9 we will present detailed applications of this approach to find optimum design values based on the life-cycle of engineering systems.

1.6.5 Optimization of Performance Over Time

Finally, management of the engineered system may involve decisions that unfold over time; that is, certain operational decisions may not be effectively made at the beginning of the operational life of the system. In this case, a sequence of decisions must be made over time and every decision may depend on the previous one. Then, at a given time, the state of the system is evaluated and an intervention is chosen, when necessary, from a set of feasible alternatives [47]. In this book, we consider the case of systems that deteriorate over time and that may require interventions to guarantee that they operate as expected. In this case, optimum decisions focus on finding the *policy* ν that maximizes the return on investment over a given time span. An *operation policy* is basically a double sequence $\nu = \{(\tau_i, \zeta_i)\}_{i \in \mathbb{N}}$ of intervention times τ_i at which the performance is improved an amount ζ_i.

In this particular case, the optimization problem can be written as

$$\max_{\circ}\{g(\mathbf{x}) = J(v_0, \nu)\} \tag{1.4}$$

where $J(v_0, \nu)$ describes the expected net-present profit (benefit-costs) that results from an operation policy ν given that the system initial state is v_0. Then, the purpose of the optimization is to find the operation policy with the maximum return. The term $J(v_0, \nu)$ in Eq. 1.4 can be written as [48]

$$J(v_0, \nu) = \mathbb{E}\left[\int_0^{t_f} G(V_u^\nu)\delta(u)du - \sum_{\tau_i < t_f} C(V_{\tau_i -}^\nu, \zeta_i)\delta(\tau_i)\right], \tag{1.5}$$

where t_f if the time at which the failure occurs, v_0 is the initial state of the system, measured in physical units (e.g., resistance), and the term $\delta(t) = e^{-\gamma t}$ corresponds to the discounting function used to evaluate the net present value. The term V_t^ν in Eq. 1.5 describes the state of the system at time t for an operation policy ν. This clearly depends on the initial condition v_0, the degradation process (e.g., shocks), and the size of all previous interventions ζ_i up to time t (i.e., operation policy) [48].

The function G can be interpreted as a utility function; thus, the first term in Eq. (1.5) corresponds to the discounted benefits; and the second term describes the discounted costs of interventions, with $C(V_{\tau_i -}^\nu, \zeta_i)$ the cost of bringing the system from level $V^{\nu}_{T_i-}$ to level $V^{\nu}_{T_i-} + \zeta$. The methods that are typically used to address this formulation are known as *dynamic programming* and include techniques such as Markov decision processes. A detailed explanation of this approach will be presented in Chap. 10, when we discuss optimal maintenance strategies.

1.7 Life-Cycle Modeling

Investment decisions for engineered systems are based on predictions about the system's future performance. Within this context, *life-cycle analysis* (LCA) is the study of a system's performance over a specific time period, frequently selected as the system's lifetime; i.e., from planning to disposal. If the study focuses on costs, it is called *life-cycle cost* analysis (LCCA). LCCA provides a framework to support long-term decisions about resource allocation related to the design, construction, and operation of infrastructure systems. LCCA focuses mainly on finding the expected discounted value of a cost–benefit relationship $Z(\mathbf{p}, \ell)$ at time $t = 0$; i.e.,

$$\mathbb{E}[Z(\mathbf{p}, \ell)] = \mathbb{E}\left[\int_0^{\ell} B(\mathbf{p}, \tau)\delta(\tau)d\tau - \sum_{i=1}^{N(\ell)} C_i(\mathbf{p}, t_i)\delta(t_i)\right] \tag{1.6}$$

where $\delta(\cdot)$ is the discount function used to compute the net present value of future gains and investments, and $\mathbf{p}$ is a vector parameter used to describe the system performance. $B(\mathbf{p}, t)$ represents the benefits derived from the existence and operation of the project and $C_i(\mathbf{p}, t)$ describes all costs incurred (e.g., failure, repair, maintenance) throughout the lifetime ℓ of the system. Note that $N(\ell)$ is the number of interventions in the time interval ℓ, and it is usually a random variable. It is worth to mention that, recently, a significant effort has been devoted to measure the life cycle of a system in terms of sustainability indicators (e.g., CO_2 emissions). In this case, the analysis is not cost-based but sustainability-based and it is called *life-cycle sustainability* analysis [36].

A central element in LCA involves making predictions about the degradation of the system. It requires a clear understanding of the physical laws that define the system behavior and the associated uncertainties. The *degradation* of an engineering artifact describes the process by which one or a set of properties lose value with time. By properties we mean not only mechanical (e.g., strength, stiffness) but any other attribute that adds value to the element (e.g., functionality, aesthetics, etc.). Degradation is a decreasing function in t; thus, if $V_t(\mathbf{p})$ represents the system's state (e.g., resistance, remaining life) at time t, there is degradation if, $V_{t+1}(\mathbf{p}) \leq V_t(\mathbf{p})$, where $\mathbf{p}$, as mentioned before, is a vector parameter of the system variables that defines its performance. Chapters 4–7 describe existing modeling tools to manage degradation problems.

Life-cycle analysis is an area of great importance in modern engineering and it involves most key elements presented and discussed in previous sections. It encompasses the need for making decisions and the uncertain performance of degrading engineering systems. Life-cycle analysis helps the efficient use of resources needed to mitigate the physical, financial, and sustainable risks associated to the degradation of large engineering projects. The book is intended to provide the basis for modeling degradation, planing optimum maintenance strategies, and evaluating the life-cycle performance of large engineering systems.

1.8 Risk and Engineering Decisions

All decisions in engineering, as in everyday life, are based on the assessment of the future consequences of a set of possible actions. Because we cannot predict the future with certainty, while we may expect our decisions to yield benefits, we do understand that they may also result in loss. In this section, we discuss the concept of risk from a variety of perspectives. We then provide a definition from a mathematical point of view and show the role that it plays into the engineering decision-making framework.

1.8.1 Interpretations and Approaches to Risk

In colloquial use, the term *risk* connotes a situation involving exposure to harm or danger. The concept of risk is used in many fields, and consequently, its precise definition and usage is dependent on context. For example, in the area of cognitive psychology [49], risk is taken to mean the "fear and dread we feel when considering a hazard" [3]. This concept of risk, also called *perceived risk* [50, 51], is an attribute associated with the characteristics of an individual and his/her worldview. In general, the public is more concerned with perceived risks than with any other type of risk (e.g., quantified risks) [3]. Although perceived risk is difficult to evaluate rigorously, and decisions based on perceived risk do not necessarily fit the framework of rational decisions (see Sect. 1.3) [52], psychologists and neuroscientists understand that it has been the basis for the survival and development of human beings. Several cognitive studies have shown that one of the main tasks of the brain is to carry out risk analyses of its environment as a way to improve decision making [53]. Within the context of perceived risk, additional considerations involve distinctions between "voluntary" and "involuntary" risks and between "individual" and "societal" risks; see for instance [54].

On the other hand, in business management, risk is understood primarily as a qualitative assessment of the possibility of financial loss due to particular threats faced by a company. These may be external threats (market conditions, competition, natural disasters, etc.) or internal threats (corporate structure, workforce dynamics). In this context, a description of threats, their consequences, and their likelihoods is very useful in deploying strategies to reduce exposure to monetary loss; they include insurance, hedging, and business reorganization. The financial sector has developed an entire and unique taxonomy of risks (e.g., capital risk, liquidity risk, geopolitical risk, sovereign risk, etc.) that are used to evaluate investment opportunities. Risk analysis and management is a major aspect in business operations.

Yet another usage of risk that often has no inherent monetization is the concept of medical risk. Any medical therapy intended to improve the well-being of the patient, whether it involves surgery, nonsurgical treatment, dispension of drugs, etc., carries the possibility (i.e., risk) that it will leave the patient worse off than if no therapy had been performed. To assess the likelihood of this type of risk, the healthcare community relies primarily on a *quantitative* assessment that arises from experimentation and observation of many previous therapeutic procedures. This assessment is obviously quite difficult, and must take into significant variability between patients, but provides the basis for medical decisions regarding choice of therapy from available alternatives.

In addition to the few specific and illustrative cases mentioned above, there are many other fields in which the term "risk" has a particular connotation. However, it is clear that the overall concept has to do with the likelihood of undesired consequences within a given context [55].

1.8.2 Mathematical Definition of Risk

In engineering, risk has a precise *quantitative* (i.e., mathematical) definition that is consistent with the framework for rational decision-making described in the previous sections. This definition of risk involves the probabilistic assessment of the outcome of an action.

For a given feasible action (or feasible set of decision variables), define X to be the *return* associated with that action. In most engineering decision problems, return is expressed in monetary terms (i.e., US$). However, depending upon the problem at hand, the "return" can be evaluated in terms of any measure of interest; e.g., *utility*. Because the outcome is a function of both the action (or control) and some random events, the return, X, is a random variable. If X takes on a positive value, we receive a *gain* (*reward* or *payoff*) for taking the action; if X takes on a negative value, we experience a *loss*. To reiterate, positive values of return are called *gains*, and negative values of return are called *losses*.

In mathematical terms, *Risk* is simply the probability law (in the form of the distribution function) of return X[5]; i.e., risk is a function R such that

$$R(a) = P(X \leq a). \quad -\infty \leq a \leq \infty \tag{1.7}$$

[5]Note that in colloquial usage, risk generally refers only to the negative values of the return function; positive values are frequently described as an *opportunity*. Despite these interpretations, in mathematical terms, and for completeness, it is most convenient to include both positive and negative returns as part of any risk analysis.

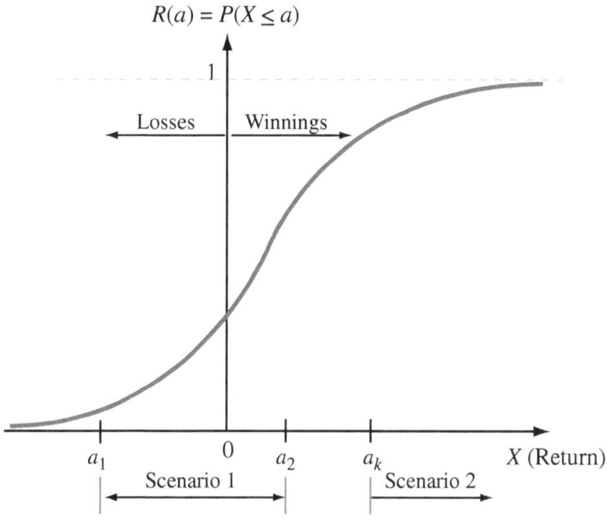

Fig. 1.2 Distribution function of the return

The probability distribution function of the return assigns a likelihood (measured between 0 and 1) to every measurable event (collection of outcomes). Thus, the specification of risk requires the evaluation of the likelihood of all possible consequences of an action. Figure 1.2 shows an example of what the risk of a particular action might look like. Note that risk can be used with advantage to make an evaluation for any scenario of interest, such as a return within the range $[a_1, a_2]$ (Fig. 1.2); or a return larger than a_k. Furthermore, when coupled with the concept of utility, this definition provides a mechanism for decision making called *risk tolerance*; for example, how much potential gain one is willing to accept to mitigate against the likelihood of a potential loss.

While mathematically rigorous, this definition of risk is also deceptively simple. Note that because risk is characterized by a probability distribution function, it is inherently subjective since it expresses the decision maker's best guess as to the likelihoods of outcomes resulting from an action. Therefore, decision makers with different information or experience may assign different distribution functions to the return. An action may almost certainly result in a loss under one decision maker's assessment of likelihood, while it may almost certainly result in a gain under another's assessment. In this sense, risk is not a physical quantity that can be weighed or measured. Rather, it is a quantitative assessment of the decision maker's model of the uncertain future.

Note also that this definition differs substantially from the widely used interpretation that risk is simply "the multiplication of the probability of occurrence by the consequence of occurrence."This notion lacks precision and is incompatible with a rational decision-making framework. Notably, it leads to conclusions that equate scenarios as different as "low-probability high consequence" with "high probability and low consequence," which are clearly different cases [51], and such usage of risk should be avoided in an engineering context.

Several extensions of the definition of risk presented in Eq. 1.7 can be made [51]; in particular, by including time (transient system states) in the analysis. Time may be included by restricting the assessment to a specific time window, $[0, t]$ for example, the probability that the return on the investment, X, falls below a by time t. Alternatively, it may be possible that the risk function changes over time due to the nature of the process.

On a final note, it is important to stress that most engineering decisions are based on assessing risks in the public interest; i.e., risks that may have an impact on society as a whole, not on a particular individual. Therefore, the approach to understanding and managing risk should be made based on a "rational" and well-grounded approach, which is consistent with the decision context.

1.9 Summary and Conclusions

This chapter presents an overview of the key elements that will be discussed in this book. As we do throughout the book, we emphasize the importance of developing dependable probabilistic models that provide evidence for making better decisions. Decisions about construction and operation (e.g., maintenance and repair) of large engineered systems depend on how we value the consequences that their performance might have on our society and future generations. This assessment can only be performed if we are able to understand and model risk; this depends greatly on how we characterize and manage the uncertainties associated with failure mechanisms. In the following chapters we will discuss all these aspects in detail and provide an insight into areas of great importance in modern engineering.

References

1. D.I. Blockley, *Engineering: A Very Short Introduction* (Oxford University Press, Oxford, 2012)
2. G.A. Hazelrigg, *Systems Engineering: An Approach to Information-Based Design* (Prentice Hall, New Jersey, 1996)
3. J.S. Nathwani, M.D. Pandey, N.C. Lind, *Engineering Decisions for Life Quality: How Safe is Safe Enough?* (Springer-Verlag, London, 2009)
4. G.A. Hazelrigg, *Fundamentals of decision making for engineers: for engineering design and systems engineering*. Independent, http://www.engineeringdecisionmaking.com/, (2012)
5. R.A. Howard, Decision analysis: applied decision theory, in *Proceedings of the Fourth International Conference on Operational Research* eds. by D. Bendel Hertz, J. Mse. International Federation of Operational Research Societies. (WIley-Interscience, 1966), 55–71
6. J. Von Neummann, O. Morgenstern, *Theory of Games and Economic Behavior*, 3rd edn. (Princeton University Press, Princeton, New Jersey, 1953)

7. P.C. Fishburn, *The Foundations of Expected Utility* (Reidel Publishing (Kluwer group), The Netherlands, 2010)
8. A.N. McCoy, M.L. Platt, Expectations and outcomes: decision-making in the primate brain. J. Comp. Physiol. A **191**, 201–211 (2005)
9. P. Glimcher, *Decisions, Uncertainty, and The Brain: The Science of Neuroeconomics* (MIT Press, Cambridge, MA, 2003)
10. R.J. Herrnstein, *The Matching Law: Papers in Psychology and Economics* (Harvard University Press, Cambridge, MA, 1997)
11. R.T. Clemen, *Making Hard Decisions: An Introduction to Decision Analysis* (Duxbury Press, Albany, NY, 1996)
12. K.T. Marshall, R.M. Oliver, *Decision Making and Forecasting with Emphasis on Model Building and Policy Analysis* (McGraw Hill, New York, 1995)
13. J.C. Hartman, *Engineering economy and the decision-making process* (Prentice Hall, New Jersey, 2007)
14. G.S. Parnell, P.J. Driscoll, D.L. Henderson, *Decision Making in Systems Engineering and Management* (Wiley, New York, 2010)
15. P. Chekland, *Systems Thinking, Systems Practice: Includes A 30-year Retrospective* (Wiley, Chichester, 1999)
16. R.L. Keeney, H. Raiffa, *Decisions with Multiple Objectives* (Cambridge University Press, Cambridge, MA, 1993)
17. C. Yoe, *Principles of Risk Analysis: Decision Making Under Uncertainty* (CRC Press—Taylor Francis, Boca Raton, 2011)
18. G.A. Holton, Defining risk. Financ. Anal. J. **60**(6), 19–25 (2004)
19. L.R. Duncan, H. Raiffa, *Games and Decisions: Introduction and Critical Survey* (Dover, New York, 1985)
20. M.H. Faber, *Statistics and Probability Theory: In Pursuit of Engineering Decision Support* (Springer-Verlag, London, 2012)
21. A.H-S. Ang, W.H. Tang, *Probability Concepts in Engineering Planning and Design: Volume II Decision Risk and Reliability* (Wiley, New York, 1984)
22. D. Kreps, *Notes on the Theory of Choice (underground classics in economics)* (Westview Press, Boulder, Colorado, 1988)
23. R. Rackwitz, A. Lentz, M.H. Faber, Socio-economically sustainable civil engineering infrastructures by optimization. Struct. Saf. **27**, 187–229 (2005)
24. E. Patte-Cornell, Discounting in risk analysis: capital vs. human safety, in *Risk, Structural Engineering and Human Error* eds. by M. Grigoriu, (University of Waterloo Press, Waterloo, Canada, 1984)
25. M.C. Weinstein, W.B. Stason, Foundation of cost-effectiveness analysis for health and medical practices. New Engl. J. Med. **296**(31), 716–721 (1977)
26. T.C. Schelling, Intergenerational discounting. Energy Policy **23**(4/5), 395–401 (1995)
27. A. Rabl, Discounting of long term costs: what would future generations prefer us to do? Ecol. Econ. **17**, 137–145 (1996)
28. S. Bayer, Generation-adjusted discounting in long-term decision-making. Int. J. Sustain. Dev. **6**(1), 133149 (2003)
29. C. Price, *Time: Discounting and Value* (Blackwell, Cambridge, MA, 1993)
30. G. Gigerenzer, R. Selten, *Bounded Rationality* (MIT Press, Cambridge, MA, 2002)
31. M.H. Faber, M.A. Maes, J.W. Baker, T. Vrouwenvelder, T. Takada, Principles of risk assessment of engineered systems, in *Proceedings of the Applications of Statistics and Probability in Civil Engineering*, edS. by J. Kanda, T. Takada, H. Furuta. (Taylor & Francis Group, London, 2007), 1–8
32. UN. Brundland Commission, *Our common future*. (UN World Commission on Environment and Development, 1987)
33. R. Rackwitz, Optimization and risk acceptability based on the life quality index. Struct. Saf. **24**, 297–331 (2002)

34. N.N. Taleb, *The Black Swan: Second Edition: The Impact of the Highly Improbable* (Random House Trade paperback, USA, 2010)
35. D. Hume, *A treatise of human nature*. Project Gutemberg e-book, www.gutemberg.org/files/4705/4705-h/4705-h.htm, Accessed 13 Aug 2015
36. J.E. Padgett, C. Tapia, Sustainability of natural hazard risk mitigation: a life-cycle analysis of environmental indicators for bridge infrastructure. J. Infrastruct. Syst. ASCE **19**(4) 395-408 (2013)
37. A. Alcorn, *Embodied energy and CO_2 coefficients for New Zealand building materials* (Center for Building Performance Research, New Zealand, 2003)
38. A.R. Pearce, J.A. Vanegas, Defining sustainability for built environments systems: an operational framework. Int. J E Technol. Manage. **2**(1–3), 94–113 (2002)
39. M. Ehrgott, *Multicriteria Optimization* (Springer-Verlag, Berlin, 2005)
40. M.S. Bazaraa, H.D. Sherali, C.M. Shetty, *Nonlinear Programming: Theory and Algorithms* (Wiley, New Jersey, 2006)
41. I. Griva, S.G. Nash, A. Sofer, *Linear and Nonlinear Optimization*, 2nd edn. (SIAM, Philadelphia, 2009)
42. R. Fletcher, *Practical Methods of Optimization* (Wiley, Cornwall, U.K., 2000)
43. S.S. Rao, *Engineering Optimization: Theory and Practice*, 3rd edn. (Wiley, New Jersey, 2009)
44. Y. Collete, P. Siarry, *Multi-objective Optimization: Principles and Case Studies* (Springer-Verlag, Berlin, 2004)
45. J.R. Birge, F. Louveaux, *Introduction to Stochastic Programming* (Springer-Verlag, New York, 1997)
46. A. Shapiro, D. Dentcheva, A. Ruszczynski, *Lectures on stochastic programming: modeling and theory* (The Society of Industrial and Applied Mathematics (SIAM) and the Mathematical Programming Society, Philadelphia, 2009)
47. S.M. Ross, *Introduction to Stochastic Dynamic Programming* (Academic Press, New York, 1983)
48. M. Junca, M. Sánchez-Silva, Optimal maintenance policy for a compound poisson shock model. IEEE Trans. Reliab. **62**(1), 66–72 (2012)
49. D. Gardner, *Risk: The Science and Politics of Fear* (McClelland and Stewart, Toronto, 2008)
50. Slovic, *The Perception of Risk* (Earthscan, Virginia, 2000)
51. S. Kaplan, J. Garrick, On the quantitative definition of risk. Risk Anal. **1**(1), 11–27 (1981)
52. D. Ariely, *Predictably Irrational: The Hidden Forces that Shape Our Decisions* (Harper Collins, New Jersey, 2008)
53. R. Llinas, *I of the Vortex: From Neurons to Self* (MIT Press, Cambridge, MA, 2002)
54. M.G. Stewart, R.E. Melchers, *Probabilistic Risk Assessment of Engineering Systems* (Chapman & Hall, Suffolk, U.K., 1997)
55. D.I. Blockley, *Engineering Safety* (McGraw Hill, New York, 1992)

Chapter 2
Reliability of Engineered Systems

2.1 Introduction

Making decisions about the design and operation of infrastructure requires estimating the future performance of systems, which implies evaluating the system's ability to perform as expected during a predefined time window. This evaluation fits within what is known as *reliability analysis*. This chapter presents an introduction to the basic concepts and the theory of reliability in engineering, which provides the foundation for constructing degradation models (see Chaps. 4–7), performing life-cycle cost analyses (see Chaps. 8 and 9), and to designing maintenance strategies (Chap. 10). In the first part of this chapter, we present some conceptual issues about reliability and a description of basic reliability approaches. The second part of the chapter, Sect. 2.7 and onward, presents an overview of reliability models and sets the basis for theory that will be used and discussed in the rest of the book.

2.2 The Purpose of Reliability Analysis

Reliability analysis is the study of how things fail. Any engineered system, be it a facility (e.g., power plant) or infrastructure component (e.g., bridge), an electro-mechanical device, a consumer product, or even a manufacturing process, is designed and built to perform a specific *function* for a *specified duration* (the *mission* of the system). Once in use, the physical properties of the system will inevitably decline, and any engineered system will eventually *fail* (i.e., be unable to perform its designated function), possibly before completion of the mission. Moreover, engineered systems are typically operated in environments that are neither controllable nor predictable, and even well-designed and constructed systems may not fulfill their intended purpose due to unforeseen or unexpected events. As technology improves and new products enter the marketplace, consumers have become accustomed to expecting

M. Sánchez-Silva and G.-A. Klutke, *Reliability and Life-Cycle Analysis of Deteriorating Systems*, Springer Series in Reliability Engineering,
DOI 10.1007/978-3-319-20946-3_2

dependable performance in the goods and services they buy and in the infrastructure developed to support their operation. Reliability analysis is the quantitative study of system failures and is an integral aspect of ensuring high-quality system performance.

As an engineering discipline, the field of reliability engages engineers of all disciplines, as well as physicists, statisticians, operations researchers, and applied probabilists. Furthermore, it encompasses a wide range of activities, which include, among others:

- collecting and analyzing data from physical and virtual experiments (*design of experiments, statistical, and simulated life testing*);
- characterizing the physical processes that lead to system failure (*physics of failure and degradation modeling*) and modeling the uncertainties that govern those failures (*probabilistic lifetime modeling*); and
- understanding the logical structure that determines the interactions and the dependencies between system components and their influence on overall system performance (*reliability systems analysis*).

The purpose of reliability analysis is not simply to describe how, when, and why systems fail, but rather to use information about failures to support decisions that improve the system's quality, safety and performance, and to reduce its cost. This aspect is especially important in areas where failures have serious consequences, for example, where public safety is involved or where significant financial investments are at stake (e.g., bridge failure). The acceptable performance of a system can be achieved in many ways; for example, through improvements in design and manufacture, and through better planning of operations (e.g., maintenance policies and warranty procedures); within this context, reliability analysis provides a quantitative foundation to support decisions that make these activities more efficient.

Reliability evaluation methods have been presented and discussed in a wide variety of applications, and many journals and books are available on the topic; see for instance [1–8]. This chapter presents some of the fundamental concepts of relaibility analysis and introduces reliability methods which are of particular importance to support of decisions about future investments (e.g., design, manufacture, operation, and maintenance). Several references have been included for the reader to find more detailed information.

2.3 Background and a Brief History of Reliability Engineering

The field of reliability analysis began in earnest after World War II, when the U.S. and Soviet militaries both began systematic studies of newly developed weapons systems with the goal of improving their operation. In subsequent years, reliability engineering permeated the military, aerospace (particularly during the "space race"), and nuclear energy sectors. These sectors were still highly regulated by governmental entities, which led to the development of many standards, specifications, and procedures that govern product development in these sectors. Driven by increasing competition and demands for high-quality and dependable consumer products, eventually, reliability analysis became widely adopted by many commercial enterprises, such automotive manufacturing, consumer electronics, software, and appliances, to name just a few. In these industries, reliability analysis remains an important part of the product development and manufacturing process. Many reliability engineering techniques, such as fault tree analysis (FTA), failure mode, effects and criticality analysis (FMECA), and root cause analysis, are commonly used in the design and planning of engineered systems. Reliability analysis has also driven the development of fatigue and wear models, crack propagation models, corrosion models, and other methods of modeling physical wear out.

Reliability of infrastructure is, to a large extent, linked with the history of structural reliability. The first papers utilizing a probabilistic approach in design and analysis of structures were published in the late 1940s by Freudenthal [9], who discussed the basic reliability problem in structural components subjected to random loading, and in the early 1950s by Johnson [10], who proposed the first comprehensive formulation of structural reliability and economical design. These papers basically set the basis for a new field in structural engineering. In the 1960s, the basic concepts of safety (e.g., safety margin and safety index) were developed by Basler [11] and Cornell [12, 13], although there were also important contributions by other researches such as Ferry-Borges [14] and Pugsley [15]. During the period from 1967 until 1974, the area of structural reliability attracted a great deal of interest in the academic community; however, its application and use in practice evolved only very slowly [3]. The work of Hasofer and Lind [16] and Veneziano [17] in the early 1970s, among others, led to the first standard in limit state format based on a probabilistic approach, the CSA [18], published in 1974. This publication was followed by development other worldwide standards, and nowadays the probabilistic approach (mostly through partial safety factors) is used in almost every code of practice. More recently, the Join Committee on Structural Safety (http://www.jcss.byg.dtu.dk/) has been working extensively to improve the general knowledge and understanding within the fields of safety, risk, reliability, and quality assurance in infrastructure design and development.

Interestingly, there are several important commercial sectors, where reliability engineering is still in a relatively nascent phase. These sectors include medical device manufacturing and food engineering. In medical device manufacturing, only relatively simple, qualitative techniques are commonly employed, and then primarily to respond to regulatory requirements. While it may appear somewhat unorthodox to consider food as an engineered system, many new methods of treating, processing, and packaging food are under development, and only very few studies on their reliability have been performed. Thus there is still a great need for engineers educated in the principles of reliability analysis among all sectors of the economy.

Despite the fact that the field of reliability now comprises a mature body of work, it is by no means a closed subject. In particular, there is still much work to be done in dealing with complex models such as those that describe the performance of large infrastructure systems. New developments in the theory and analysis of random processes have appeared that lend themselves particularly well to the performance analysis of infrastructure systems. At the same time, the increasing scrutiny of the financial performance of massive infrastructure projects, and the socio-technical aspects of project execution and operation [19, 20], has demanded advanced reliability models to support both public and private investment decisions.

2.4 How do Systems Fail?

Before describing a mathematical framework for reliability analysis, it is valuable to develop a simple conceptual model for the "natural history" of an engineered object (e.g., a bridge, an electronic circuit, or an engine). A newly produced engineered object is imbued during manufacture with an initial physical capacity/resistance, commonly referred to as the *nominal life* of the object. Nominal life is measured in terms of a physical quantity (or indeed, a *vector* of physical quantities) whose units will be referred to as "life units." Because of variations in materials, manufacturing or construction processes, etc., the nominal life of an engineered object is taken to be a random quantity. In the object's operating environment, the physical capacity/resistance of the object declines through the process of *degradation*; see Chap. 4. Conceptually, degradation is the process of stripping out life units from the object over time; it can be described mathematically as a random process that measures the life units removed from the object over time. At any point in time, the *remaining life* (or *remaining capacity/resistance*)[1] is defined to be the difference between the nominal life and the accumulated degradation up to that time. When the remaining life declines to zero (or reaches a minimum performance threshold), the object fails. The time at which failure occurs is referred to as the *lifetime* of the system.

Two examples, shown in Fig. 2.1, illustrate the concepts of nominal life and deterioration. The first example considers an incandescent lightbulb. The lightbulb contains a filament that converts electrical energy into light energy (photons) and heat. Over time, the heat causes a reduction in the material of the filament (tungsten atoms detach from the filament), and eventually, when the filament becomes thin enough, detachment results in the loss of the electrical conductivity, and the bulb fails. In this example, we use the initial width of the filament as the nominal life of the bulb. Deterioration is the process by which the width of the filament is reduced during use. The second example is the case of a bridge located in a seismic region. In this example, nominal life is measured in terms of the bridge's initial structural capacity (e.g., stiffness). Degradation can then be described by two mechanisms, one related to the weakening of the structure due to steady wear out over time, the second related to sudden decreases in capacity as a result of earthquakes of various magnitudes. As a result of these two phenomena, the structural capacity is reduced over time until it reaches a minimum performance threshold that defines system failure.

This characterization of failure is useful for several reasons. It suggests that in most systems there are two distinct and independent factors that determine system

[1] Throughout the book the terms "remaining life" and "remaining capacity/resistance" will be used interchangeably.

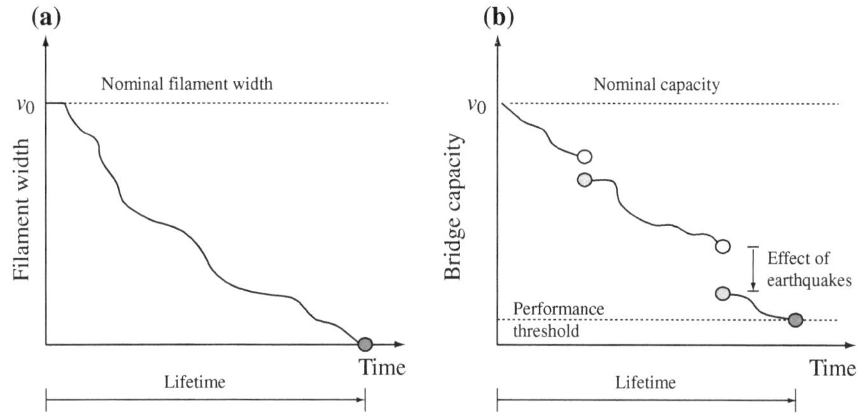

Fig. 2.1 Sample path of degradation of two systems over time: **a** the filament thickness of a light bulb; and **b** a bridge structural capacity

lifetime, namely the manufacturing or construction process that establishes nominal life (resistance/capacity), and the operating conditions and environmental processes that govern deterioration. Thus it is natural to study these factors separately. There are a variety of tools for studying the distribution of nominal life (resistance/capacity) via manufacturing process variability (e.g., quality control) and static design reliability estimation (see for instance Sect. 2.8). On the other hand, studying system degradation involves generally more elaborated stochastic process models such as Markov chains, Brownian motion, compound Poisson processes, and Lévy processes; some of them will be discussed in Chaps. 3–5 and 7. The characterization of failure in terms of these two factors (manufacturing and environment) is central to evaluating the system lifetime as a random variable.

2.5 The Concept of Reliability

The widely used and general accepted definition of reliability, and one which will be adopted in this book, is the following:

> The reliability of a system[2] is the likelihood that it will perform its required functions under stated conditions for a specified period of time.

Note that for any given situation, it is necessary to define exactly what is understood by the terms used above. Thus, unavoidably, engineering judgement is required in defining essential concepts such as "required functions," "stated conditions," and "specified period of time"; these make up the *mission* of the system. Furthermore,

---

[2]In this book, we will use also the terms *system, device or component* as the object of a reliability study. Most of the concepts and theory presented here are applicable to a wide range of objects, therefore, the term *system* is used as a general description of the object of study.

the notion that the system "performs its required functions" suggests the need to distinguish clearly between two possible system operating states, namely "satisfactory" and "not satisfactory" (i.e., failed).

The definition of reliability presented above also introduces the need to measure a "likelihood," and hence, it rests on the mathematical foundations of probability theory as the means by which reliability is characterized. Taking the system's *lifetime* to be its operating time, the definition of reliability above can be rephrased as follows:

> The reliability of a system is the probability that the system's lifetime exceeds a specific period of time (e.g., its mission time).

Finally, in terms of a system's performance indicator, an alternate (equivalent) definition of reliability is:

> The reliability of a system is the probability that the system's performance indicator remains above a predefined threshold within a specific period of time.

In the definition of reliability based on a system performance indicator (e.g., resistance measure), the threshold is the minimum value above which the system is deemed to operate successfully. This threshold is a very important concept in engineering design and is frequently referred to as the *limit state*: the value of a performance measure below which a system fails to perform its function satisfactorily.

The limit state concept has been used extensively as a design and operating criterion in mechanical problems and especially in various civil engineering fields such as soil mechanics, pavements, and structures. Although different limit states can be defined, there are two of particular importance, which will be used throughout this book: *ultimate* and *serviceability* limit states. Ultimate limit states describe the system's condition beyond which its operation is unacceptable, for instance partial or total structural instability, structural collapse, attainment of the maximum resistance (for some components or the entire system) or unacceptable deterioration. On the other hand, serviceability limit states allow for the system to perform below the expectations but without failure, for instance, excessive deformations, vibration or noise, or esthetic degradation.

2.6 Risk and Reliability

Reliability is often associated with the terms "risk" and "risk analysis"; nevertheless, risk and reliability are different concepts. The field of risk analysis differs from reliability engineering in that it takes a broader approach to threats and their consequences. Risk analysis is a process of collecting evidence of possible unwanted future scenarios (consequences of detrimental outcomes) throughout the system's life cycle; therefore, both qualitative and quantitative analysis are important. The results from reliability analysis can be used as evidence in risk analysis. In risk analysis, aspects such as the socioeconomic evaluation of consequences, communication, management, and policy are very important. Frequently, probabilistic risk analysis (PRA), which is commonly taken as a systematic evaluation of the likelihoods of some consequences, is seen as subsumed in reliability analysis. However, although they might sometimes look similar, there are some important differences in the fundamentals of both approaches.

In this book, we will mention the term risk marginally; our focus is only on the theoretical aspects of reliability, as described in the following sections. Further reading on the conceptual aspects of risk analysis and its relationship with reliability can be found in [7, 21, 22]

2.7 Overview of Reliability Methods

Although there are many ways of approaching reliability, the selection of any strategy cannot be detached from the decision problem. This means that the analysis should balance relevance and precision so that the results become meaningful evidence for the decision. The selection of the approach that best suits the decision problem depends on the knowledge and understanding of the performance of the system, as well as on aspects such as the availability and quality of information, and the resources available.

The traditional way to classify reliability methods groups them in four levels based on the extent of information that is used [3, 5]. Thus, *level* I methods use one characteristic value of each uncertain parameter. It is basically a non-probabilistic approach and a generalized version of the *safety factor* commonly used in engineering design. In *level* II methods, random variables are described by two parameters (e.g., mean and variance), and they are usually assumed to be normally distributed. Furthermore, in these models the reliability problem is described by a simple limit state function. The reliability index presented in Sect. 2.8.1 is a case in point. The third category, *level* III methods, focuses on estimating the probability of failure, which requires information about the joint distribution of all uncertain parameters. This level also includes system reliability problems and transient (time-dependent models) analysis. Finally, *level* IV methods combine reliability models with information about the context, for example, cost-benefit analysis, life-cycle cost analysis, failure consequences, operation policies (maintenance and intervention strategies) and so on. Within this context, most of this book is about *level* IV reliability methods.

2.8 Traditional Structural Reliability Assessment

2.8.1 Basic Formulation

It is quite common in the civil engineering literature (cf. [3, 5, 8, 23]) to assess structural reliability in a static sense by comparing the (random) capacity/resistance (strength) of the system to the load/demand (stress) placed on the system. In the literature, this approach, also termed *interference theory* [24] or the *basic reliability problem* [5], is most useful during the design phase, when physical models for determining the system capacity may be available.

In this case, the system is deemed to fail when the demand (e.g., load) exceeds the capacity (e.g., resistance) of the system. Thus, if we define a random variable C to be the capacity (with density f_C) and D to be the demand of the system (with density f_D), the limit state in this formulation is $C - D = 0$, where $C - D$ is the so-called *safety margin*. By definition, the reliability R of the system is given by

$$R = P(C > D) = P(C - D > 0) \tag{2.1}$$

If we further assume that C and D are independent and nonnegative random variables; then,

$$R = \int_{-\infty}^{\infty} f_D(x)\left(\int_{x}^{\infty} f_C(y)dy\right)dx, \tag{2.2}$$

which can also be written as

$$R = \int_{-\infty}^{\infty} f_D(x)[1 - F_C(x)]dx = \int_{-\infty}^{\infty} F_D(y) f_C(y)dy \tag{2.3}$$

Example 2.1 Consider a system subjected to a demand, which is assumed to be log-normally distributed. Three demand cases are considered. The density of all three possible demand functions (with the same mean but different COV) and the distribution of the resistance are shown in Fig. 2.2. The system's capacity (i.e., ability to accommodate the demand) is also assumed to follow a log-normal distribution with mean $\mu_C = 15$ and coefficient of variation $COV_C = 0.2$. Compute the reliability of the system.

The reliability of the system can be computed using Eq. 2.3:

$$R = \int_{-\infty}^{\infty} F_D(y) f_C(y)dy \tag{2.4}$$

For the particular case of lognormal demand and resistance, there is a close form solution; i.e.,

$$R = 1 - \Phi\left[-\frac{\ln\left(\frac{\mu_C}{\mu_D}\sqrt{\frac{1+COV_D^2}{1+COV_C^2}}\right)}{\sqrt{\ln[(1 + COV_D^2)(1 + COV_C^2)]}}\right] \tag{2.5}$$

where Φ is the normal standard distribution and $COV_{X_i} = \sigma_{X_i}/\mu_{X_i}$. Then, for the data used in this example, the reliability values for the three cases considered are: $R_{(\mathrm{COV}=0.1)} = 0.961$, $R_{(\mathrm{COV}=0.2)} = 0.926$, and $R_{(\mathrm{COV}=0.3)} = 0.89$. These results

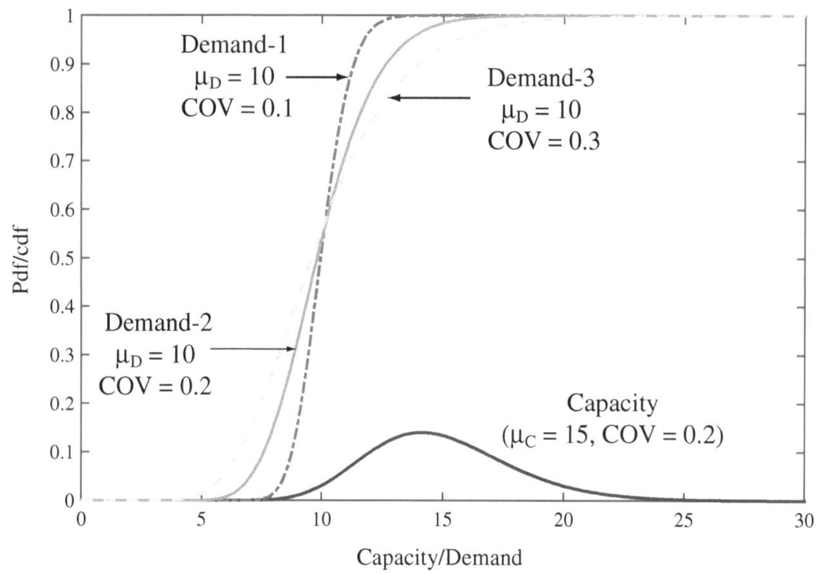

Fig. 2.2 Density function of the capacity and distribution function of the demand

show that larger variability implies larger failure probabilities and, therefore, smaller reliability values.

Let us now consider the special case where C and D in Eq. 2.3 are independent and normally distributed random variables. Let us further define $Z = C - D$, which is also normally distributed with parameters $\mu_Z = \mu_C - \mu_D$ and $\sigma_Z^2 = \sigma_C^2 + \sigma_D^2$; the density of Z is shown in Fig. 2.3. Then, the limit state can be defined as $Z = 0$. For this particular case, the reliability can be computed as:

$$R = \int_0^\infty f_Z(z)dz = 1 - \Phi\left(\frac{0 - \mu_Z}{\sigma_Z}\right) = 1 - \Phi(-\beta) \quad (2.6)$$

where $\beta = \mu_Z/\sigma_Z$ is called *safety* or *reliability index* [5]. The index β is a central concept in structural reliability. It is frequently used as a surrogate of failure probability and is widely used as a criteria for engineering design. For example, typical safety requirements for standard civil infrastructure (e.g., bridge design [25]) use $\beta \approx 3.5$–4.0 as an acceptable performance criteria [25].

2.8.2 Generalized Reliability Problem

Often, the formulation of the reliability problem (limit state) in terms of capacity, C, and demand, D, alone (Eq. 2.3) is not feasible, or it is incomplete because additional

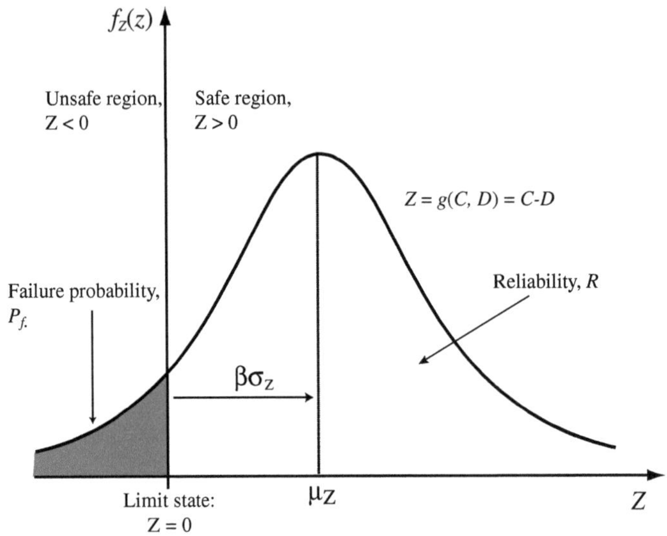

Fig. 2.3 Definition of the *reliability index* for the case of two normal random variables

information needs to be considered. In these cases, it may be of interest to describe the reliability problem in terms of a set of basic variables: $\mathbf{X} = \{X_1, X_2, \ldots, X_n\}$. In this n-dimensional variable space, the limit state $g(\mathbf{X}) = 0$ separates the safe $(g(\mathbf{X}) > 0)$ and failure $(g(\mathbf{X}) \leqslant 0)$ regions. The function $g(\mathbf{X}) = 0$ is a measure of a specific system performance condition based on a set of random variables $\mathbf{X}$ and other parameters that are not random.

Thus, a general form of Eq. 2.3 can be written as,

$$R = P(g(\mathbf{X}) > 0) = \int \cdots \int_{g(\mathbf{X})>0} f_{\mathbf{X}}(\mathbf{x}) d\mathbf{x} \tag{2.7}$$

where $f_{\mathbf{X}}(\mathbf{x})$ is the joint probability density function of the n-dimensional vector $\mathbf{X}$ of basic variables. Note that neither the resistance nor the demand are explicitly mentioned in this formulation. Equation 2.7 is usually referred to as the *generalized reliability problem* [5].

The solution of Eq. 2.7 is not always an easy task. For instance, there may be a large number of variables involved, the limit state function may not be explicit (i.e., it cannot be described by a single equation), or the solution cannot be found either analytically or numerically. Then, several alternative approaches have been proposed to solve Eq. 2.7; they can be grouped in:

- analytical solutions (e.g., direct integration) or numerical methods;
- simulation methods (e.g., Monte Carlo); or
- approximate methods (e.g., FORM/SORM)

Solving Eq. 2.7 by direct integration or through numerical methods is possible using specialized software such as Matlab®, Mathcad®, or Mathematica®. However, in most cases, this is only possible for simple mechanical problems with few variables and known probability distributions. Therefore, alternative approaches such as simulation and approximate methods have been proposed; they will be briefly discussed in the following subsections.

2.8.3 Simulation

As problems become complex, simulation appears as a good option to estimate reliability. Consider a system whose performance is defined by a set of random variables $\mathbf{X} = \{X_1, X_2, \ldots, X_n\}$ with joint probability density function $f_{\mathbf{X}}(\mathbf{x})$. Let us define an indicator function $I[\cdot]$ such that $I[\mathbf{x}] = 0$ for $g(\mathbf{X}) \leq 0$ (failure) and $I[\mathbf{x}] = 1$ for $g(\mathbf{X}) > 0$ (not failure). Then, the reliability can be estimated as the expected value of the indicator function; this is,

$$R = \int \cdots \int I[\mathbf{x}] f_{\mathbf{X}}(\mathbf{x}) d\mathbf{x} \tag{2.8}$$

The unbiased estimator of the reliability is:

$$R \approx \frac{1}{N} \sum_{i=1}^{N} I[\mathbf{x}] = \frac{N_F(g(\mathbf{x}) > 0)}{N} \tag{2.9}$$

where N is the number of simulations and $N_F(g(\mathbf{x}) > 0)$ is the number of cases in which the system has not failed.

Although simulation is a very valuable tool, it should be used with care. For instance, an aspect that requires special attention is the case of correlated variables. For correlated normal random variables, methods such as the Cholesky decomposition can be used [8, 23]; for arbitrary correlated variables, there are other methods available; e.g., see [5, 26]. Furthermore, defining the number of simulations necessary to obtain a dependable solution is also a difficult task. It clearly depends on the actual result; for example, if the failure probability is estimated to be about 10^{-4}, the number of simulations required should be larger than 10^4. Although several statistical models have been proposed to select the number of simulations [8]; the best approach consists of drawing the expected value and the variance of the result as function of the number of simulations; in this case, the solution is reached at convergence.

Clearly the computational cost of simulation is a central issue. The computational cost grows with the number of variables and the complexity of the limit state function. Then, in order to reduce the number of simulations several *variance reduction* techniques have beenproposed. Among the most used are importance sampling, directional simulation, the use of antithetic variables and stratified sampling [5, 27]. Recently, due to the sustained growth of computational capabilities, enhanced simulation methods have gained momentum. Some examples are subset simulation [28, 29], enhanced Monte Carlo simulation [30], methods that use a surrogate of the limit state function based on polynomial chaos expansions and kriging [31, 32], and statistical learning techniques [33].

2.8.4 Approximate Methods

There are some widely used methods to approximate the solution of Eq. 2.7 out of which the most popular is called *First-Order Second Moment* (FOSM) approach. In this case, the information about the distribution of the variables is discarded and only the first two moments are considered. When the information about the distributions is retained and included in the analysis, this method changes the name to *Advanced First-Order Second Moment* (AFOSM). In these case, the limit state, i.e., $g(\cdot) = 0$ is approached using Taylor series facilitating the evaluation. When the method uses a first-order approximation, the method is called *First-Order Reliability Method* (FORM); and when it is based on a second-order approximation it is referred to as *Second-Order Reliability Method* (SORM). Both FORM and SORM are widely used in practical engineering problems [5, 34].

Both FORM and SORM are carried out in the standard or normalized variable space (i.e., $U_i = (X_i - \mu_{X_i})/\sigma_{X_i}$). In FORM, the reliability index, β (see Sect. 2.8.1), is calculated as the minimum distance from the origin to the first-order approximation (using Taylor series) of limit state function [5] (Fig. 2.4). Then, FORM consists on solving the following optimization problem:

$$\begin{aligned} &\text{Minimize } \sqrt{\mathbf{U} \cdot \mathbf{U}^T} \\ &\text{subject to } g(X_1, X_2, \ldots, X_n) = 0 \end{aligned} \tag{2.10}$$

where $\mathbf{X} = \{X_1, X_2, \ldots, X_n\}$ defines the space of the original variables; and $\mathbf{U} = \{U_1, U_2, \ldots, U_n\}$ is the set of normalized independent variables.

Frequently, the limit state function is not linear. In this cases, FORM can be used only to approximate the solution and the quality of the results depends on the nonlinearity of the limit state function g (Fig. 2.4); i.e., as g becomes highly nonlineal the FORM approximation is less accurate. SORM is an alternative to deal with this problem since it uses a second-order approximation to the limit state function; however, the mathematical complexity of the solution increases significantly for high-dimensional variable problems. Another important difficulty of this approach arises when the random variables are not normally distributed. In this case, FORM cannot be applied directly. To manage this problem, Fiessler and Rackwitz [35] proposed a solution that approximate the tail of nonnormal distributions to normal distributions; this method has been used widely used with rather good results.

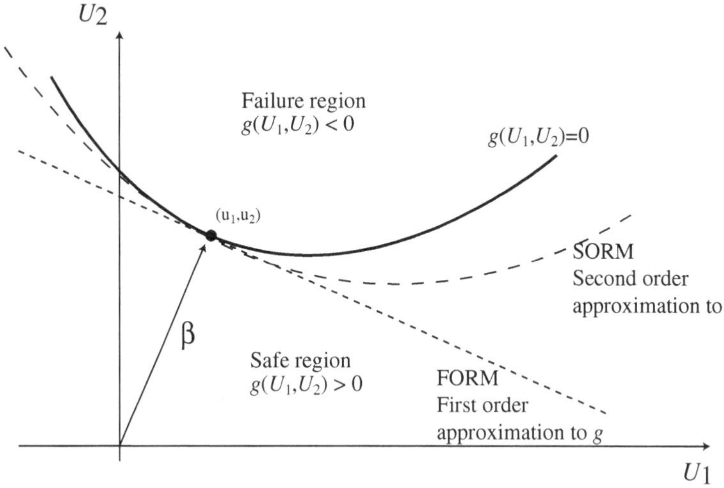

Fig. 2.4 Definition of the *reliability index* as the distance to the limit state function for the case of two random variables

The details of these methods are beyond the scope of this book and have been widely discussed elsewhere; e.g., [3, 5, 8, 23, 36].

2.9 Notation and Reliability Measures for Nonrepairable Systems

The static approach shown above lends itself very well for design studies and when the mission length of the system is fixed in advance. However, the primary focus of this book is on systems that evolve over time and which have an indeterminate mission length. Thus, it is important to distinguish between systems that are nonrepairable (that is, they are abandoned after a failure occurs), and systems that can be maintained operational through some external actions. In the latter, the system may experience a sequence of failures, repairs, replacements, and other maintenance activities.

The purpose of this section is to introduce the notation and basic notions of reliability that will be used later on in the book. Initially, we consider the case of a system that terminates upon failure, but in the later sections, we will extend this framework to include repairable systems. For these systems, we require a somewhat more general (although completely consistent) approach. These definitions are all quite standard and can be found in many reliability texts; e.g., [1, 2, 37–39].

2.9.1 Lifetime Random Variable and the Reliability Function

The study of reliability revolves around the idea that the time at which a system fails cannot be predicted with certainty. We define the *lifetime*, or *time to failure* (these are equivalent concepts) as a nonnegative random variable L, measured in units of time and described by its cumulative distribution function:

$$F_L(t) = P(L \leq t), \qquad t \in [0, \infty] \tag{2.11}$$

We will typically assume that the lifetime is continuous, and thus has density f_L, where

$$f_L(t) = \frac{dF_L(t)}{dt}. \tag{2.12}$$

When the context is clear, we will drop the subscript and refer to the distribution function of the lifetime simply as F; with density f.

The reliability of the system at time t, $R(t)$, is defined as the probability that the system is operational at time t; i.e.,

$$R(t) = P(L > t) = 1 - F(t) = \overline{F}(t) \tag{2.13}$$

Clearly, the reliability function $R(\cdot)$ is simply the complement of the distribution function of the lifetime evaluated at time t. Also known as the *survivor function*, $R(t)$ represents the probability that the system operates satisfactorily up to time t. Then, it follows that

$$R(t) = 1 - \int_0^t f(\tau)d\tau = \int_t^{\infty} f(\tau)d\tau \tag{2.14}$$

and the density of the time to failure can be expressed in terms of the reliability as:

$$f(t) = -\frac{d}{dt}R(t) \tag{2.15}$$

2.9.2 Expected Lifetime (Mean Time to Failure)

The mean system lifetime (also known as mean time to failure or MTTF) is simply the expectation of L; i.e.,

$$\mathbb{E}[L] = MTTF = \int_0^{\infty} \tau f(\tau)d\tau. \tag{2.16}$$

Because the lifetime is a nonnegative random variable, the *MTTF* can be expressed (using integration by parts) in terms of the reliability function as

$$MTTF = \int_0^\infty R(\tau)d\tau. \tag{2.17}$$

2.9.3 Hazard Function: Definition and Interpretation

The (unconditional) probability of failure of a device in the time interval $[t_1, t_2]$ is given by $F(t_2) - F(t_1)$ (or $R(t_1) - R(t_2)$). To compute the (conditional) probability of failure of a device in a certain time interval *given* that the device is working at the beginning of the time interval involves the concept of the *hazard function*, also called the *hazard rate*, $h(t)$. The hazard function can be interpreted as the instantaneous failure rate (i.e., failure in the next small instant of time) of a system of age t; in terms of conditional probability

$$h(t)\Delta t \approx P(L \le t + \Delta t \mid L > t), \tag{2.18}$$

for small values of Δt. Therefore, the hazard function $h(t)$ is defined by

$$\begin{aligned} h(t) &= \lim_{\Delta t \to 0} \frac{P(L \le t + \Delta t | L > t)}{\Delta t} \\ &= \lim_{\Delta t \to 0} \frac{P(t < L \le t + \Delta t)}{\Delta t \; P(L > t)} \\ &= \frac{f(t)}{R(t)} \end{aligned} \tag{2.19}$$

Consequently, the cumulative hazard function, denoted by Λ, is defined by:

$$\Lambda(t) = \int_0^t h(s)ds. \tag{2.20}$$

It is easy to show that [6]

$$\Lambda(t) = -\ln\{R(t)\}, \tag{2.21}$$

or put differently,

$$R(t) = \exp\left\{-\int_0^t h(s)ds\right\} = \exp\{-\Lambda(t)\}. \tag{2.22}$$

This relationship establishes the link between the cumulative hazard function, i.e., $\Lambda(t)$, and the reliability function. Inserting Eq. 2.22 in 2.19 and solving for $f(t)$, we can also obtain an expression for the lifetime density in terms of the hazard function:

$$f(t) = h(t)\exp\{-\Lambda(t)\}. \tag{2.23}$$

A constant hazard function ($h(t) \equiv \lambda$ for all t and some $\lambda > 0$) holds if and only if the lifetime L has an exponential distribution with parameter $\lambda > 0$; i.e.,

$$f(t) = \lambda e^{-\lambda t} \tag{2.24}$$

and the reliability function can be expressed as

$$R(t) = e^{-\lambda t} \tag{2.25}$$

Exponentially distributed lifetimes have the "memoryless" property; that is, failures are neither more likely early in a system's life nor late in a system's life, but are in some sense "completely" random.

The hazard function has been used to study the performance of a wide variety of devices [6]. Generally, the hazard function will vary over the life cycle of the system, particularly as the system ages. A conceptual description of the hazard function that proves useful for some engineered systems is the so-called "bathtub" curve shown in Fig. 2.5.

The bathtub curve proposes an early phase, characterized by a decreasing hazard function (i.e., DFR), that reflects early failures due to manufacturing quality or design defects. This phase is commonly termed the *infant mortality* phase and is followed by a period of constant hazard, where failures are due to random external factors,

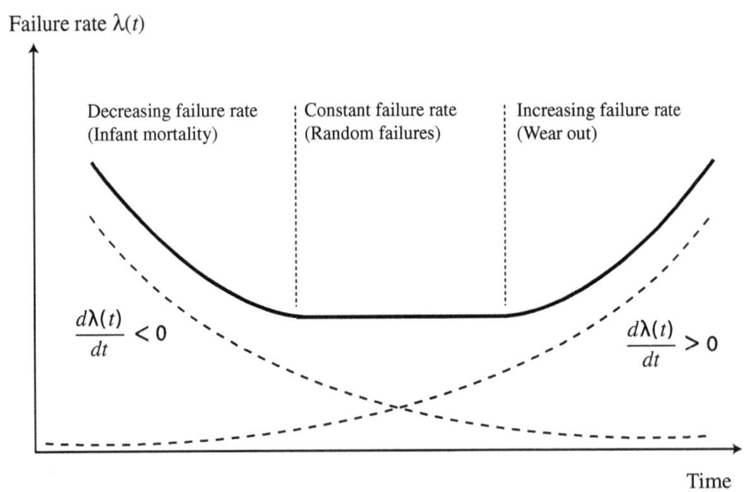

Fig. 2.5 Time-dependent failure rate: the *bathtube* curve

such as high vibrations, over-stresses, unexpected changes in temperature, and other extreme conditions. Finally, if units from the population remain in use long enough, the failure rate begins to increase as materials wear out and degradation failures occur at an ever increasing rate (i.e., IFR); this is known as the *wear out failure* period. Wear out is the result of aging due to, for instance, fatigue or depletion of materials (such as lubrication depletion in bearings).

Despite the fact that the bathtub curve is presented and discussed in almost all reliability books, some caveats on its practical applicability are in order. Its use as a conceptual device may be appropriate for some product *populations*, and in particular, the decreasing hazard part of the curve corresponds to the elimination through failure of relatively weaker members of the population (i.e., those of poor quality). There has been little published empirical evidence for the bathtub curve as a general model for the hazard function over a product's life, and a number of authors [40–42] have cautioned against its indiscriminate use in practice.

Statistical information about failure rates is usually fitted to a probability model. The numerical methods used for this purpose can be found elsewhere [4, 6, 23].

2.9.4 Conditional Remaining Lifetime

Another important concept in reliability analysis is the conditional remaining life distribution $H(t|x)$, defined as follows (Fig. 2.6):

$$H(t|x) = P(L \leq x + t | L > x) = \frac{F(x+t) - F(x)}{1 - F(x)}, \qquad t, x \geq 0 \tag{2.26}$$

where L is the time to failure with distribution $F(t)$, and $H(t|x)$ is a conditional distribution, which can be interpreted as the distribution of the remaining life of a system of age x. If L is continuous, with density f, the conditional remaining life density is given by

$$h(t|x) = \frac{f(x+t)}{1 - F(x)}, \tag{2.27}$$

which is basically the density function of the time to failure truncated in x. The mean of this distribution gives the conditional expected remaining life $\mathbb{E}[L|x]$ of a system of age x:

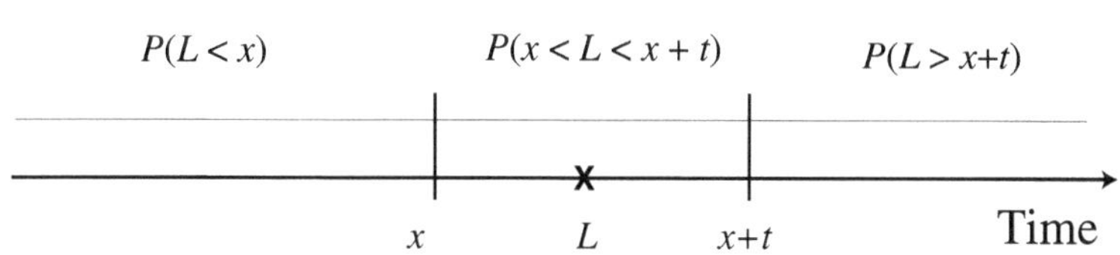

Fig. 2.6 Conditional remaining life

$$\mathbb{E}[L|x] = \mathbb{E}[L - x|L > x] = \int_0^\infty (1 - H(\tau|x))d\tau = \int_0^\infty \tau h(\tau|x)d\tau, \quad (2.28)$$

where the last equality holds if the lifetime distribution is continuous.

Example 2.2 According to field reports, the mean time to failure of a specific type of component was found to be $\mu = 12$. Because there is not clear information about the distribution of the time to failure, it is required to compute the basic reliability quantities for the following three distributions: lognormal (mean $\mu = 12$ and COV $= 0.25$), uniform [43, 44], and exponential with $\lambda = 1/12$.

Equation 2.19 was used to evaluate the hazard rate for the three distributions; the results are shown in Fig. 2.7. Note that for the particular and important case of the exponential distribution:

$$h(t) = \frac{f(t)}{1 - F(t)} = \frac{\lambda \cdot \exp(-\lambda t)}{\exp(-\lambda t)} = \lambda = \frac{1}{12}. \quad (2.29)$$

which is time-independent and reflects the memoryless property of the exponential distribution. The corresponding reliability functions were evaluated using Eq. 2.25, for $T_0 = 0$, the results are presented in Fig. 2.7b.

On the other hand, the conditional survival probability density (Eq. 2.27) for a value of $x = 3$ is shown in Fig. 2.8a. Note that the x-axis represents the time t after $x = 3$; for instance $h(t = 5|x = 3)$ means the density at a time $t = 8$. Finally, the evolution of the conditioned survival density function, for various x and for the lognormal case only, is presented in Fig. 2.8b. It can be observed that larger values of x shift the function to the left. This is caused by the fact that as x becomes larger, $1 - F(x)$ becomes smaller.

2.9.5 Commonly Used Lifetime Distributions

Among the most commonly used distribution functions in reliability and survival analysis are the exponential (described above), Weibull, lognormal, and gamma (although this list is by no means complete; for a more comprehensive list see) [45]. These distributions can be represented as special cases of the *generalized gamma* family. The generalized gamma is a three-parameter distribution; its density and cumulative distribution functions are given below [45]:

$$f(t; \theta, \beta, \kappa) = \frac{\beta}{\Gamma(\kappa)\theta} \left(\frac{t}{\theta}\right)^{\kappa\beta - 1} e^{-(t/\theta)^\beta}, \qquad t > 0 \quad (2.30)$$

$$F(t; \theta, \beta, \kappa) = \Gamma_1 \left[\left(\frac{t}{\theta}\right)^\beta ; \kappa \right]. \quad (2.31)$$

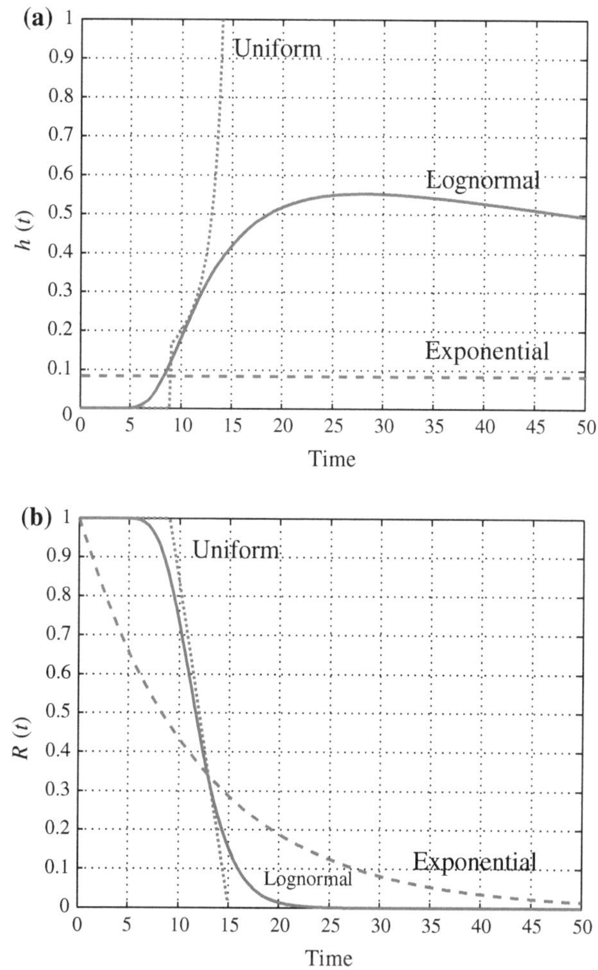

Fig. 2.7 **a** Failure rate and **b** reliability function for the three distributions

where $\theta > 0$ is a scale parameter, and $\beta > 0$ and $\kappa > 0$ are shape parameters; Γ is the gamma function and Γ_1 is the incomplete gamma function; i.e.,

$$\Gamma(\kappa) = \int_0^{\infty} z^{\kappa-1} e^{-z} dz, \quad z > 0 \tag{2.32}$$

$$\Gamma_1(z; \kappa) = \frac{\int_0^z y^{\kappa-1} e^{-y} dy}{\Gamma(\kappa)}, \quad z > 0. \tag{2.33}$$

Table 2.1 shows the parameter selection for the special cases of the generalized gamma mentioned above.

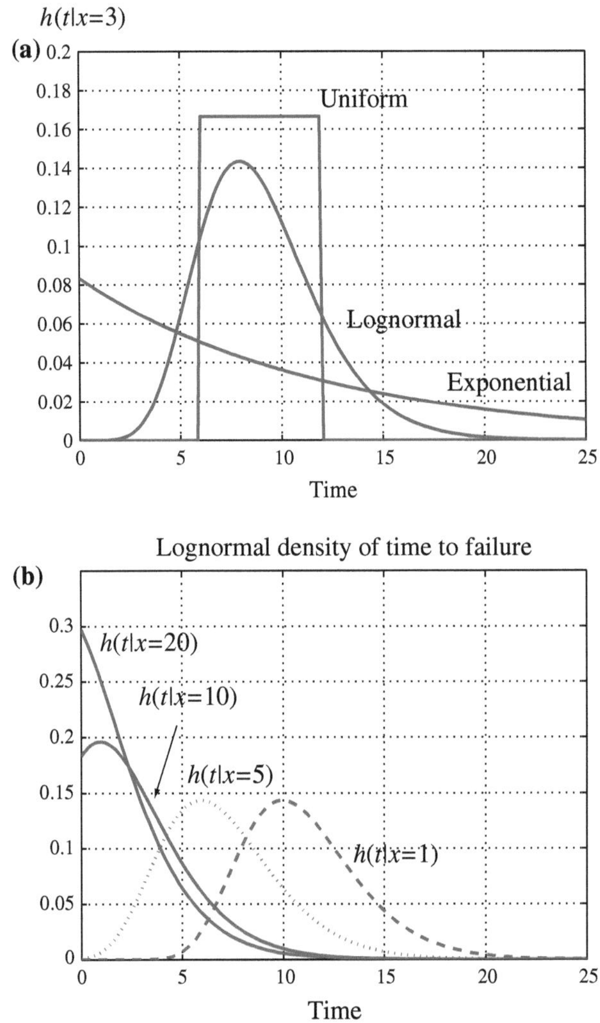

Fig. 2.8 Conditional density function for **a** $x = 3$ and all three failure time distributions; and **b** for $x = \{1, 5, 10, 20\}$ and the lognormal failure time distribution

2.9.6 Modeling Degradation to Predict System Lifetime

Based on the discussion in Sect. 2.4, L is realized when the degradation accumulated by the system meets or exceeds its nominal life (or more generally, the performance threshold or limit state); see Fig. 2.9.

To formalize this idea, let us define Y as a positive random variable that measures nominal capacity of a system (in physical units); i.e., initial capacity. Let us further define $V(t)$ to be a system performance indicator at time t; for example, the structural

Table 2.1 Special cases of the generalized gamma distribution [45]

Parameters of the generalized gamma	Distribution	$F(t)$
$\beta = 1$	Gamma (θ, κ)	$\Gamma_1\left(\frac{t}{\theta}; \kappa\right)$
$\kappa = 1$	Weibull $(\ln(\theta), 1/\beta)$	$1 - \exp\left(-\left(\frac{t}{\ln(\theta)}\right)^{1/\beta}\right)$
$\beta = 1; \kappa = 1$	Exponential (θ)	$1 - \exp(-\theta t)$
$\kappa \to \infty$	Lognormal	$\Phi\left(\frac{\ln(t) - \frac{\ln(\theta)+\ln(\kappa)}{\beta}}{1/(\beta\sqrt{\kappa})}\right)$

capacity of a bridge after t years. To allow generality, we will henceforth refer to $V(t)$ simply as "remaining capacity" of the system at time t and $D(t)$ as the *total* degradation by time t. Then, if the remaining capacity decreases over time as a result of the process of degradation, the random variable that describes the systems lifetime can be viewed as the length of time required for the remaining capacity to reach a threshold k^*, with $k^* \leq Y$. Therefore, for $t \geq 0$,

$$V(t) = \max(Y - D(t), k^*) \tag{2.34}$$

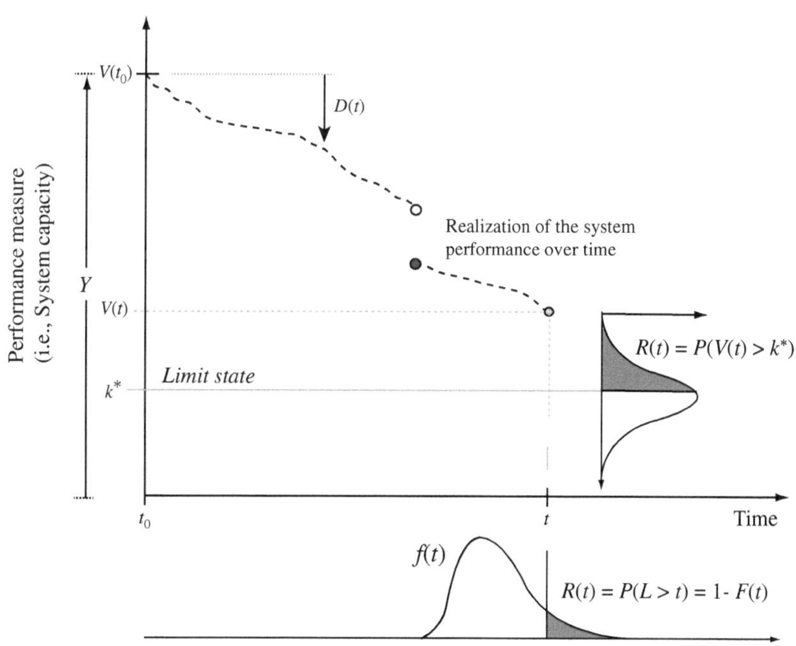

Fig. 2.9 Illustration of the definition of reliability

and

$$L = \inf\{t \geq 0 : V(t) \leq k^*\}, \tag{2.35}$$

or equivalently,

$$L = \inf\{t \geq 0 : D(t) \geq Y - k^*\}. \tag{2.36}$$

where k^* is the minimum performance threshold for the system to operate successfully; i.e., *limit state* (see Fig. 2.9). So we can interpret the device lifetime L as a first passage time of the total degradation process to a random threshold $Y - k^*$. As we mentioned earlier, this characterization allows, at least conceptually, for us to model the fact that random environmental effects "drive" system degradation. However, we should note at the outset that first passage problems are, in general, somewhat difficult to analyze for general degradation processes. The later chapters of this book will be devoted to these types of problems.

Note also that the relationship between reliability evaluated in terms of the system life, L, and as a static condition at a given point in time t is shown also in Fig. 2.9; this complementarity can be observed as well in Eqs. 2.35 and 2.36.

2.10 Notation and Reliability Measures for Repairable Systems

The previous section presented notation and reliability measures for systems consisting of a single lifetime; that is, systems that are abandoned upon failure. Most systems of interest, however, are not discarded (or replaced) upon failure, but rather made operational again by some type of maintenance or repair. Maintenance activities may be scheduled *prior* to failure as well (preventively), in an attempt to avoid failures at inopportune times (see Chap. 10). Repairable systems are studied with a variety of outcomes in mind, such as to minimize overall life-cycle costs, to develop effective inspection/maintenance strategies, to estimate warranty costs, and to decide when an aging system should be replaced (completely overhauled) rather than simply repaired. A sample path of a reparable system is shown in Fig. 2.10.

We will assume that failures render the system inoperable for a random amount of time during which the repair (or replacement) is made. In the simplest case, we might consider a sequence of successive lifetimes $\{L_1, L_2, \ldots\}$ and a sequence of repair times $\{R_1, R_2, \ldots\}$, where each lifetime is followed by a repair time.

Let us define the *system state* at time t, $Z(t)$, as operational ($Z(t) = 1$) or failed ($Z(t) = 0$); then we can define *point availability* $A(t)$ as the probability that the system is operational at time t. That is,

$$A(t) = P(Z(t) = 1) = P(V(t) > 0). \tag{2.37}$$

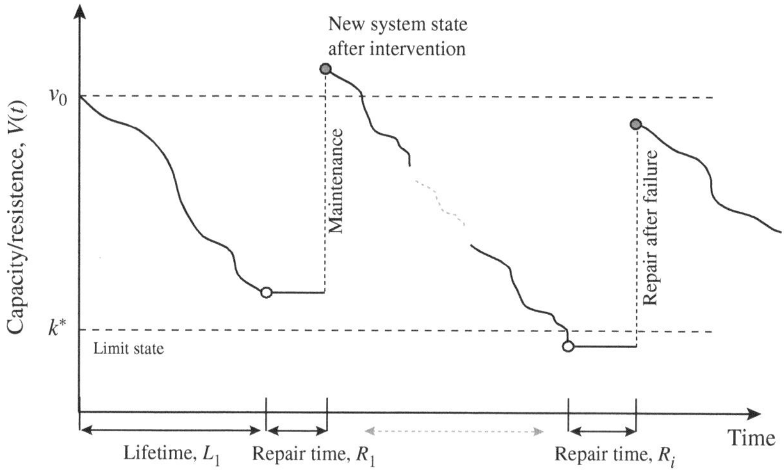

Fig. 2.10 Sample path of repairable system

Let us make note of the obvious—namely, that point availability is a time-dependent quantity that will typically depend on the *initial conditions*, that is, what is going on at the origin.

In addition to point availability, we will also be interested in the *limiting* availability A; i.e.,

$$A = \lim_{t \to \infty} A(t). \tag{2.38}$$

In order to work with limiting availability, we will first need to make sure that this quantity exists. For the models we will work with, the limiting availability will typically also be a *stationary* availability; that is, for certain initial conditions, the limiting availability will describe the time-dependent availability for all t. Later in the book, we will discuss the problem of availability in more detail. Moreover, we will make some assumptions about the probability laws associated with lifetimes and repair times in order to calculate availability.

2.11 Summary and Conclusions

Reliability, the probability that the system performs as conceived, is a key concept in the design and operation of any engineered system. In structures and infrastructure, reliability methods have been traditionally classified in four levels (I to IV) depending of their complexity when modeling uncertainty; and according to the type and extent of informationused in the analysis. Reliability models can be organized also based on the relevance of the information that they provide for the decision making process.

Overall decisions about the performance of the system use models based on failure observations. On the other hand, decisions about specific system components require models that carefully describe their performance in time. In this chapter, we discussed and presented existing models to manage these types of problems. Since the theoretical aspects presented here have been widely discussed elsewhere, the chapter is intended only as a conceptual summary of the main ideas and techniques behind reliability modeling.

References

1. R.E. Barlow, F. Proschan, *Mathematical theory of reliability* (Wiley, New York, 1965)
2. T.J. Aven, U. Jensen, *Stochastic Models in Reliability*. Series in Applications of Mathematics: Stochastic Modeling and Applied Probability, vol. 41 (Springer, New York, 1999)
3. H.O. Madsen, S. Krenk, N.C. Lind, *Methods of Structural Safety* (Prentice Hall, Englewood Cliffs, 1986)
4. J.R. Benjamin, C.A. Cornell, *Probability, Statistics, and Decisions for Civil Engineers* (McGraw Hill, New York, 1970)
5. R.E. Melchers, *Structural Reliability-Analysis and Prediction* (Ellis Horwood, Chichester, 1999)
6. E.E. Lewis, *Introduction to Reliability Engineering* (Wiley, New York, 1994)
7. M.G. Stewart, R.E. Melchers, *Probabilistic Risk Assessment of Engineering Systems* (Chapman & Hall, Suffolk, 1997)
8. A. Haldar, S. Mahadevan, *Probability, Reliability and Statistical Methods in Engineering Design* (Wiley, New York, 2000)
9. A.M. Freudenthal, The safety of structures. Trans. ASCE **112**, 125–180 (1947)
10. A.I. Johnson, *Strength, Safety and Economical Dimensions of Structures*, vol. 22 (Statens Kommitte for Byggnadsforskning, Meddelanden, Stockholm, 1953)
11. E. Basler, Analysis of structural safety. In *Proceedings of the ASCE Annual Convention*, Boston MA, June 1960
12. C.A. Cornell, Bounds on the reliability of structural systems. ASCE-J. Struct. Div. **93**, 171–200 (1967)
13. C.A. Cornell, Probability-based structural code. J. Am. Concr. Inst. (ACI) **66**(12), 974–985 (1969)
14. J. Ferry-Borges, Implementation of probabilistic safety concepts in international codes, *Proceedings of the International Conference on Structural Safety and Reliability* Verlag, Dusseldorf, Aug 1977, pp. 121–133
15. A. Pugsley, *The Safety of Structures* (Edward Arnold, London, 1966)
16. A.M. Hasofer, N.C. Lind, Exact and invariant second moment code format. ASCE J. Eng. Mech. Div. **100**, 111–121 (1974)
17. D. Veneziano, Contributions to second moment reliability theory. Research Report R-74-33, Department of Civil Engineering, MIT, Cambridge, MA, 1974
18. Canadian Standard Association (CSA), Standards for the design of cold-formed steel members in buildings. CSA-S-136, Canada, 1974
19. D. Paez-Pérez, M. Sánchez-Silva, A dynamic principal-agent framework for modeling the performance of infrastructure. Eur. J. Oper. Res. (2016) (in press)
20. D. Paez-Pérez, M. Sánchez-Silva, Modeling the complexity of performance of infrastructure (2016) (under review)

21. D.I. Blockley, *Engineering Safety* (McGraw Hill, New York, 1992)
22. T. Bedford, R. Cooke, *Probabilistic Risk Analysis: Foundations and Methods* (Cambridge University Press, Cambridge, 2001)
23. A.S. Nowak, K.R. Collins, *Reliability of Structures* (McGraw Hill, Boston, 2000)
24. K.C. Kapur, L.R. Lamberson, *Reliability in Engineering Design* (Wiley, New York, 1977)
25. M. Ghosn, B. Sivakumar, F. Moses, Infrastructure planning handbook: planning engineering and economics. NCHRP Report 683: Protocols for Collecting and Using Traffic Data in Bridge Design. National Academy Press (National Academy of Science), Washington, 2011
26. P-L. Liu, A. Der Kiuregian. Optimization algorithms for structural reliability analysis. Report UCB SESM-86 09, Department of Civil Engineering, University of California at Berkeley, 1986
27. S.M. Ross, *Simulation*, 4th edn. (Elsevier, Amsterdam, 2006)
28. S.K. Au, J. Beck, Estimation of small failure probabilities in high dimensions by subset simulation. Prob. Eng. Mech. **16**(4), 263–277 (2001)
29. S.K. Au, Reliability-based design sensitivity by efficient simulation. Comput. Struct. **83**, 1048–1061 (2005)
30. A. Naes, B.J. Leira, O. Batsevych, System reliability analysis by enhanced monte carlo simulation. Struct. Saf. **31**, 349–355 (2009)
31. B. Sudret, Global sensitivity analysis using polynomial chaos expansions. Reliab. Eng. Syst. Saf. **93**, 964–979 (2008)
32. B. Sudret, Meta-models for structural reliability and uncertainty quantification. In *Proceedings of the 5th Asian-Pacific Symposyum on Structural Reliability and its Applications—Sustainable infrastructures*, ed. by K.K. Phoon, M. Beer, S.T. Quek, S.D. Pang (Reserch Publishing, Chennai, 2012), Singapore, 23–25 May 2012
33. J.E. Hurtado, *Structural Reliability: Statistical Learning Perspectives* (Springer, New York, 2004)
34. A. Haldar, S. Mahadevan, *Reliability Assessment Using Stochastic Finite Element Analysis* (Wiley, New York, 2000)
35. R. Rackwitz, B. Fiessler, Structural reliability under combined random load sequences. Struct. Saf. **22**(1), 27–60 (1978)
36. M. Sánchez-Silva, Introducción a la confiabilidad y evaluacin de riesgos: teoría y aplicaciones en ingeniera. Segunda Edición (Ediciones Uniandes, Bogotá, 2010)
37. E. Çinlar, *Introduction to Stochastic Processes* (Prentice Hall, New Jersey, 1975)
38. M. Finkelstein, *Failure Rate Modeling for Risk and Reliability* (Springer, New York, 2008)
39. I.B. Gerstbakh, *Reliability Theory with Applications to Preventive Maintenance* (Springer, New York, 2000)
40. G.-A. Klutke, P.C. Kiessler, M.A. Wortman, A critical look at the bathtube curve. IEEE Trans. Reliab. **52**(1), 125–129 (2003)
41. D. Kececioglu, F. Sun, *Environmental Stress Screening: Its Quantification, Optimization, and Management* (Prentice Hall, New York, 1995)
42. W. Nelson, *Applied Life Data Analysis* (Wiley, New York, 1982)
43. A.H-S. Ang, W.H. Tang, *Probability Concepts in Engineering: Emphasis on Applications to Civil and Environmental Engineering* (Wiley, New York, 2007)
44. S. Asmussen, F. Avram, M.R. Pistorius, Russian and American put options under exponential phase-type levy models. Stoch. Process. Appl. **109**, 79–111 (2004)
45. W.Q. Meeker, L.A. Escobar, *Statistical Methods for Reliability Data* (Wiley, New York, 1998)

Chapter 3
Basics of Stochastic Processes, Point and Marked Point Processes

3.1 Introduction

The study of the dynamic performance of engineered systems subject to uncertainty requires the use of tools from stochastic processes. Although stochastic processes have been used extensively in many disciplines (e.g., see [1–4]), this chapter will focus on the the mathematical background that supports the models presented later in the book. The topics of stochastic processes presented in this chapter include definition of point processes, basic theorems, renewal theory, and regenerative processes. Not all theory about stochastic processes presented in this book is included in this chapter; some additional concepts and formalisms are presented and discussed in the following chapters when appropriate. This chapter is not intended as a comprehensive review, and several references are included for the reader to explore some of the topics in more detail.

3.2 Stochastic Processes

Stochastic processes are used in most modern engineering disciplines to model the dynamics of physical processes that evolve over time according to random phenomena. It is common in reliability and life-cycle engineering to model actual physical degradation as well as maintenance activities using stochastic processes. In this section we present a general definition and basic properties of stochastic processes, before providing specific degradation-related stochastic models in succeeding sections.

M. Sánchez-Silva and G.-A. Klutke, *Reliability and Life-Cycle Analysis of Deteriorating Systems*, Springer Series in Reliability Engineering,
DOI 10.1007/978-3-319-20946-3_3

3.2.1 Definition

Definition 1 A stochastic process is an indexed family of random variables $\mathscr{X} = \{X(t), t \in \Gamma\}$ all defined on a common probability space $(\Omega, \mathscr{F}, P)$.

The index set Γ may be countable, e.g., $\Gamma = \mathbb{N} = \{0, 1, 2, \ldots\}$, in which case the process is a *discrete parameter* process, or uncountable, e.g., $\Gamma = \mathbb{R}_+ = [0, \infty)$, in which case the process is a *continuous parameter* process. It is quite common, especially in engineering applications, to think of the index $t \in \Gamma$ representing time, and the random variable $X(t)$ to represent the *state* of the process at time t. The set in which the random values $X(t), t \in \Gamma$ take values is called the *state space* of the stochastic process. In engineering applications, we will always take the state space to be a Euclidean space.

A note on notation: we will generally use script characters as a concise way to describe the family of random variables (e.g., $\mathscr{X} = \{X(t), t \in \mathbb{R}\}$ or $\mathscr{T} = \{T_n, n \in \mathbb{N}\}$).

A *sample path* of a stochastic process is simply a realization of the process; that is, an observation of the entire sequence of random variables in the process for a given outcome (sample point). For example, if we let $X(t)$ be the number of customers present in a service system at time t, a sample path of the process $\mathscr{X} = \{X(t), t \in \mathbb{R}\}$ is shown in Fig. 3.1; note that here we label the vertical axis as $X(t; \omega)$ to remind the reader that the values are for the particular sample point ω.

In order to employ stochastic processes to make predictions, we must build (or determine from assumptions) the *probability law* or equivalently, the *distribution* of the process (see Appendix). In its most general form, the probability law of a

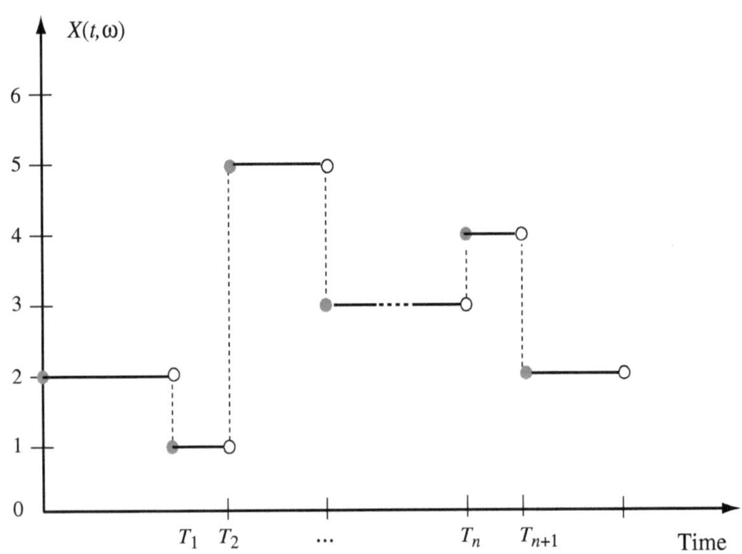

Fig. 3.1 Sample path of $\mathscr{X}$

stochastic process is determined by all possible *finite* joint probabilities of random variables of the process; that is, probabilities of the form

$$P(X(s_1) \in A_1, X(s_2) \in A_2, \ldots, X(s_k) \in A_k) \tag{3.1}$$

for any k and any $s_i \in \Gamma$ and $A_i \in \mathscr{F}$, with $i = 1, \ldots, k$.

A stochastic process is said to be *stationary* if its probability law is invariant to shifts along the time axis; that is, for all $k, \delta, s_1, \ldots, s_k$,

$$P(X(s_1) \in A_1, \ldots, X(s_k) \in A_k) = P(X(s_1+\delta) \in A_1, \ldots, X(s_k+\delta) \in A_k) \tag{3.2}$$

The joint probabilities in Eq. 3.1 allow us to evaluate (predict) any property of interest about the stochastic process such as marginal and conditional probabilities, as well as limiting distributions and properties such as stationarity. As one might imagine, determining the joint probabilities in (3.1) is no easy task. In order to achieve tractable results, we will generally need to make assumptions that simplify the structure of dependencies between the random variables of the process. While perhaps restricting their applicability, such assumptions will, however, lead to useful model and manageable properties that engineers can apply in a variety of complex settings.

3.2.2 Overview of the Models Presented in this Chapter

In this chapter we present an overview of stochastic processes that are relevant and frequently used in modeling degradation and failure. We first provide a very general, but appropriately formal, description of an important class of stochastic processes known as point processes (along with their associated counting processes) and the tools used to analyze them. We will then expand the underlying description of time-dynamics to include additional random information, leading to the idea of a marked (or compound) point process. In subsequent sections we discuss specific assumptions that lead to Poisson processes and renewal processes. These processes form the basis of important processes in modeling degradation and maintenance activities, namely compound Poisson processes and alternating renewal processes, which are presented in this chapter. Additional stochastic processes used in modeling degradation, namely Markov chains, gamma and Lévy processes, are discussed in Chaps. 5–7.

Our intention here is to provide the basic notation and mathematical framework for the models developed in succeeding chapters for degradation, failure, and repair. In our exposition, we wish not only to summarize the properties of these processes but also to provide some context for when particular models are appropriate or useful to describe degradation, failure, and repair. This section is not intended to be a comprehensive treatment of stochastic processes, and for additional background in stochastic processes the reader is highly recommended to visit the elementary texts of [3] or [4] or the more advanced research monographs of [5–9].

3.3 Point Processes and Counting Processes

Suppose we observe some (randomly occurring) phenomenon over time, e.g., the times at which a device or piece of equipment fails, or the arrivals of customers to a service station. As time goes on, we obtain a collection of points (a "point pattern") that denote occurrences of the phenomenon. Point processes are stochastic models that aim to characterize the probabilistic behavior of these point patterns.

Point process models are widely used in all domains of engineering (as well as many fields of science), in applications as varied as modeling electrical pulses, demands for products, traffic at a web site, security breaches at a port of entry, lightening strikes that may instigate wildfires, defects on a semiconductor wafer, etc. While we generally think of points evolving over time, we may also consider the distribution of points in some geographical space as well. In the field of reliability engineering, they are particularly relevant to modeling system failures over time, as well as modeling shocks that may cause damage to a system. Point processes are also embedded in more complicated stochastic processes, such as the times at which a stochastic process reaches a given threshold value, or point processes with associated "marks" or jump sizes at event occurrences.

3.3.1 *Simple Point Processes*

A point process describes a random distribution of points in a topological state space (which may represent time, two- or three-dimensional geographical space, or something more abstract). Typically, we think of the points representing the times of occurrences of a particular phenomenon or object of interest; if we start observing the processes at time 0, the state space is $\mathbb{R}_+$. Points may also represent the locations of factories that may produce airborne pollutants (state space $\mathbb{R}^2$ or the location of stars in a galaxy (state space $\mathbb{R}^3$). In this section, we limit ourselves to point processes on $\mathbb{R}_+$, and we will generally think of the points as the epochs of a specific phenomenon such as a failure or a repair, but it is important to keep in mind that point processes can model "spatial" processes as well.

In what follows, we will assume an underlying probability space $(\Omega, \mathscr{F}, P)$, as described in the Appendix (see section A.2). A point process has the following definition:

Definition 2 A simple point process $\mathscr{T} = \{T_n, n \in \mathbb{N}\}$ is an ordered sequence of nonnegative random variables $0 = T_0 < T_1 < \cdots$ denoting the locations (or times) of the points.

We make some simplifying assumptions to ensure that our point processes are well behaved. First, we will assume that points occur one at a time; that is, two or more occurrences cannot happen simultaneously. If this assumption holds, we say that the process is *orderly*, so that for any t, there is either one point at t or no points at t.

We will formalize this property in the Poisson process section. Further, we assume that any finite interval of time can contain only finitely many occurrences (so that $\sup_n T_n = \infty$).

A point process has an associated *counting process* that provides an equivalent characterization.

Definition 3 A counting process is a stochastic process $\mathcal{N} = \{N(t), t \geq 0\}$ on $0 \leq t < \infty$ with $N(0) = 0$ and $N(t) < \infty$ for each $t < \infty$, whose sample paths are piecewise constant, right continuous, and have jumps (at random times) of size 1.

The random variable $N(t) - N(s)$ for $s < t$ is called an *increment* of $\mathcal{N}$, and it counts the number of jumps of the process in the interval $(s, t]$. A counting process and its associated point process are related in the following way (Fig. 3.2):

$$N(t) = \max\{n \geq 0 : T_n \leq t\} = \sum_{n=1}^{\infty} 1_{\{T_n \leq t\}},$$

where 1_B is the indicator random variable, i.e.,

$$1_B(\omega) = \begin{cases} 1, & \text{if } \omega \in B \\ 0, & \text{if } \omega \notin B \end{cases}.$$

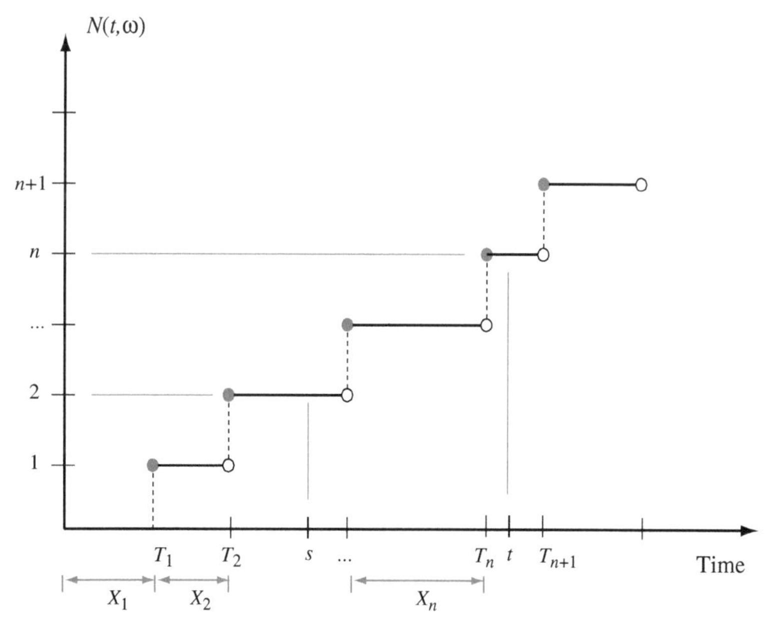

Fig. 3.2 Sample path of a counting process

It also follows that

$$\{N(t) \geq n\} = \{T_n \leq t\} \quad \text{and} \quad \{N(t) = n\} = \{T_n \leq t < T_{n+1}\}.$$

Figure 3.2 presents a typical sample path of a counting process; it includes the point process $\mathscr{T}$ and the inter-event time process $\mathscr{X}$.

A point process is typically characterized by its *(conditional) intensity function.* To define the conditional intensity function, we must introduce the concept of the "history" $\mathscr{H}(t)$ of a point process. Informally, by the history of a point process at time t, we mean information revealed by the process in $[0, t]$; that is, the realization of all random variables associated with the point process up to (and including) time t. Formally, we define the history (in terms of the counting process) as

$$\mathscr{H}(t) = \sigma\{N(s), 0 \leq s \leq t\}, \tag{3.3}$$

where σ denotes the smallest σ-algebra with respect to which the random variables under consideration are measurable (see Appendix A for further details, but for us, the informal description of the history will be adequate to explain the idea of the point process intensity).

Now the conditional intensity of a point process can be defined as follows:

Definition 4 The conditional intensity $\lambda(t|\mathscr{H}(t))$ of a point process is given by

$$\lambda(t|\mathscr{H}(t)) = \lim_{\delta \to 0} \frac{P(N(t+\delta) - N(t^-) = 1|\mathscr{H}(t^-))}{\delta} \tag{3.4}$$

The conditional intensity of the point process measures the likelihood that the process has a point "at" time t given the past pattern of points (the history) up to (but not including) time t.

The conditional intensity function is also called the *hazard function* or, in some cases, the *rate* of the point process. In general, it is a complicated stochastic process, because future points may depend in a very complex way on past points. In some special cases, however, it can be a constant (Poisson process), a deterministic function (nonhomogeneous Poisson process), or a random variable (renewal process).

Finally, we will often be interested in the *inter-event time process* process of a point process, denoted by $\mathscr{X} = \{X_n, n \geq 1\}$, where

$$X_1 = T_1,$$
$$X_n = T_n - T_{n-1}, \quad n = 2, 3, \ldots$$

Clearly, the event times determine the inter-event times, and vice versa; thus the inter-event time process gives us yet another way to characterize the point process. Since these three ways of characterizing the distribution of points in time are essentially equivalent (although clearly, each process has different properties), much of the literature refers to each of these processes colloquially as a *point process*.

3.3.2 Marked Point Processes

Beyond considering only the time of a random occurrence of events over time, in many situations we may be interested in capturing additional information about the occurrence. For instance, in models of shock degradation, we may think of shocks occurring at random times, each inflicting a random amount of damage on the system (see Fig. 3.3), so that we are interested in both the time of the shock and its magnitude. In a queueing context, we may think of an arrival to a service system bringing along a request for a random amount of service. We may handle such situations using a *marked point process*, which is defined as follows:

Definition 5 Let $\mathscr{T} = \{T_n, n = 0, 1, 2, \ldots\}$ be a point process, and let $\mathscr{M} = \{M_n, n = 0, 1, 2, \ldots\}$ be a sequence of random variables taking values in a mark space $\mathbb{M}$. Then a marked point process $\{(T_n, M_n), n = 0, 1, 2, \ldots\}$ is the ordered sequence consisting of the time points T_n and their associated marks M_n.

Depending on the context, we can think of the mark M_n as an additional description of the event occurring at time T_n, for example, as the size of the shock occurring at time T_n, or the repair cost associated with the failure occurring at time T_n. For marked point processes, we have to adjust our definition of the associated counting process to include information about the mark. We do this by defining a counting process for each subset $A \subset \mathbb{M}$ by

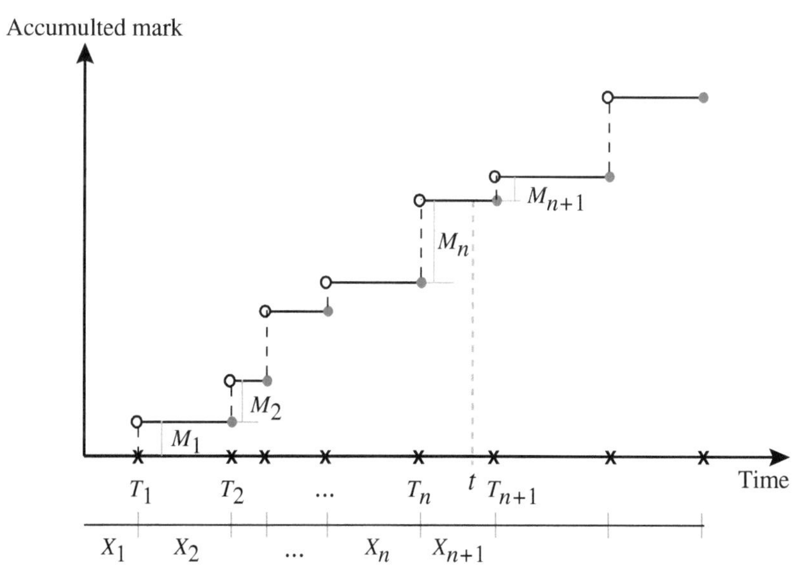

Fig. 3.3 Sample path of a marked point process

$$N^A(t) = \sum_{n=1}^{\infty} 1_{\{M_n \in A\}} 1_{\{T_n \le t\}} \tag{3.5}$$

Thus the counting process $\mathscr{N}^A = \{N^A(t), t \geq 0\}$ counts the number of points up to time t whose marks fall in the subset A, and we can think of a family of counting processes $\{\mathscr{N}^A, A \subset \mathbb{M}\}$ that conveys the same information as the marked point process $\{(T_n, M_n), n = 0, 1, 2, \ldots\}$.

Both simple and marked point processes are widely used in modeling device lifetimes and in models of maintained systems. As an example, suppose we model the degradation process as a marked point process where the times represent shocks that affect a system and the marks represent the amount of damage incurred (capacity/resistance or in general life units removed) at a shock. We might then model the shock process as a marked point process, and we would be interested in the time at which the amount of damage exceeds the nominal life. Or consider a maintained system, where occurrence times represent times of failures, repairs, or preventive replacements. A model for this system might involve a complicated point process. Before developing such models, we introduce the Poisson process (and its variants). We will see that the Poisson process is useful, but somewhat restrictive. We will next introduce the renewal process and present a particularly useful variant in maintenance modeling, the alternating renewal process.

3.4 Poisson Process

The Poisson process is one of the simplest and most widely used point processes in engineering applications. The Poisson process has been used to model arrivals to a service system (it plays a central role in the development of queueing theory), solar flares, radioactive decay, material flaws, accidents on a roadway, among many other phenomena.

The Poisson process can be defined equivalently in several different ways. We begin with a completely qualitative definition, from which the quantitative properties of the process can be derived. In fact, the qualitative and quantitative definitions are equivalent. We state most of the important properties of the Poisson process without proof; proofs and derivations are available in any standard textbook on stochastic processes (c.f. [3, 4]).

Definition 6 A Poisson process is a counting process $\mathscr{N}$ with the following properties:

(i) $N(0) = 0$.
(ii) Nonoverlapping increments are independent, i.e., for any $t, s \geq 0$, the distribution of $N(t+s) - N(t)$ is independent of $\{N(u), u \leq t\}$.

(iii) The process has *stationary* increments, i.e., the distribution of $N(t+s) - N(s)$ is the same, for all t and any $s \geq 0$.
(iv) The process is orderly, i.e., $\lim\limits_{h\to 0} \frac{P(N(h)>1)}{h} = 0$, or equivalently $P(N(h) > 1) = o(h)$.

To move from this completely *qualitative* definition of the Poisson process to a characterization of its probability law, first note that the assumptions that $\mathscr{N}$ has stationary, independent increments imply that

$$\begin{aligned} P(N(t+s)=0) &= P(N(s)=0, N(t+s)-N(s)=0) \\ &= P(N(s)=0)P(N(t+s)-N(s)=0) \\ &= P(N(s)=0)P(N(t)=0) \end{aligned}$$

As the exponential function is the only nonzero continuous function that satisfies this expression, we have

Lemma 7 *Let* $\{N(t), t \geq 0\}$ *be a counting process that has stationary, independent increments, and suppose that, for all* $t > 0$, *we have that* $0 < P(N(t) = 0) < 1$. *Then for any* $t \geq 0$,

$$P(N(t)=0) = e^{-\lambda t}$$

for some $\lambda > 0$.

This lemma and orderliness imply that for the Poisson process,

$$P(N(h)=0) = 1 - \lambda h + o(h),$$

and

$$P(N(h)=1) = \lambda h + o(h).$$

From this result we obtain the distribution of $N(t)$.

Theorem 8 *Let* $\{N(t), t \geq 0\}$ *be a Poisson process (as defined in Definition 6) with* $0 < P(N(t) = 0) < 1$, *for all* $t > 0$. *Then*

$$P(N(t)=n) = \frac{e^{-\lambda t}(\lambda t)^n}{n!}$$

for some $\lambda > 0$ *and all* $t \geq 0$.

Outline of the proof: using the properties above, we have

$$
\begin{aligned}
P(N(t+h)=n) &= \sum_{l=0}^{n} P(N(h)=l, N(t+h)-N(h)=n-l) \\
&= \sum_{l=0}^{n} P(N(h)=l)P(N(t)=n-l) \\
&= P(N(h)=0)P(N(t)=n) + P(N(h)=1)P(N(t)=n-1) \\
&\quad + \sum_{l=2}^{n} P(N(h)=l)P(N(t)=n-l) \\
&= (1-\lambda h + o(h))P(N(t)=n) + (\lambda h + o(s))P(N(h)=n-1) + o(h)
\end{aligned}
$$

From here, we can develop a differential equation for $P(N(t)=n)$ as follows:

$$
\begin{aligned}
\frac{dP(N(t)=n)}{dt} &= \lim_{h\to 0} \frac{P(N(t+h)=n) - P(N(t)=n)}{h} \\
&= \lim_{h\to 0} \frac{-\lambda h P(N(t)=n) + \lambda h P(N(t)=n-1) + o(h))}{h} \\
&= -\lambda P(N(t)=n) + \lambda P(N(t)=n-1),
\end{aligned}
$$

for $n = 1, 2, \ldots$. Coupled with the initial probability in Eq. 3.7, this system of equations can be solved recursively to yield Eq. 3.8.

Corollary 9 *The expectation of $N(t)$ is given by*

$$\mathbb{E}[N(t)] = \lambda t, \quad t \geq 0. \tag{3.6}$$

The parameter λ in the equation above is called the *rate* or *intensity* of the Poisson process; it is the conditional intensity defined in equation Eq. 3.4. In the case of the Poisson process, the conditioning history is irrelevent because of independent increments, and the conditional intensity is simply a deterministic constant. It will also be useful for what follows to note that $\mathbb{E}[N(t)]$ can be written as

$$\mathbb{E}[N(t)] = \int_0^t \lambda du. \tag{3.7}$$

3.4.1 Inter-event Times and Event Epochs of the Poisson Process

Let $\{N(t), t \geq 0\}$ be a Poisson counting process, and for $i > 0$, let us denote the time of the i-th event by T_i, with $T_0 := 0$. Further, let the i-th inter-event time be $X_i := T_i - T_{i-1}$. In this section, we study the processes $\{X_i, i = 1, 2, \ldots\}$ and $\{T_i, i = 1, 2, \ldots\}$. We begin with the following characterization of $\{X_i, i = 1, 2, \ldots\}$.

Theorem 10 *The sequence $X_1, X_2, \ldots$ is a sequence of independent, identically distributed exponential random variables with parameter λ (mean $1/\lambda$).*

This result should come as no big surprise. After all, the assumptions of stationarity and independent increments essentially means that the process has *no memory*. That is, from any point on, the process is independent of what happened in the past (independent increments) and also has the same distribution as the process starting at the origin (stationarity). Since the process has no memory, the exponential interarrival times are expected.

With this characterization of the inter-event time process $\{X_i, i \geq 1\}$ we can easily characterize the point process of *event times* $\{T_i, i \geq 0\}$; thus, we have

$$T_0 = 0$$

$$T_n = \sum_{i=1}^{n} X_i, \quad n \geq 1.$$

therefore, it follows that the distribution of T_n is the distribution of the sum of n independent exponential random variables, each with parameter λ. This distribution is known as the gamma distribution with parameters n and λ. (For integer n, such a gamma distribution is also known as an Erlang distribution). The pdf of T_n is given by

$$f_{T_n}(t) = \lambda e^{-\lambda t} \frac{(\lambda t)^{n-1}}{(n-1)!}, \quad t \geq 0 \tag{3.8}$$

An alternate way to derive the distribution of T_n is to note that

$$\{T_n \leq t\} = \{N(t) \geq n\} \tag{3.9}$$

and hence

$$F_{T_n}(t) = P(T_n \leq t) = P(N(t) \geq n) = \sum_{j=n}^{\infty} e^{-\lambda t} \frac{(\lambda t)^j}{j!} \tag{3.10}$$

Differentiating this expression leads to the pdf given in Eq. 3.8. To summarize, we have the following result for the point process $\{T_i, i \geq 0\}$.

Theorem 11 *If $T_0 = 0$ and T_n has a gamma distribution with parameters n and λ for $n = 1, 2, \ldots$; then, T_i and T_{i+1} are related by*

$$T_{i+1} = T_i + X_{i+1},$$

where X_{i+1} is independent of $T_0, T_1, \ldots, T_i$.

3.4.2 Conditional Distribution of the Arrival Times

If we know the number of events that happened in a given time interval (say $N(t) = n$), we may be interested in knowing something about when those events occurred. Then, in this section we compute the probability distribution of the arrival times, given that we know the number of arrivals.

In order to compute this conditional distribution, let us begin with an easy case: $N(t) = 1$. Then, conditional distribution of T_1 is given by (using first principles and properties of the Poisson process):

$$\begin{aligned} F_{T_1}(u|N(t) = 1) &:= P(T_1 \leq u|N(t) = 1) \\ &= \frac{P(T_1 \leq u, N(t) = 1)}{P(N(t) = 1)} \\ &= \frac{P(N(u) = 1, N(t) - N(u) = 0)}{P(N(t) = 1)} \\ &= \frac{\lambda u e^{-\lambda u} e^{-\lambda(t-u)}}{\lambda t e^{-\lambda t}} \\ &= \frac{u}{t}, \qquad 0 \leq u \leq t \end{aligned}$$

This result says that, given that one event has occurred in the interval $[0, t]$, the time of occurrence of the event is uniformly distributed on $[0, t]$. It follows that

$$\mathbb{E}[T_1|N(t) = 1] = \frac{t}{2}. \tag{3.11}$$

Generalizing this result when n events are observed in the time interval $[0, t]$, we have the following result.

Theorem 12 *Let $\{N(t), t \geq 0\}$ be a Poisson process with rate λ. Given that $N(t) = n$, the n arrival times $(T_1, T_2, \ldots, T_n)$ have the conditional density*

$$f(t_1, t_2, \ldots, t_n|N(t) = n) = \frac{n!}{t^n}, \qquad 0 < t_1 < t_2 < \cdots < t_n. \tag{3.12}$$

Note: The conditional distribution given above is the distribution of the *order statistics* of a random sample of n uniformly distributed random variables on $[0, t]$. The order statistics are relevant here because the T_i are (by definition) ordered, i.e., $0 \leq T_i \leq T_2 \leq \ldots \leq T_n$.

Corollary 13

$$\mathbb{E}[T_k|N(t) = n] = \frac{kt}{n+1}.$$

Finally, in this section we state another property of the Poisson Process; again, this property is conditioned on the number of events by time t.

Theorem 14 *Let $\{N(t), t \geq 0\}$ be a Poisson process with rate λ, and suppose that we are given that $N(t) = n$ for some fixed t. Then we have*

$$P(N(u) = i | N(t) = n) = \binom{n}{i} \left(\frac{u}{t}\right)^i \left(1 - \frac{u}{t}\right)^{n-i}, \quad i = 0, 1, \ldots, n \quad 0 < u < t \tag{3.13}$$

That is, given $N(t) = n$, the number of events that have occurred by time u is binomial with parameters n and u/t.

3.4.3 Nonhomogeneous Poisson Process

We can generalize the Poisson process discussed in the sections above somewhat. If we relax the assumption of independent increments, much of the structure of the process is lost. However, we can relax the assumption of stationarity by allowing the number of points in an interval to depend on both the length and the location of the interval. Thus, we have the following definition:

Definition 15 The counting process $\{N(t), t \geq 0\}$ is called a nonhomogeneous (or nonstationary) Poisson process with rate function $\lambda(t), t \geq 0$ if

(i) $N(0) = 0$.
(ii) $\{N(t), t \geq 0\}$ has independent increments.
(iii) $P(N(t+h) - N(t) \geq 2) = o(h)$.
(iv) $P(N(t+h) - N(t) = 1) = \lambda(t)h + o(h)$.

Note that in the case of the nonhomogeneous Poisson process, the rate (intensity) $\lambda(t)$ is a deterministic *function* of t. If let

$$m(t) = \int_0^t \lambda(u)du, \tag{3.14}$$

then, the following theorem gives the distribution of $N(t+u) - N(t)$.

Theorem 16 *If $\{N(t), t \geq 0\}$ is a nonhomogeneous Poisson process with rate function $\lambda(t)$, then*

$$P(N(t+u) - N(t) = n) = e^{-(m(t+u)-m(t))} \frac{(m(t+u) - m(t))^n}{n!}, \quad n = 0, 1, 2, \ldots \tag{3.15}$$

The theorem above states that the increments of the nonhomogeneous Poisson counting process still have a Poisson distribution, but now the rate of the Poisson distribution depends not only on the *length* of the increment, but also on *where* the increment starts.

Corollary 17 *The expectation of $N(t+s) - N(t)$ is given by*

$$\mathbb{E}[N(t+s) - N(t)] = m(t+s) - m(t), \quad t, s \geq 0. \tag{3.16}$$

where $m(t)$ is as defined in Eq. 3.14.

If $\{T_n, n = 1, 2, \ldots\}$ are the arrival times of the nonhomogeneous Poisson process, then, from the above theorem and the independent increments property, we have the following conditional probability:

$$P(T_{n+1} - T_n > t \,|T_1, \ldots, T_n) = e^{-[m(T_n+t)-m(T_n)]}, \quad t \geq 0. \tag{3.17}$$

Thus the density of the interarrival time $X_n = T_{n+1} - T_n$ conditioned on $T_1, \ldots, T_n$ is given by

$$f_{T_{n+1}-T_n}(t\,|T_1, \ldots, T_n) = f_{T_{n+1}-T_n}(t\,|T_n) = \lambda(T_n + t)e^{-[m(T_n+t)-m(T_n)]}, \quad t \geq 0. \tag{3.18}$$

Nonhomogenous Poisson processes are clearly a natural way of modeling degradation processes associated to either increasing or decreasing failure rate (i.e., IFR/DFR). Typical examples include aging in most large civil infrastructure systems (e.g., due to corrosion or creep), fatigue in pavement of metal structures, moisture damage, etc.

3.4.4 Compound Poisson Process

Compound Poisson processes are marked point process whose events occur over time according to a Poisson process and whose marks are independent, identically distributed (*iid*) random variables (see Sect. 3.3.2). Formally, we have the following definition:

Definition 18 A compound Poisson process is a stochastic process $\mathscr{X} = \{X(t), t \geq 0\}$ of the form

$$X(t) = \sum_{i=1}^{N(t)} Y_i, \qquad t \geq 0, \tag{3.19}$$

where $\{N(t), t \geq 0\}$ is a Poisson process, and the sequence $\{Y_i, i = 1, \ldots, \}$ is a sequence of *iid* random variables, independent of $\{N(t), t \geq 0\}$.

If the common distribution function of the jump sizes is G, and the Poisson process $\{N(t), t \geq 0\}$ has rate λ, then the distribution of the increments is given by

$$\begin{aligned}
P(X(t) - X(s) \leq y) \\
&= \sum_{k=0}^{\infty} P(X(t) - X(s) \leq y | N(t) - N(s) = k) P(N(t) - N(s) = k) \\
&= 1 \cdot P(N(t) - N(s) = 0) \\
&\quad + P(\sum_{k=1}^{\infty} P(Y_1 + \cdots + Y_k \leq y) P(N(t) - N(s) = k) \\
&= e^{-\lambda(t-s)} + \sum_{k=1}^{\infty} G_n(x) \frac{\lambda(t-s)^k}{k!} e^{-\lambda(t-s)} dx, \qquad (3.20)
\end{aligned}$$

where G_n is the n-fold convolution of G with itself. Similarly, the moment generating function $M_{X(t)}(u)$ of $X(t)$ has the form

$$\begin{aligned}
M_{X(t)}(u) &= \mathbb{E}[e^{uX(t)}] \\
&= \sum_{k=0}^{\infty} \mathbb{E}[e^{u(Y_1 + \cdots + Y_k)}] \frac{\lambda(t)^k}{k!} e^{-\lambda(t)} \\
&= \sum_{k=0}^{\infty} (\mathbb{E}[e^{u(Y_1)}])^k \frac{\lambda(t)^k}{k!} e^{-\lambda(t)} = \sum_{k=0}^{\infty} (M_{Y_1}(u))^k \frac{\lambda(t)^k}{k!} e^{-\lambda(t)} \\
&= e^{\lambda t (M_{Y_1}(u) - 1)}. \qquad (3.21)
\end{aligned}$$

The mean and variance of the compound Poisson process are then given by

$$\mathbb{E}[X(t)] = \lambda t \mathbb{E}[Y_1] \qquad (3.22)$$

$$Var[X(t)] = \lambda t \mathbb{E}[Y_1^2]. \qquad (3.23)$$

Compound Poisson processes are commonly used in modeling degradation due to shocks that occur at random times with random sizes (see Chap. 5).

3.5 Renewal Processes

Renewal processes are point processes that generalize the Poisson process assumption that inter-event times are exponentially distributed, while maintaining the assumption that they are independent. Renewal processes have advantages over the Poisson process for modeling systems that are completely replaced upon failure as, unlike the Poisson process, they allow for the time to failure to incorporate some notion of aging. Renewal processes, however, do not possess independent increments, so that their analysis is somewhat more complicated. Nonetheless, they are widely used to model maintained systems that are, at some point, replaced and restarted [10]. In Chap. 8 we will use renewal models to deal with systems systematically reconstructed.

Again, our interest in this section is to introduce notation and basic properties; proofs are generally omitted but can be found in most common books on stochastic processes (e.g., [3, 4]).

3.5.1 Definition and Basic Properties

We define a *renewal process* from its inter-event times as follows:

Definition 19 A renewal process $\mathscr{N}$ is a counting process whose inter-event times $\{X_i, i = 1, 2, \ldots\}$ comprise a sequence of independent, identically distributed non-negative random variables.

We take F to be the common distribution function of the inter-event times, and we will often refer to the renewal process by either its counting process $\mathscr{N} = \{N(t), t \geq 0\}$, where $N(t) = \sup\{n : T_n \leq t\}$, or by its inter-event time sequence $\{X_i, i = 1, 2, \ldots\}$.

Definition 20 A renewal process $\{X_n, n \geq 1\}$ with $P(X_1 < \infty) = 1$ is called a persistent (nonterminating) renewal process. If $P(X_1 < \infty) < 1$ then we have a transient (terminating) renewal process.

For our purposes, unless otherwise stated, we will consider *persistent* renewal processes. To avoid trivialities, we will also assume that $P(X_1 > 0) > 0$; this condition ensures that X_1 has a mean $\mathbb{E}[X_1] =: \mu > 0$ (keep in mind that it may be $+\infty$).

Now let us interpret X_i in a point process context as the time between the $i - 1$ and the ith event. For $n = 0, 1, 2, \ldots$, if let

$$T_0 = 0, \quad \text{and} \quad T_n = X_1 + X_2 + \ldots + X_n,$$

then T_n is the time, measured from the origin, at which the n-th event occurs. Because the process "regenerates" at the time of an event (that is, the future looks statistically identical when viewed at any event time), we refer to the events as *renewals*. As a direct consequence of the strong law of large numbers,

$$\lim_{n\to\infty} \frac{T_n}{n} = \mu \qquad \text{a.s.} \tag{3.24}$$

and since we assume $\mu > 0$, T_n must approach infinity as n approaches infinity. Thus T_n must be less than or equal to t for at most a finite number of values of n, and hence an infinite number of renewals cannot occur in a finite time.

The random variable $N(t)$ denotes the number of renewals by time t. Then, based on the assumptions made regarding the inter-event times, we have the following theorem.

Theorem 21 *$N(t)$ is a random variable with finite moments of all orders, i.e.,*

(i) $P(N(t) < \infty) = 1$,
(ii) $\mathbb{E}[N(t)^k] < \infty, \quad k = 1, 2, \ldots$.

A couple of observations are in order. First, note that even though $N(t) < \infty$ for each (finite) t, it is true that, with probability 1, $N(\infty) = \lim_{t\to\infty} N(t) = \infty$, since

$$\begin{aligned} P(N(\infty) < \infty) &= P(X_n = \infty \text{ for some } n) \\ &= P(\cup_{n=1}^{\infty}\{X_n = \infty\}) \le \sum_{n=1}^{\infty} P(X_n = \infty) = 0. \end{aligned}$$

Second, as the following example indicates, the fact that $N(t)$ is finite does not necessarily imply that $\mathbb{E}[N(t)]$ is finite (this is a good example to remember!):

Example 3.3 Let Y be a random variable with $P(Y = 2^n) = (\frac{1}{2})^n, n \ge 1$. Now

$$P(Y < \infty) = \sum_{n=1}^{\infty} P(Y = 2^n) = \sum_{n=1}^{\infty} \left(\frac{1}{2}\right)^n = 1.$$

But

$$\mathbb{E}[Y] = \sum_{n=1}^{\infty} 2^n P(Y = 2^n) = \sum_{n=1}^{\infty} 2^n \left(\frac{1}{2}\right)^n = \infty.$$

We have already shown that $N(\infty) = \lim_{t\to\infty} N(t) = \infty$. Of interest, too, is the time average *rate* of renewals in $[0, t]$, $N(t)/t$. For this, we have the following theorem:

Theorem 22 (Strong Law for Renewal Processes) *With probability 1,*

$$\frac{N(t)}{t} \to \frac{1}{\mu} \quad \text{as } t \to \infty. \tag{3.25}$$

Proof Since $T_{N(t)} \leq t < T_{N(t)+1}$, we have

$$\frac{T_{N(t)}}{N(t)} \leq \frac{t}{N(t)} < \frac{T_{N(t)+1}}{N(t)}, \tag{3.26}$$

where $T_{N(t)}$ is the time of the last renewal prior to time t and $T_{N(t)+1}$ is the time of the first renewal after time t. For each sample point ω, $T_{N(t)}(\omega)/N(t,\omega)$ runs through precisely the same values as $t \to \infty$ as does $T_n(\omega)/n$ as $n \to \infty$, and since $N(t) \to \infty$ and $T_n/n \to \mu$ *a.s.*, it follows that $T_{N(t)}/N(t) \to \mu$ *a.s.* as $t \to \infty$ as well. Furthermore,

$$\frac{T_{N(t)+1}}{N(t)} = \Big(\frac{T_{N(t)+1}}{N(t)+1}\Big)\Big(\frac{N(t)+1}{N(t)}\Big) \to \mu \cdot 1 = \mu \; a.s. \qquad \text{as } t \to \infty,$$

and therefore, since $\frac{t}{N(t)}$ is caught between two random variables, both of which converge to μ

$$\frac{t}{N(t)} \to \mu \qquad \text{as } t \to \infty, \tag{3.27}$$

and the result follows. □

It is important to note that the strong law for renewal processes states that the *time* averages $N(t,\omega)/t$ converge to $1/\mu$ for each sample path ω. Much of renewal theory concerns the behavior of the ensemble (or statistical) average $\mathbb{E}[N(t)]/t$, and the ensemble average near a particular point t, $\mathbb{E}[N(t+\alpha) - N(\alpha)]/\alpha$. We will see later that for renewal processes, all three averages coincide in the limit (as $t \to \infty$). This most important property forms the basis of the *ergodic* property of renewal processes. The practical implications of these results are significant.

3.5.2 Distribution of $N(t)$

The distribution of $N(t)$ can be obtained using the important relationship between $N(t)$ and T_n, namely:

$$\{N(t) \geq n\} \iff \{T_n \leq t\}; \tag{3.28}$$

that is, there have been at least n renewals by time t if and only if the nth renewal occurs before or at time t. This observation leads directly to the following theorem.

Theorem 23 *The distribution of $N(t)$ is given by*

$$P(N(t) = n) = F_n(t) - F_{n+1}(t), \qquad n \geq 0, \tag{3.29}$$

where

$$F_0(t) = 1,$$
$$F_1(t) = F(t),$$
$$F_n(t) = \int_0^t F(t-u)dF_{n-1}(u), n = 2, 3, \ldots,$$

that is, F_n is the n-fold convolution of F with itself.

Example 3.4 The Erlang case.
Let $f(x) = e^{-\lambda x}\lambda^p x^{p-1}/(p-1)!,\ 0 \le x < \infty$. Then

$$f_n(x) = \frac{e^{-\lambda x}\lambda^{np}x^{np-1}}{(np-1)!} \tag{3.30}$$

and

$$\begin{aligned} F_n(x) &= \int_0^x e^{-\lambda x}\lambda^{np}\frac{y^{np-1}}{(\lambda p-1)!}dy \\ &= 1 - e^{-\lambda x}\sum_{j=0}^{np-1}\frac{(\lambda x)^j}{j!} \qquad n \ge 1 \end{aligned}$$

Hence

$$\begin{aligned} P(N(t) = n) &= F_n(t) - F_{n+1}(t) \\ &= e^{-\lambda t}\sum_{j=np}^{np+p-1}\frac{(\lambda t)^j}{j!} - e^{\lambda t}\sum_{j=0}^{np-1}\frac{(\lambda t)^j}{j!} \\ &= e^{-\lambda t}\sum_{j=np}^{np+p-1}\frac{(\lambda t)^j}{j!}, \quad n = 0, 1, 2, \ldots. \end{aligned}$$

While an analytic expression for the distribution of $N(t)$ is difficult to obtain for arbitrary inter-renewal distribution F, for small values of t, the distribution of $N(t)$ can be approximated using Theorem 23 and ignoring terms in the sum for large n. For larger values of t, we can use transform methods to obtain an expression for distribution of $N(t)$. Recall that the Laplace transform of a nondecreasing function G with $G(x) = 0$ for $x < 0$ is given by

$$\mathscr{L}(G) = G^*(s) = \int_0^\infty e^{-sx}G(x)dx \tag{3.31}$$

whenever the integral exists. Now if G is the distribution function of a nonnegative random variable that has density g, we have (integration by parts)

$$G^*(s) = \frac{1}{s}\, g^*(s). \tag{3.32}$$

Also, if G_n is the n-fold convolution of a nonnegative random variable G, then

$$G_n^*(s) = (sG^*(s))^n \tag{3.33}$$

Since $N(t)$ is a discrete nonnegative random variable, we can define its *probability generating function (pgf)* as follows:

$$G(t, z) = \sum_{n=0}^{\infty} P(N(t) = n) z^n. \tag{3.34}$$

Then, we have the following expression for $G(t, z)$:

Lemma 24

$$G(t, z) = 1 + (z - 1) \sum_{n=1}^{\infty} z^{n-1} F_n(t). \tag{3.35}$$

Proof Substituting $P(N(t) = n)$ from Theorem 23 yields

$$\begin{aligned} G(t, z) &= \sum_{n=0}^{\infty} (F_n(t) - F_{n+1}(t)) z^n \\ &= \sum_{n=0}^{\infty} F_n(t) z^n - \sum_{n=0}^{\infty} F_{n+1}(t) z^n \\ &= F_0(t) z^0 + z \sum_{n=1}^{\infty} F_n(t) z^{n-1} - \sum_{n=1}^{\infty} F_n(t) z^{n-1} \\ &= 1 + (z - 1) \sum_{n=1}^{\infty} F_n(t) z^{n-1}. \end{aligned}$$

□

Furthermore, let $\mathscr{L}(G(s, z)) = G^*(s, z) = \int_0^\infty e^{-st} G(t, z) dt$ be the Laplace transform of $G(t, z)$; then,

Theorem 25

$$G^*(s, z) = \frac{1 - sF^*(s)}{s(1 - zsF^*(s))}. \tag{3.36}$$

Proof

$$
\begin{aligned}
G^*(s,z) &= \int_0^\infty e^{-st}\big(1+(z-1)\sum_{n=1}^\infty F_n(t)z^{n-1}\big)dt \\
&= \frac{1}{s}+(z-1)\sum_{n=1}^\infty z^{n-1}F_n^*(s) \\
&= \frac{1}{s}+(z-1)F^*(s)\sum_{n=1}^\infty z^{n-1}(sF^*(s))^{n-1} \\
&= \frac{1}{s}\big[1+\frac{s(z-1)F^*(s)}{1-zsF^*(s)}\big] \\
&= \frac{1-sF^*(s)}{s(1-zsF^*(s))}.
\end{aligned}
$$

□

Corollary 26 *When $F(x)$ is the distribution function of an absolutely continuous random variable with density function $f(x)$,*

$$G^*(s,z)=\frac{1-f^*(s)}{s(1-zf^*(s))}.$$

Example 3.5 The exponential case.
Let $F(x)=1-e^{-\lambda x},\quad 0\le x<\infty$. Then $f(x)=\lambda e^{-\lambda x}$ and $f^*(s)=\frac{\lambda}{\lambda+s}$. Now

$$
\begin{aligned}
G^*(s,z) &= \frac{1-f^*(s)}{s(1-zf^*(s))} = \frac{1-\frac{\lambda}{\lambda+s}}{s(1-\frac{z\lambda}{\lambda+s})} \\
&= \frac{1}{s+\lambda(1+z)},
\end{aligned}
$$

which implies

$$
\begin{aligned}
G(t,z) &= e^{-\lambda(1-z)t} = e^{\lambda(z-1)t} \\
&= e^{-\lambda t}\sum_{n=0}^\infty \frac{(\lambda tz)^n}{n!}
\end{aligned}
$$

so

$$P(N(t)=n)=\frac{e^{-\lambda t}(\lambda t)^n}{n!}.$$

A renewal process with exponentially distributed inter-renewal times is, of course, the *Poisson process*.

For the case of density functions that have rational Laplace transforms, inversion techniques exist that can, in principle, produce the distribution of $N(t)$. In general, however, the distribution of $N(t)$ is difficult to obtain. For large t, we can approximate the distribution of $N(t)$ using a Central Limit Theorem; the proof is somewhat technical and can be found in [3].

Theorem 27 (Central Limit Theorem for Renewal Processes) *If both the mean* μ *and the variance* σ^2 *of the inter-renewal times are finite, then*

$$\lim_{t\to\infty} P\Big(\frac{N(t)-t/\mu}{\sigma\sqrt{t/\mu^3}} < y\Big) = \frac{1}{\sqrt{2\pi}}\int_{-\infty}^{y} e^{-x^2/2}dx = \Phi(y). \tag{3.37}$$

where Φ *is the distribution function of the standard normal.*

This result is quite useful as it allows us to approximate the distribution of $N(t)$ for large values of t.

3.5.3 The Renewal Function and the Elementary Renewal Theorem

The *renewal function* $m(t) := \mathbb{E}[N(t)]$ has been the subject of extensive study and research. Much of renewal theory is concerned with the properties of this function, particularly it's asymptotic behavior as $t \to \infty$. This section develops those properties. In some cases, the proofs are relatively straightforward, and we present them here. However, some results are beyond the scope of this book, and we provide only a rough sketch of the proof. Further details are available in [3, 4].

We begin with an exact expression for $m(t)$ in terms of the distribution function F of inter-event times.

Theorem 28

$$m(t) = \sum_{n=1}^{\infty} F_n(t).$$

Proof

$$\begin{aligned} m(t) = \mathbb{E}[N(t)] &= \sum_{n=1}^{\infty} nP(N(t)=n) = \sum_{n=1}^{\infty} n(F_n(t) - F_{n+1}(t)) \\ &= \sum_{n=1}^{\infty} F_n(t). \end{aligned}$$

Note that the finiteness of $m(t)$ was established in Theorem 21 under the assumption that $F(0) < 1$. □

While the expression above appears quite simple, in practice the renewal function is generally difficult to calculate, even for moderately large t. The Elementary Renewal Theorem provides the asymptotic behavior of the expected rate of renewals. The proof of the theorem involves the concept of stopping times and uses a very important result known as Wald's equation, topics beyond the scope of the book, so we state the theorem without proof.

Theorem 29 (Elementary Renewal Theorem)

$$\frac{m(t)}{t} \to \frac{1}{\mu} \quad \text{as } t \to \infty. \tag{3.38}$$

The Elementary Renewal Theorem states that the statistical (ensemble) average number of renewals in $[0, t]$ is proportional to t for large values of t, a result that is intuitively appealing. It is reasonable to conjecture that a similar statement holds for the average number of renewals in an interval $(t, t+\alpha]$ as $t \to \infty$ for fixed α. In fact, the conjecture holds for *continuous* (nonlattice) inter-renewal distributions. A lattice distribution is a discrete probability distribution whose probability is concentrated on a set of points of the form $a+nd$, $n = 0, 1, \ldots, d > 0$; the period of the distribution is the largest number d for which this holds. For example, if a random variable takes on values 3, 6, and 12, the random variable is lattice with period 3. A little care must be observed in taking the limit for lattice distributions because there will be "gaps" where no renewals can occur. This result is due to David Blackwell; the proof is surprisingly complicated, and no simple proof has yet emerged.

Theorem 30 (Blackwell's Theorem)

1. *If F is not lattice, then*

$$m(t+\alpha) - m(t) \to \frac{\alpha}{\mu} \quad \text{as } t \to \infty \tag{3.39}$$

   for all $\alpha \geq 0$.
2. *If F is lattice with period d, then*

$$E[\text{Number of renewals at } nd] \to \frac{d}{\mu} \quad \text{as } n \to \infty \tag{3.40}$$

3.5.4 Renewal-Type Equations

Much of renewal theory involves studying the properties of solutions to certain integral equation of the form

$$g(t) = h(t) + \int_0^t g(t-u)dF(u), \quad t \geq 0, \tag{3.41}$$

or in convolution form,

$$g = h + g * F. \tag{3.42}$$

Here $h(t)$ is a known function and $g(t)$ is an unknown function, often, in our context, a time-dependent probability or expectation. Such an equation is called a *renewal-type equation*, and these equations have been well studied in analysis. Renewal equations are generally constructed using conditioning arguments.

The following theorem gives a renewal-type equation satisfied by the renewal function:

Theorem 31 *The renewal function $m(t)$ satisfies*

$$m(t) = F(t) + \int_0^t m(t-u)dF(u), \quad t \geq 0.$$

Proof if we define

$$m(t) = \begin{cases} 0 & \text{on } \{X_1 > t\}, \\ 1 + m(t-u) & \text{on } \{X_1 = u \leq t\}. \end{cases}$$

then,

$$\begin{aligned} m(t) &= 0 \cdot \overline{F}(t) + \int_0^t (1 + m(t-u))dF(u) \\ &= F(t) + \int_0^t m(t-u)dF(u), \quad t \geq 0. \end{aligned}$$

□

Example 3.6 (Adapted from [11]) One instance in which it is possible to obtain an analytical solution for the renewal equation is when the distribution of interarrival times is uniform on (0, 1). In this case, and for $t < 1$, the renewal function becomes:

$$m(t) = t + \int_0^t m(t-x)dx = t + \int_0^t m(u)du \quad \text{by making } u = t - x \tag{3.43}$$

By taking the derivative, this equation becomes:

$$m'(t) = 1 + m(t) \tag{3.44}$$

Furthermore, by making $h(t) = 1 + m(t)$, we obtain $h'(t) = h(t)$. The solution of this differential equation leads to $h(t) = Ke^t$; which can be used to obtain the following expression for $m(t)$:

$$m(t) = Ke^t - 1 \tag{3.45}$$

Then, since $m(0) = 0$, then $K = 1$ and we get the final expression for $m(t)$:

$$m(t) = e^t - 1 \qquad \text{for} \quad 0 \le t \le 1 \tag{3.46}$$

Solutions to renewal equations are characterized in the following theorem:

Theorem 32 *If h is bounded and vanishes for $t < 0$, the solution to the renewal-type equation is given by $g = h + m * h$, or equivalently,*

$$g(t) = h(t) + \int_0^t h(t-u)dm(u).$$

Proof (Kao, p. 102)[4]: Suppose that the inter-renewal distribution has density f, so that the renewal-type equation can be written as

$$g(t) = h(t) + \int_0^t g(t-u)f(t)dt.$$

Then the renewal function has density

$$m'(t) = \sum_{n=1}^{\infty} f_n(t),$$

where $f_n(t)$ is the n-fold convolution of f with itself. The Laplace transform of the renewal-type equation is given by

$$g^*(s) = h^*(s) + g^*(s)f^*(s).$$

From this expression, it follows that

$$\begin{aligned} g^*(s) &= \frac{h^*(s)}{1 - f^*(s)} = h^*(s)\big[1 + f^*(s) + (f^*(s))^2 + \cdots\big] \\ &= h^*(s) + h^*(s)m'^*(s) \end{aligned}$$

and the result follows by inverting the last expression. □

We will now present some examples of renewal-type equations that arise naturally in the study of renewal processes.

Example 3.7 We know already that the renewal function varies as t/μ for large t. We can refine this a bit by studying the difference

$$g(t) = m(t) - \frac{t}{\mu}. \tag{3.47}$$

Note that g satisfies the renewal equation

$$g = h + g * F \tag{3.48}$$

therefore, $h(t)$ satisfies

$$h(t) = \frac{1}{\mu}\int_t^{\infty} \overline{F}(u)du - \overline{F}(t). \tag{3.49}$$

Example 3.8 Let $U(t)$ be the time since the last renewal before time t in a renewal process; that is, let $U(t) = t - T_{N(t)}$. $U(t)$ is known as the *backward recurrence time* or *age* of the renewal process at time t. For fixed x, let $g(t) = P(U(t) > x)$. Then g satisfies the renewal equation

$$g = h + g * F$$

where

$$h(t) = \overline{F}(t)1_{(x,\infty)}(t).$$

Example 3.9 Let $K(t)$ be the length of time from time t until the next renewal occurs in a renewal process; $K(t) = T_{N(t)+1} - t$. $K(t)$ is called the *forward recurrence time* or *excess life*. For fixed x, let $g(t) = P(K(t) > x)$; $g(t)$ satisfies the renewal equation $g = h + g * F$, where

$$h(t) = \overline{F}(t + x).$$

3.5.5 Key Renewal Theorem

While the time-dependent behavior of solutions to renewal-type equations is often difficult to obtain, we can analyze the asymptotic behavior of these solutions using the so-called *Key Renewal Theorem*. The proof of this theorem requires that the function $h(t)$ be *directly Riemann integrable*, that is, that the upper and lower Riemann sums, defined, respectively, by

$$\underline{s} = a\sum_{n=1}^{\infty} \underline{m}_n(a) \quad \text{and} \quad \overline{s} = a\sum_{n=1}^{\infty} \overline{m}_n(a) \tag{3.50}$$

where $\underline{m}_n(a)$ and $\overline{m}_n(a)$ are, respectively, the infimum and the supremum of $h(t)$ on the interval $(n-1)a \le t \le na$, are finite and tend to the same limit as $a \to 0$. A function h is directly Riemann integrable on $[0, \infty]$ if it is integrable over every finite interval $[0, a]$ and if $\overline{s} < \infty$ for some a (then automatically $\overline{s} < \infty$, for all a). Direct Riemann integrability ensures that $h(t)$ does not oscillate wildly as $t \to \infty$.

The following proposition lists some useful results for identifying directly Riemann integrable functions:

Proposition 33 *Let h be a nonnegative function. Then*

(i) h is directly Riemann integrable if it is continuous and vanishes outside a finite interval.
(ii) if h is bounded and continuous, h is directly Riemann integrable if and only if $\overline{s} < \infty$ for some $a > 0$.
(iii) if h is monotone nonincreasing, h is directly Riemann integrable if and only if h is Riemann integrable.

Proof see Çinlar [2]. □

We are now in a position to state the Key Renewal Theorem, which characterizes the asymptotic behavior of the solutions to renewal-type equations.

Theorem 34 *Key Renewal Theorem. If the inter-renewal distribution is not lattice, and if $h(t)$ is any directly Riemann integrable function on $t \geq 0$, then if $\mu < \infty$,*

$$\lim_{t\to\infty} \int_0^t h(t-u)dm(u) = \frac{1}{\mu}\int_0^\infty h(u)du,$$

where

$$m(x) = \sum_{n=1}^{\infty} F_n(x)$$

Furthermore, if $\mu = \infty$, then

$$\lim_{t\to\infty} \int_0^t h(t-u)dm(u) = 0.$$

It can be shown that the Key Renewal Theorem and Blackwell's Theorem (Theorem 30) are equivalent. We do not provide the proof here, but it can be found in [12].

Using the Key Renewal Theorem (hereafter abbreviated KRT), we can evaluate the limit as $t \to \infty$ of the quantities for which we obtained renewal-type equations in Examples 3.6–3.8, as well as other such quantities.

Example 3.10 Consider $g(t) = m(t) - t/\mu$ in Example 3.6. Employing the KRT, we obtain (using integration by parts)

$$\lim_{t\to\infty} m(t) - \frac{t}{\mu} = \frac{\sigma^2 - \mu^2}{2\mu^2},$$

where $\sigma^2 = Var[X_i]$.

Example 3.11 Consider $g(t) = P(U(t) > x)$ in Example 3.7. Employing the KRT, we obtain

$$\lim_{t\to\infty} P(U(t) > x) = \frac{1}{\mu}\int_x^\infty \overline{F}(u)du.$$

Example 3.12 Consider $g(t) = P(K(t) > x)$ in Example 3.8. Employing the KRT, we obtain

$$\lim_{t\to\infty} P(K(t) > x) = \frac{1}{\mu}\int_x^\infty \overline{F}(u)du.$$

3.5.6 Alternating Renewal Processes and the Distribution of $T_{N(t)}$

An alternative approach to developing renewal-type equations by conditioning on X_1 is to condition on $T_{N(t)}$ instead. This approach, pioneered by Ross (1993), leads directly to an expression whose asymptotic behavior can be examined via the Key Renewal Theorem. This section presents Ross' approach and introduces the idea of an *alternating* renewal process, a construct that turns out to be quite useful in analyzing renewal processes.

Lemma 35 *The distribution of $T_{N(t)}$ is given by*

$$P(T_{N(t)} \le x) = \overline{F}(t) + \int_0^x \overline{F}(t-u)dm(u). \tag{3.51}$$

Proof

$$\begin{aligned}
P(T_{N(t)} \le x) &= \sum_{n=0}^{\infty} P(T_n \le x, T_{n+1} > t)\\
&= \overline{F}(t) + \sum_{n=1}^{\infty} P(T_n \le x, T_{n+1} > t)\\
&= \overline{F}(t) + \sum_{n=1}^{\infty}\int_0^\infty P(T_n \le x, T_{n+1} > t | T_n = u)dF_n(u))\\
&= \overline{F}(t) + \sum_{n=1}^{\infty}\int_0^x \overline{F}(t-u)dF_n(u)\\
&= \overline{F}(t) + \int_0^x \overline{F}(t-u)d\big(\sum_{n=1}^{\infty} F_n(u)\big)\\
&= \overline{F}(t) + \int_0^x \overline{F}(t-u)dm(u).
\end{aligned}$$

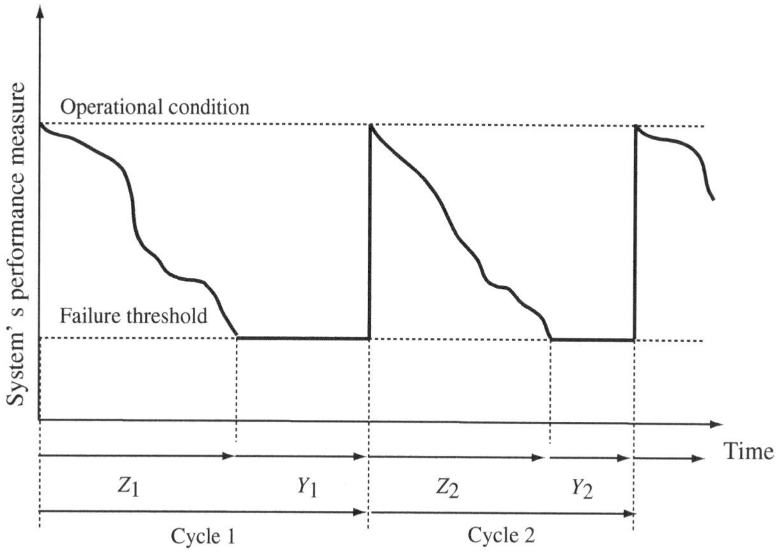

Fig. 3.4 Sample path of alternating renewal process

The interchange of integration and summation is justified because all terms are nonnegative. □

Now consider a system that can be in one of two states, either *on* or *off*. The system starts on, and it remains on for a length of time Z_1; it then goes off and remains off for a length of time Y_1. The system is then on again for a length of time Z_2, then off for a length of time Y_2, and so on. We refer to the time between the starts of two successive on times as a *cycle* (Fig. 3.4).

We assume that $\{Z_i, i \geq 1\}$ is an *iid* sequence with common distribution function H, that $\{Y_i, i \geq 1\}$ is also an *iid.* sequence with common distribution function G, and that the random pairs $\{(Z_i, Y_i), i \geq 1\}$ are *iid*. We do, however, allow Z_i and Y_i to be dependent; that is, within a cycle, the lengths of the on and off times may depend on each other. If $P(t)$ is the probability that the system is on at time t; then, we have the following result.

Theorem 36 *If* $\mathbb{E}[Z_n + Y_n] < \infty$, *and* F *is nonlattice, then*

$$\lim_{t\to\infty} P(t) = \frac{\mathbb{E}[Z_n]}{\mathbb{E}[Z_n] + \mathbb{E}[Y_n]}. \tag{3.52}$$

Proof Define renewal epochs for this process as the times at which the system goes on. Conditioning on the time of the last renewal prior to time t, we have

$$\begin{aligned} P(t) &= P(\text{on at } t | T_{N(t)} = 0) P(T_{N(t)} = 0) \\ &+ \int_0^\infty P(\text{on at } t | T_{N(t)} = u) dP(T_{N(t)} \leq u). \end{aligned}$$

Now

$$P(\text{on at } t|T_{N(t)} = 0) = P(Z_1 > t|Z_1 + Y_1 > t) = \frac{\overline{H}(t)}{\overline{F}(t)},$$

and, for $0 < u < t$,

$$P(\text{on at } t|T_{N(t)} = u) = P(Z_{N(t)+1} > t - u|Z_{N(t)+1} + Y_{N(t)+1} > t - u) = \frac{\overline{H}(t-u)}{\overline{F}(t-u)},$$

hence,

$$P(t) = \frac{\overline{H}(t)}{\overline{F}(t)}\overline{F}(t) + \int_0^t \frac{\overline{H}(t-u)}{\overline{F}(t-u)}\overline{F}(t-u)dm(u) = \overline{H}(t) + \int_0^t \overline{H}(t-u)dm(u).$$

Since $\overline{H}(t)$ is nonnegative, nonincreasing, and Riemann integrable, we can apply the Key Renewal Theorem to this last expression to obtain

$$\lim_{t\to\infty} P(t) = \frac{1}{\mu}\int_0^\infty \overline{H}(u)du = \frac{\mathbb{E}[Z_n]}{\mathbb{E}[Z_n] + \mathbb{E}[Y_n]}.$$

□

Example 3.13 To see the usefulness of the alternating renewal process approach, consider a renewal process $\{X_i, i \geq 1\}$ with distribution function F and mean μ, and say the system is "on" at time t if the backward recurrence time at time t is less than x (for fixed x) and "off" otherwise. That is, the process is "on" for the first x units of a renewal interval and "off" the remaining time. Then, the "on" time in a cycle is $min(x, X)$ and,

$$\lim_{t\to\infty} P(U(t) \leq x) = \frac{\mathbb{E}[\min(x, X)]}{\mathbb{E}[X]} = \frac{1}{\mu}\int_0^\infty P(\min(x, X) > u)du = \frac{1}{\mu}\int_0^x \overline{F}(u)du,$$

which agrees with Example 3.10.
Similarly, if we say the system is "off" the last x units of the cycle and "on" otherwise, we can conclude that

$$\lim_{t\to\infty} P(K(t) \le x) = \lim_{t\to\infty} P(\text{off at } t) = \frac{\mathbb{E}[\min(x, X)]}{\mathbb{E}[X]} = \frac{1}{\mu}\int_0^x \overline{F}(u)du,$$

which agrees with Example 3.11.
Finally, consider the random variable $X_{N(t)+1} = T_{N(t)+1} - T_{N(t)} = U(t) + V(t)$. $X_{N(t)+1}$ represents the length of the renewal interval that contains t. To compute the distribution function of $X_{N(t)+1}$, let an on–off cycle correspond to a renewal interval, and say that the on time in the cycle is the total cycle time if that time is greater than x and zero otherwise. Then, provided F is not lattice,

$$\lim_{t\to\infty} P(X_{N(t)+1} > x) = \frac{1}{\mu}\mathbb{E}[\text{on time in cycle}] = \frac{1}{\mu}\mathbb{E}[X|X > x]P(X > x) = \frac{1}{\mu}\int_x^\infty u dF(u),$$

or equivalently,

$$\lim_{t\to\infty} P(X_{N(t)+1} \le x) = \frac{1}{\mu}\int_0^x u dF(u). \tag{3.53}$$

This result is the so-called *inspection paradox*; it demonstrates that, asymptotically, the interval containing t does not have the same statistical properties as an arbitrary inter-renewal interval; indeed, it is stochastically larger than an arbitrary interval.

3.6 Summary and Conclusions

In this chapter, we reviewed basic concepts of stochastic process that will be of great use for modeling deteriorating systems in the subsequent chapters. We first discussed the conceptual aspects and theoretical foundations of point process. Special emphasis and detailed discussion was provided for Poisson processes. Due to the importance for systematically reconstructed systems (see Chaps. 8 and 9), renewal theory was also reviewed.

References

1. L. Takacs, *Stochastic processes* (Wiley, New York, 1960)
2. E. Çinlar, *Introduction to stochastic processes* (Prentice Hall, New Jersey, 1975)
3. S.M. Ross, *Stochastic processes*, 2nd edn. (Wiley, New York, 1996)
4. E.P.C. Kao, *An introduction to Stochastic Processes* (Duxbury Press, Belmont, 1997)
5. T.J. Aven, U. Jensen. *Stochastic Models in Reliability*. Series in Applications of Mathematics: Stochastic Modeling and Applied Probability, vol. 41, (Springer, New York, 1999)
6. T.R. Fleming, D.P. Harrington, *Counting Processes and Survival Analysis* (Wiley, New York, 1991)
7. P. Bremaud, *Point Processes and Queues* (Springer, New York, 1981)
8. S.N. Ethier, T.G. Kurtz, *Markov Processes: Characterization and Convergence* (Wiley, New York, 1986)
9. S.P. Meyn, R.L. Tweedie, *Markov Chains and Stochastic Stability* (Springer-Verlag, London, 1993)
10. R. Rackwitz, Optimization—the basis of code making and reliability verification. Structural Safety **22**(1), 27–60 (2000)
11. S. Ross, *Introduction of Probability Models* (Academic Press, San Diego, CA, 2007)
12. S. Resnick, *Adventures in Stochastic Processes* (Birkhauser, Boston, 1992)

Chapter 4
Degradation: Data Analysis and Analytical Modeling

4.1 Introduction

A central element in life-cycle modeling of engineered systems is the appropriate understanding, evaluation, and modeling of degradation. In this chapter we first provide a formal definition and a conceptual framework for characterizing system degradation over time. Afterward, we discuss the importance of actual field data analysis and, in particular, we present a conceptual discussion on data collection. We also present briefly the basic concepts of regression analysis, which might be considered the first and simplest approach to constructing degradation models. Regression analysis will be used later to obtain estimates of the parameters of degradation models. As an example, the special case of estimating the parameters of the gamma process (see Chap. 5) is presented. This chapter is not intended as a comprehensive discussion on degradation data analysis, as this topic has been widely studied in a variety of different research fields, and many tools and procedures are available for modeling degradation data. If the reader is interested, some of the most relevant references with respect to failure data in engineering problems are [1, 2].

Finally, the discussion presented in Chaps. 1–3, which has provided motivation for the study of engineered systems subject to failure, as well as an overview of the mathematical background in stochastic processes, will serve as the foundation for modeling degradation analytically. In the last part of the chapter, and as an introduction to the rest of the book, we provide a conceptual framework for characterizing system degradation over time and define the appropriate random variables that will be used later. We discuss the general properties of progressive and shock degradation mechanisms, which are illustrated with several examples of physical degradation in various engineering fields. This chapter is intended as a conceptual and general discussion of degradation before we present specific analytical degradation models in detail in Chaps. 5 through 7.

M. Sánchez-Silva and G.-A. Klutke, *Reliability and Life-Cycle Analysis of Deteriorating Systems*, Springer Series in Reliability Engineering,
DOI 10.1007/978-3-319-20946-3_4

4.2 What Is Degradation?

When an engineered system is put into use, physical changes to the system occur over time. These changes may be the result of internal processes, for instance, natural changes in material properties, or external processes, such as environmental conditions and operating stresses. Regardless of the cause, these changes may result, over time, in a reduced capacity of the system to perform its intended function.

We measure the capacity of a system by one or more physical quantities that serve as performance measures, such as the inter-story drift of a building, the vibrational signature of a bridge, or the tread depth of a tire. By the term *degradation* (or equivalently, *deterioration*), we mean

> the decrease in capacity of an engineered system over time, as measured by one or more performance indicators.

Thus degradation is a process that describes the loss of system capacity over time. We make a distinction in this book between the definition of degradation given above and the actual physical processes that result in the decline in capacity. As noted in [3], what we define as degradation above is in reality only the observable damage produced by a number of different physical processes that may, themselves, be unobservable. For example, in the case of concrete bridge decks, physical changes due to corrosion, cracking and spalling, load related fatigue, and so on [4] occur over time as a result of exposure and system use; the processes related to these phenomena are typically not directly observable. However, these processes all manifest themselves through changes in performance measures, and the latter is what we refer to as degradation. In this sense, theoretical and empirical models of the physical processes that result in system damage are quite valuable (and in some cases, critical) in developing effective models of degradation. Ben-Akiva and Ramaswamy [3] pioneered an approach to this problem using *latent* variables or processes, a concept that was first introduced in social sciences to model those characteristics that are not easily measurable or directly observable in a population [5]. While several attempts have been made to link the physical changes observed in the system to the system's capacity to perform its function [3, 6–8], these procedures are generally quite data intensive and suffer from computational limitations; nevertheless, this remains an open and very important problem in all aspects of engineering. However, we will not address this issue directly, and our main concern will be with the characterization of degradation as the reduction of the system capacity over time.

In engineering practice, system capacity is often characterized by an *index* or *rating* that is intended to combine a number of performance indicators into a single measure that represents the system state. Examples of such indices include the Present Serviceability Index (PSI) in pavement management, the Utah Bridge Deck Index (UBDI) for concrete bridge deck management, [9–13]. While these indices do serve as a guide for determining whether the system performance at a given time is acceptable, they have little *predictive* value [14], which is crucial to supporting operational and maintenance decisions. In this book, we will study predictive models for degradation that incorporate inherent randomness due to such factors as material variability, changes in operating conditions, and variable environmental factors.

4.3 Degradation: Basic Formulation

We desire a formulation of system performance over time that explicitly incorporates randomness in the design, manufacture, and operation of the system. To that end, let us assume that a new system (device or component) is placed in operation at time 0, and let V_0 be a positive random variable that measures the initial capacity of the system (also referred to as the *nominal life*). The nominal life of a system is generally determined by the system's design and manufacturing, and is independent of the operating conditions once the system is placed in service. Let $D(t)$ be a random variable that measures the accumulated degradation by time t, and let $V(t)$ be the *remaining* capacity of the system at time t. The remaining capacity at time t will simply be the nominal life decreased by the accumulated degradation up to time t, provided the system remains operational at time t; that is,

$$V(t) = V_0 - D(t), \tag{4.1}$$

Conceptually, failure occurs when the remaining life declines to zero; however, for our purposes, it will be useful to define performance states characterized by remaining life falling below a prespecified critical value [15] known as a *limit state*. Many maintenance and intervention models are based on *control-limit* policies that call for a particular action once a limit state is entered. A particularly important limit state, which will be widely used in this book, corresponds to a minimum performance level (here designated by k^*). Once this limit state is reached, the system will be removed from service (see Fig. 4.1), or replaced. We refer to this state as the *failure limit state*; even though a structure may still be minimally operational past this state, its continued use will pose unacceptable risks, and for all intents and purposes, it will be considered to have "failed" and will require complete replacement. The selection of k^* is usually obtained based on experience; frequently, $k^* = 0$ but in some cases it is reasonably to assume that $k^* > 0$.

Once the limit state k^* has been defined, we can revise our expression for remaining life as follows:

$$V(t) = \max(V_0 - D(t), k^*). \tag{4.2}$$

The system *lifetime* can then be defined as

$$L = \inf\{t \geq 0 : V(t) \leq k^*\}, \tag{4.3}$$

or equivalently,

$$L = \inf\{t \geq 0 : D(t) \geq V_0 - k^*\}. \tag{4.4}$$

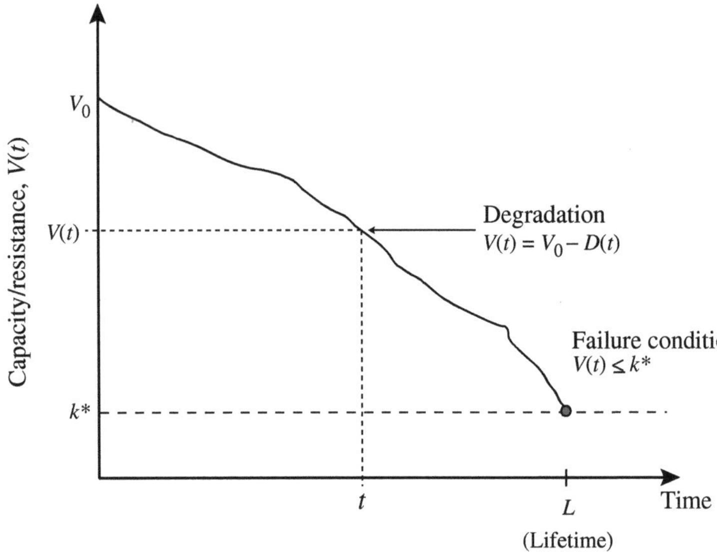

Fig. 4.1 Basic formulation of degradation

Note that we can interpret the device lifetime L as the *first passage* time for the total degradation process $\{D(t), t \geq 0\}$ to reach $V_0 - k^*$.

Other limit states may similarly be defined that correspond to acceptable performance levels determined, for instance, by a regulatory agency; i.e., a *serviceability limit state*. These states may indicate the need for a preventive intervention or maintenance but might not require complete replacement of the system, and again, the intervention times will be determined as first passage times to a limit state.

If the system is systematically maintained (repaired preventively and/or at times of failure), we can define system *availability* at time t as

$$A(t) = P(V(t) \geq k^*), \quad t \geq 0. \tag{4.5}$$

Based on models developed to describe nominal life and degradation over time, we are interested in estimating such quantities as:

- the probability distribution of capacity of the system at time t and, if it exists, in the limit as $t \to \infty$;
- the first passage time distribution for the capacity to fall below a prespecified threshold level; and
- the system availability at time t and, if it exists, the limiting system availability (this is of particular importance in cases where the system is systematically reconstructed—see Chap. 8).

4.4 Degradation Data

This book is concerned with models that characterize system degradation; that is, models that describe the deterioration in system performance over time. To "calibrate" these models, to estimate model parameters and to validate model performance requires the collection of data on actual system behavior. Data collection involves the structured gathering of empirical observations of systems, either under controlled, experimental conditions, or under uncontrolled operating conditions. Because it is often difficult to observe the physical changes that accompany degradation directly and continuously, we often monitor surrogates for these physical changes, or alternately, we may monitor some system performance indicator over time. While our main focus in this book is on model development, in this section we present an overview of the nature, problems, and challenges of collecting and analyzing data to characterize degradation.

4.4.1 Purpose of Data Collection

One of the primary objectives of data collection in modeling degradation is to predict the time until the system reaches particular operational states. The types of data gathered for infrastructure degradation and reliability studies, and the methods used for their analysis, can be generally aggregated into two main directions. The first involves the direct study of the time at which system failure occurs. The vast majority of reliability studies are related to failure time estimation for systems that are replaced upon failure (so-called non-repairable systems). Common experimental techniques involve placing statistically identical items on test under operating conditions (or accelerated operating conditions) and observing the time of failure of each item. Because not all items may have failed by the end of the study period, failure time studies typically involve *censored* data. In these cases, precise failure times are not known, but the censored observations provide a lower bound on the actual failure time. This is particularly true in the case of infrastructure components that are designed to be highly reliable and have life spans of several decades. Statistical methods for dealing with censored data have a long history in the field of survival analysis and life testing; some further reading on this topic can be found in [16–18].

When modeling failure times, data is used to estimate parameters of a (positive) lifetime random variable. Common distributions used in modeling time to failure include the exponential, gamma, Weibull, lognormal, and several other less common distributions (e.g., inverse Gaussian, Birnbaum–Saunders, Gompertz–Makeham). A number of references are available for the statistical properties of these distributions, including [1, 2, 19]. For reliability prediction, moment-based parameters, such as the mean and variance of lifetime, are often not of primary interest. Rather, engineers may be more interested in estimating quantiles of the lifetime or (similarly) failure probabilities for given (fixed) mission lengths. The choice of distribution to fit often involves the phase of life that is of interest, as determined by the shape of the hazard function, and many techniques have been developed that address modeling the hazard rate directly as a linear or polynomial function; cf. [20].

A second direction for data collection and analysis in degradation modeling involves situations where actual physical changes that lead to deterioration of system performance can be measured. Examples include material fatigue induced by crack formation and propagation, material removal due to wear or thermal cycling, corrosion, and fracture. If direct measurements of these processes can be made over time, the analyst often has more information available that may allow modeling of the actual failure mechanism. In cases where actual degradation processes are not observable, it may still be possible to observe a performance measure that acts as a surrogate for degradation, for instance, decreasing power output of an electronic device over time. Techniques for modeling degradation paths over time are quite complex, and necessarily employ analytical models of specific physical failure mechanisms. These models generally involve the effects of stressors such as temperature, duty cycle, vibration, humidity on the material properties of a system. In contrast to direct measurement of failure times, these degradation models are often used to predict when the measured degradation (or its performance surrogate) reach a threshold that results in failure. Variability due to the initial material properties (manufacturing process) as well as actual operating conditions leads to the attainment of the failure threshold, and hence this approach can also lead to estimation of the lifetime distribution; some additional information on this approach can be found in [21].

Whether working with failure time observations or with observations of degradation or performance, highly reliable systems and those that are designed for long mission lengths may require *accelerated testing*. In accelerated testing, the level or intensity of stressors are magnified beyond what normal operating conditions would dictate in order to induce premature degradation or failure. There is a great body of work related to accelerated testing; suffice it to say that the design and analysis of accelerated tests for failure prediction is quite complicated and involves a great deal of engineering judgement.

4.4.2 Data Collection Challenges

As technology evolves toward more precise and less expensive data acquisition systems, modeling the degradation process of engineered systems should become a common practice. Today it is possible to install sensors and smart chips to measure and record data about the system performance over the life of an engineered device. In some areas, this practice belongs to the area of *system health monitoring* and materials state information. This information is used to carry out real-time monitoring and for prognostic purposes. Thus, the next generation of reliability field data will be richer in information and as the cost of technology drops, cost/benefit ratios decrease and applications spread to different practical problems [22].

Future data will also come from the development of better accelerated tests. These will require new lab techniques and methods to incorporate the main sources of uncertainty that are found in the field like load demands, temperature, humidity, material oxidation, etc., [1]. In this field, scale models and testing facilities such as the geotechnical centrifuge [23] have been used extensively.

Furthermore, the development of analytical tools to replicate actual experimental data is an area of research that is gaining a lot of attention. Frequently, simulations are used in situations where experiments are not feasible for practical or ethical reasons. The main questions associated to this issue are related to the assumptions, the validity and the conditions required for a simulation so that it can serve as a surrogate for an experiment. Thus, simulation techniques should guarantee that the results are as reliable as the results of an analogous experiment [24]. Further discussions on this topic can be found in [25–28].

4.5 Construction of Models from Field Data

The selection of the best degradation model is guided by both field data and some understanding of the mechanical laws that describe the system performance. If there is information about the physics that drive the behavior of the system, the mechanical performance can be expressed in the form of a differential equation, or a system of differential equations with some randomness that can be associated to, for instance, the model parameters (e.g., rate, material properties) [1]. A classic example is the case of fatigue of materials expressed in terms of the crack growth rate; thus, degradation can be described as:

$$\frac{da(t)}{dt} = C \times [\Delta K(a)]^m$$

where C and m are constants, and $a(t)$ is the crack size; and ΔK is the range of the stress intensity factor, i.e., the difference between the stress intensity factor at maximum and minimum loading $\Delta K = K_{max} - K_{min}$, where K_{max} and K_{min} are the maximum and minimum stress intensity factor respectively [29]. Another example is the automobile tire wear (wear rate), which is modeled as: $D(t)/dt = C$; where C is a constant. The selection of the best mechanical model depends upon the physics of the problem at hand and it is a topic that is not in the scope of this book.

Sometimes the complexity of the degradation problem makes it hard to find a unique mathematical formulation and the only information available is field data. In these cases, the only option is to make inferences from failure time observations or from data about the system condition at different points in time. The former provides information about the lifetime distribution, while the latter can be used to model and understand the system performance over time; information that can be used later to build a mechanistic model of the degradation process (see Chaps. 5–7). In this section, we will briefly mention the basic concepts of regression analysis, which can be interpreted as the most basic degradation model; literature about regression analysis is abundant, but some useful information can be found in [30, 31].

4.6 General Regression Model

Let us assume that the degradation path of the system consists of a vector of field measurements $\{y_1, y_2, \ldots, y_m\}$ made at discrete points in time $\{t_1, t_2, \ldots, t_m\}$, which reveal the actual condition of the system. Let us also assume that the system performance is characterized by a model denoted by $D(t) = y'(t)$; e.g., target degradation model (see Fig. 4.2).

Then, the relationship between actual data and the model at time t_i can be written as:

$$y(t_i) = y'(t_i) + \epsilon(t_i); \quad i = 1 \cdots m \tag{4.6}$$

where $\epsilon(t_i) = y(t_i) - y'(t_i)$ is a measure of the error (residual) at time t_i and is usually modeled as a random variable normally distributed; i.e., $N(0, \sigma_\epsilon)$. The form of $y'(t)$ is obtained from a mechanical model or can be selected arbitrarily. For example, several commonly used models for degradation are shown in Table 4.1; where $\mathbf{B} = \{\beta_0, \beta_1, \ldots \beta_k\}$ is a set of parameters that fully characterize the model. For example, if it has a linear form, $y(t_i, \mathbf{B}) = \beta_0 + \beta_1 t_i + \epsilon(t_i)$. In practice, it is

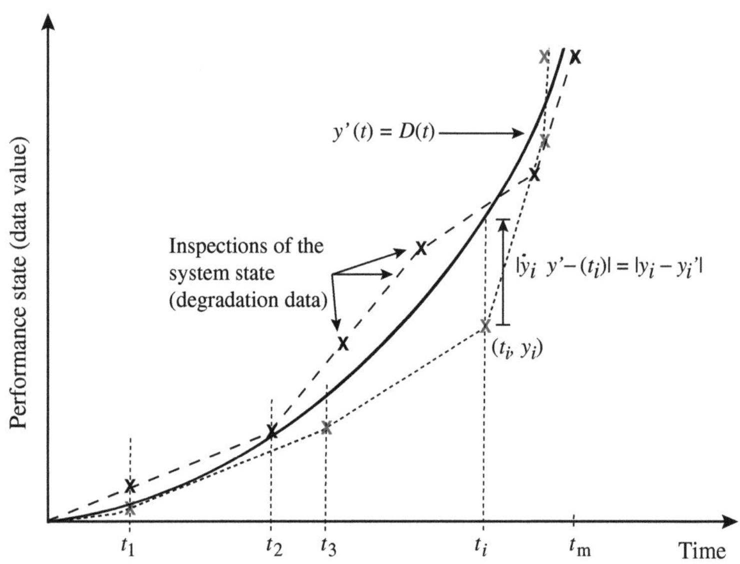

Fig. 4.2 Description of the general degradation model

Table 4.1 Common regression models useful to describe degradation

Regression type	$y'(t, \mathbf{B})$
Linear	$\beta_0 + \beta_1 t$
Exponential	$\beta_1 \cdot e^{\beta_2 t}$
Power	$\beta_1 \cdot t^{\beta_2}$
Logarithmic	$\beta_0 + \beta_1 \cdot \ln(t)$
Logistic	$\frac{\beta_1}{1+\beta_2 \exp(-\beta_3 t)}$
Gompertz	$\beta_1 \cdot \beta_2^{\beta_3^t}$
Lloyd-Lipow	$\beta_0 - \beta_1 t^{-1}$

usually assumed that the set of parameters $\mathbf{B}$ are independent of ϵ, and that σ_ϵ is constant [1]. It is important to stress that although frequently a predefined model for $y'(t)$ is selected, occasionally, the form of degradation is unknown and, therefore, nonparametric regression techniques are required to analyze the data.

Due to the inherent variability of the problem, the set of parameters $\mathbf{B}$ are uncertain, which leads to possible different degradation paths with the same general trend. For example, Fig. 4.3 shows the measurements of the crack size in a fatigue test of an Alloy-A [32], which is a standard degradation process in materials subjected to repeated loads. In this figure every curve represents the result of a specimen built and tested under the same conditions. It can be observed that there is some important variability in the results.

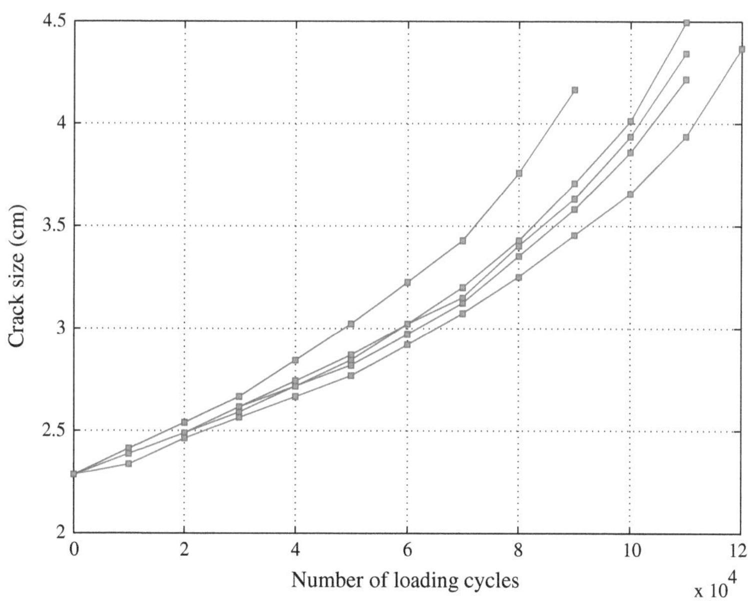

Fig. 4.3 Fatigue crack data of Alloy-A (Data reported in Lu and Meeker, 1993 [32])

In more complex structures this degradation process is more difficult to evaluate and the uncertainties more difficult to quantify. For example, in the area of asphalt pavements the surface of the top asphalt course is permanently exposed to the combined action of traffic loading and climatic effects. Among the different phenomena affecting the functionality and durability of these materials, asphalt oxidation is recognized in as one of the most relevant weather-related deterioration processes. Oxidative hardening is defined as the process by which the asphalt binder present in the mixture becomes stiffer as a consequence of its chemical reaction with the oxygen present in the air. The main consequence of this chemical process is that the mixture becomes more fragile, which in turn makes it more susceptible to undergo fracture. This is of particular concern during low temperature seasons where this condition can promote the appearance of cracks at the surface, affecting the overall serviceability, functionality, and durability of the pavement structure. Figure 4.4 presents the increase in the expected normalized dynamic modulus (i.e., the increase in modulus with respect to the modulus at the moment of opening the pavement to traffic) during the initial 5 years of a pavement. Note that there is a significant variability, depending upon the construction process, i.e., different compaction levels leading to different air void contents, and on other aspects such as the chemical kinetics that describes the coupled effect of oxidative hardening and the mechanical viscoelastic response of the material [33].

These examples show the importance of quantifying the randomness of **B**, which is clearly problem-related and can be described by a multivariate normal distribution

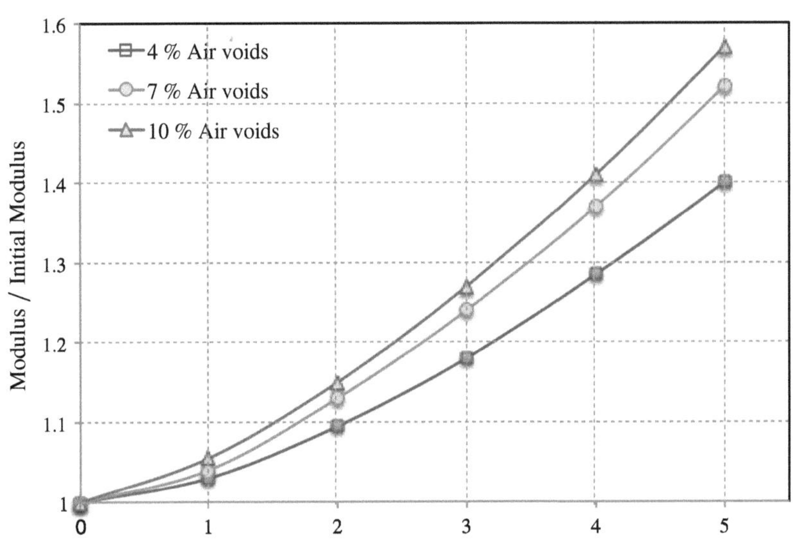

Fig. 4.4 Increase in modulus of asphalt mixtures with different air void content as a consequence of oxidative hardening (modified after Caro et al. [33])

with mean vector $\mu_{\mathbf{B}}$ and covariance matrix $\Sigma_{\mathbf{B}}$ (see Meeker and Escobar [1]). Finally, and for completeness, the analysis should also take into account the set of parameters $\mathbf{p}$ that are important to describe the process but are not necessarily random; for instance, the geometry. Then, Eq. 4.6 can be rewritten as:

$$y(t_i) = y'(t_i, \mathbf{B}, \mathbf{p}) + \epsilon(t_i); \quad i = 1 \cdots m; \quad j = 1 \cdots k \tag{4.7}$$

4.7 Regression Analysis

Finding the best degradation model requires identifying the function $y'(t)$ (we drop $\mathbf{p}$ for now) and the parameters $\mu_{\mathbf{B}}$ and $\Sigma_{\mathbf{B}}$. Thus, a regression has the following form:

$$\mathbb{E}[Y|t] = y'(t, \hat{\mathbf{B}}) \tag{4.8}$$

where $\hat{\mathbf{B}}$ is the best estimator of the vector parameter $\mathbf{B}$. For example, for the case of a linear regression: $y'(t, \hat{\mathbf{B}}) = \hat{\beta}_0 + \hat{\beta}_1 t$. The function $y'(t)$ is obtained by evaluating various models (e.g., see Table 4.1) and selecting the one with the least cumulative error; this error is evaluated as:

$$\Delta^2 = \sum_{i=1}^{n} (y_i - y'_i)^2; \quad i = 1, 2, \ldots, n, \tag{4.9}$$

where y'_i is the value of the proposed model and y_i the value of the actual data point at time t_i ($i = 1, \ldots, m$ data points). Frequently, the error is also evaluated in terms of what is called the *mean square error* (MSE) of the regression:

$$\text{MSE} = \frac{1}{n} \sum_{i=1}^{n} (y_i - y'_i)^2; \quad i = 1, 2, \ldots, n, \tag{4.10}$$

The error term in Eq. 4.7 it is usually assumed to have a constant variance, i.e., $\epsilon \approx N(0, \sigma_\epsilon^2 = \text{constant})$. However, if there is significant variation in the degrees of scatter of the control variable (i.e., data value at an inspection time), the conditional variance of the regression equation will not be constant and $\epsilon \approx N(0, \sigma_\epsilon^2 = q(t))$. In these cases, Eq. 4.9 needs to be evaluated as [31]:

$$\Delta^2 = \sum_{i=1}^{n} w_i (y_i - y'_i)^2; \quad i = 1, 2, \ldots, n, \tag{4.11}$$

where w_i is a weight assigned to the data such that data points in regions of small conditional variance (i.e., small σ_ϵ^2) should carry higher weights than those in regions with larger conditional variance. These weights are assigned inversely proportional to the conditional variance [31]; i.e.,

$$w_i = \frac{1}{\sigma^2_{\epsilon(t_i)}(y'(t_i))^2} \tag{4.12}$$

The estimation of the parameters of the regression, i.e., **B** (Eq. 4.7), can be obtained by minimizing Δ^2 in Eq. 4.9 or 4.11; i.e.,

$$\min_{\mathbf{B}} \sum_{i=1}^{n} (y_i - y'(t_i, \mathbf{B}))^2; \qquad \min_{\mathbf{B}} \sum_{i=1}^{n} w_i (y_i - y'(t_i, \mathbf{B}))^2 \tag{4.13}$$

This method is usually referred to as the *method of least squares*. It is important to mention that Eqs. 4.9 and 4.10 should be modified if there is some correlation between the observation times and the data values [34]; however, this is not usually the case in degradation problems.

There is a vast amount of literature available about regression analysis; conceptual discussions on particular aspects as well as specific examples in Civil engineering problems and calculation details can be found in, for example, [31, 34, 35]. In the following two subsections we will briefly summarize some important aspects of linear and nonlinear regression models. The case of multivariate regression models will not be discussed here but the details can be found in [31, 35].

4.7.1 *Linear Regression*

The case of linear regression $y'(t, \mathbf{B}) = \beta_0 + \beta_1 t$ has been widely studied and the derivation of the estimative for the parameters β_1 and β_2 can be obtained using the method of least squares. Let's consider a sample of observed data pairs of size n, i.e., $\{(t_1, y_2), (t_2, y_2), \ldots, (t_n, y_n)\}$, where, for example, t_i is the time at which the system is inspected and y_i the result of the inspection in terms of a given performance measure. Then, the parameters of the regression equation can be obtained analytically by solving Eq. 4.13 where $y'(t)$ has a linear form:

$$\min_{\mathbf{B}} \sum_{i=1}^{n} (y_i - y'(t_i, \mathbf{B}))^2 = \min_{\{\beta_0, \beta_1\}} \sum_{i=1}^{n} (y_i - \beta_0 - \beta_1 t_i)^2 \tag{4.14}$$

Then, computing the derivative of Eq. 4.14 with respect to the parameters and equating to 0, leads to (for the case of constant variance) [31]:

$$\hat{\beta}_0 = \frac{1}{n} \sum_{i=1}^{n} y_i - \frac{\hat{\beta}_1}{n} \sum_{i=1}^{n} t_i = \bar{y} - \hat{\beta}_1 \bar{t} \tag{4.15}$$

$$\hat{\beta}_1 = \frac{\sum_{i=1}^{n} y_i t_i - n\bar{y}\bar{t}}{\sum_{i=1}^{n} t_i^2 - n\bar{t}^2} = \frac{\sum_{i=1}^{n} (t_i - \bar{t})(y_i - \bar{y})}{\sum_{i=1}^{n} (t_i - \bar{t})^2}, \tag{4.16}$$

where $\bar{y}$ and $\bar{t}$ are the corresponding sample means, and n is the sample size. Therefore, the least-squares regression equation is:

$$\mathbb{E}[y|t, \hat{\mathbf{B}}] = \hat{\beta}_0 + \hat{\beta}_1 t \tag{4.17}$$

4.7.2 Nonlinear Regression

In most degradation problems the functional regression among variables (e.g., time and performance measure) is not always linear; on the contrary, frequently it shows nonlinear trends. The basic idea of nonlinear regression is the same as that of linear regression; the main difference is that the prediction equation $y'(t)$ (Eq. 4.7) depends nonlinearly on one or more unknown parameters. For instance, $y'(t) = \beta_0 + t/(1 + \beta_1)^2$; also some typical examples are shown in Table 4.1. It is important to stress that the definition of nonlinearity actually relates to the unknown parameters and not to the relationship between the covariates and the response. A comprehensive review of nonlinear regression models and many practical examples can be found in [30, 36, 37].

Frequently, nonlinear regression models are constructed from expressions linear in the parameters. For example (dropping **B** for now);

$$y'(t) = \beta_0 + \beta_1 g(t) \tag{4.18}$$

where $g(t)$ is a nonlinear function of t. A common model that follows this approximation is the polynomial regression, which can be written as follows:

$$y'(t) = \beta_0 + \beta_1 t + \beta_2 t^2 + \beta_3 t^3 + \cdots + \beta_n t^n \tag{4.19}$$

whose parameters can be computed using the least-squares method described above. Another important example of transforming a nonlinear function into a linear expression is the following: consider the nonlinear function $y'(t) = \beta_0 \exp(\beta_1 t)$; then, by taken logarithm in both sides we get that $\ln y'(t) = \ln \beta_0 + \beta_1 t$ and the regression equation can be computed as:

$$\mathbb{E}[\ln y'|t] = \ln \beta_0 + \beta_1 t \tag{4.20}$$

Example 4.14 In asphalt pavements, fatigue is a critical failure mechanism. Consider two asphalt mixtures subjected to a standard fatigue test[1] [38] and whose results are shown in Table 4.2.

---

[1]Data obtained from the Materials lab in the Department of Civil & Environmental Engineering at Los Andes University—Fatigue tests that follow the norm UNE-EN-12697-24:2006+A1 [38].

Table 4.2 Fatigue data of two asphalt mixtures

Asphalt Mix-1		Asphalt Mix-2	
N. Cycles ($\times 10^6$)	Def. ($\times 10^{-3}$m/m)	N. Cycles ($\times 10^6$)	Def. ($\times 10^{-3}$m/m)
0.072	0.15	0.09	0.165
0.72	0.09	0.504	0.09
2.448	0.06	1.44	0.057
0.072	0.15	0.078	0.165
0.792	0.09	0.504	0.09
2.448	0.06	1.368	0.057
0.096	0.15	0.072	0.165
0.900	0.09	0.648	0.09
2.160	0.06	1.512	0.057
0.084	0.15	0.054	0.165
0.972	0.09	0.576	0.09
2.160	0.06	1.584	0.057

Based on this information we can construct a degradation model via regression analysis. The fatigue curve can be described by the following equation:

$$\log(N) = C - m\log(S) \tag{4.21}$$

where N is the number of cycles to failure at an stress/strain amplitude S; and C and m are constants to be determined. Note that rearranging Eq. 4.21 we get

$$NS^m = C \tag{4.22}$$

which is usually referred to as the $S - N$ relationship. Equation 4.21, is a nonlinear regression, which can be expressed as a linear regression. Note that Eq. 4.21 can be expressed also as: $\log(S) = \alpha - \beta\log(N)$. Then, using the least-squares method, the estimates of the regression coefficients for the first asphalt mixture are $\hat{\alpha}_1 = -2.5291$ and $\hat{\beta}_1 = -0.2620$; and for the second asphalt mixture: $\hat{\alpha}_2 = -2.1199$ and $\hat{\beta}_2 = -0.3406$. This leads to the regression degradation model shown in Fig. 4.5. Furthermore, the fatigue formulation, Eq. 4.22, for both mixes becomes:

$$NS^{\hat{m}_1} = \hat{C}_1 = NS^{-3.817} = 9.653 \tag{4.23}$$

$$NS^{\hat{m}_2} = \hat{C}_2 = NS^{-2.936} = 6.224 \tag{4.24}$$

where $m = 1/\beta$ and $C = \alpha/\beta$.

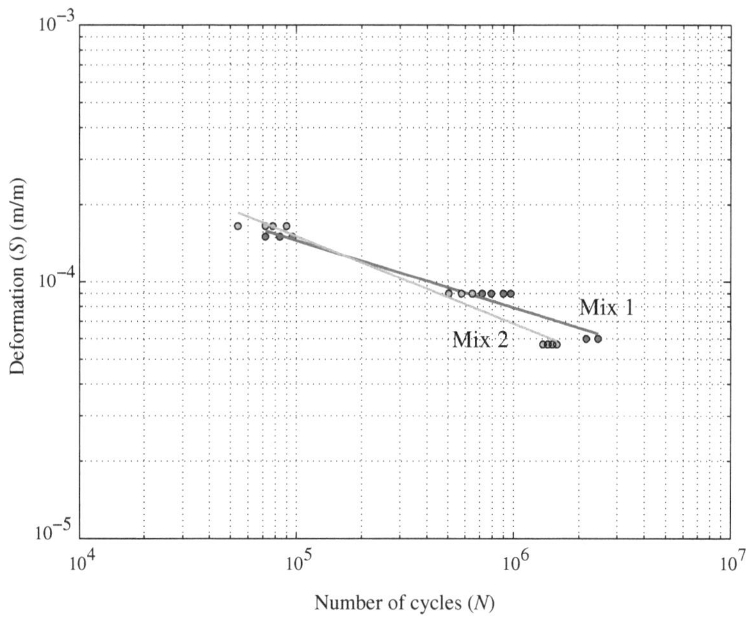

Fig. 4.5 Asphalt fatigue degradation model based on experimental data

4.7.3 Special Case: Parameter Estimation for the Gamma Process

Data analysis is essential to build any model, and degradation is not an exception. In Chap. 3 we presented the basics of the most important models that we will later develop in more detail in Chaps. 5–7. Among them, there is one particular case that is particularly important, i.e., the gamma process. It is used mostly to model progressive degradation since it is somewhat an improvement over rate-based models (see Sect. 4.9.2). The gamma process will be discussed in more detail in Sect. 5.5.1. In this section, we will present an approximation, described in [39], to find the parameters of the gamma process (i.e., the scale u, and shape v, parameters) from empirical data. For this task, we would present the results obtained by using two main methods: Moment Matching (MM) and Maximum Likelihood (ML).

The MM and the ML methods can be used also in other models described later such as when obtaining the parameters for phase-type distributions (Chap. 6). Some references will be given when necessary.

4.7.4 Moment Matching Method

Let us define the target degradation model as $D(t) = y'(t)$ (Sect. 4.6). Furthermore, consider that the underlying degradation process is represented by a gamma process (see Eq. 5.50 in Sect. 5.5.1) with scale parameter u and shape parameter $v(t)$. Then we can use the MM method to define the parameters of the gamma process that describe $D(t)$.

The expected value and variance of the accumulated deterioration at time t (i.e., calendar time), $D(t)$, with $t \geq 0$ are:

$$\mathbb{E}[D(t)] = \frac{v(t)}{u} \quad \text{and} \quad Var[D(t)] = \frac{v(t)}{u^2}. \tag{4.25}$$

The expected deterioration function can take any form depending of the problem at hand; however, as discussed later in Sect. 4.9.2, it is reasonable to assume a power law for the expected deterioration at time t, $v(t)$, [39]; i.e., $v(t) = ct^b$, for some constants $c > 0$ and $b > 0$. This kind of relationship is often present in many practical applications [9, 13].

For the particular case in which the exponent b of the power law is known, the nonstationary gamma process can be transformed into a stationary gamma process by making the following time transformation. Since $z = t^b$ then $t = z^{1/b}$ [39], and therefore the expected value and the variance in Eq. 4.25 become:

$$\mathbb{E}[D(t)] = \frac{cz}{u} \quad \text{and} \quad Var[D(t)] = \frac{cz}{u^2}. \tag{4.26}$$

which result in a stationary gamma process with respect to the transformed time z.

Suppose now that the set $\{y_0, y_1, \ldots, y_n\}$ are the results from inspections taken at times $\{t_0, t_1, \ldots, t_n\}$. Then, the transformed inspection times can be computed as: $z_i = t_i^b$ with $i = 0, 1, 2, \ldots, n$; and the transformed times between inspections can be defined as $w_i = t_i^b - t_{i-1}^b = z_i - z_{i-1}$. This means that the deterioration increment, $\Delta_i = D(t_i) - D(t_{i-1})$, has a gamma distribution with shape parameter cw_i and scale parameter u for all i. The corresponding observations of Δ_i are given by: $\delta_i = y_i - y_{i-1}$. Then, the estimators $\hat{c}$ and $\hat{u}$ from the method of moments are given by [13]:

$$\frac{\hat{c}}{\hat{u}} = \frac{\sum_{i=1}^{n} \delta_i}{\sum_{i=1}^{n} w_i} = \frac{y_n}{z_n} = \frac{y_n}{t_n^b} \tag{4.27}$$

$$\frac{\hat{c}}{\hat{u}^2}\left(t_n^b - \frac{\sum_{i=1}^{n} w_i^2}{t_n^b}\right) = \sum_{i=1}^{n}\left(\delta_i - \frac{y_n}{t_n^b} w_i\right)^2, \tag{4.28}$$

Note that the first equation involves the sum of the observed damage increments, which leads to the total damage observed, i.e., y_n, which occurs at time t_n (i.e., total time). In other words, the last observation is enough to fit the first moment, as it contains the information from all the previous damage increments.

4.7.4.1 Maximum Likelihood

The method of maximum likelihood estimates c and u by maximizing the log-likelihood function of the observed damage increments $\delta_i = y_i - y_{i-1}$. As these are independent, their joint density can be defined as $f_{\Delta_1,\ldots,\Delta_n}(\delta_1,\ldots,\delta_n)$, which is simply the product of the individual gamma densities,

$$f_{\Delta_i}(\delta_i) = \frac{u^{\Delta v_i}\delta_i^{\Delta v_i - 1}}{\Gamma(\Delta v_i)}\exp(-u\delta_i) \tag{4.29}$$

where $\Delta v_i = v(t_i) - v(t_{i-1}) = c(t_i^b - t_{i-1}^b)$, for $i = 1,\ldots,n$.

Then, the likelihood of the observed degradation increments takes the form:

$$\begin{aligned} l(\delta_1,\ldots,\delta_n|c,u) &= \prod_{i=1}^{n} f_{\Delta_i}(\delta_i) \\ &= \prod_{i=1}^{n} \frac{u^{c(t_i^b - t_{i-1}^b)}}{\Gamma(c(t_i^b - t_{i-1}^b))}\delta_i^{c(t_i^b - t_{i-1}^b)-1}\exp(-u\delta_i). \end{aligned} \tag{4.30}$$

A system of equations is obtained by evaluating the partial derivatives of the log-likelihood function of the degradation increments with respect to c and u. Then, the estimatives $\hat{c}$ and $\hat{u}$ can be solved from [13]:

$$\hat{u} = \frac{\hat{c}t_n^b}{y_n}, \tag{4.31}$$

$$t_n^b \log\left(\frac{\hat{c}t_n^b}{y_n}\right) = \sum_{i=1}^{n-1}(t_{i+1}^b - t_i^b)\{\psi(\hat{c}(t_{i+1}^b - t_i^b)) - \log\delta_i\}, \tag{4.32}$$

where $\psi(x)$ is the digamma function, defined as the derivative of the logarithm of the gamma function: $\psi(x) = \frac{d\log\Gamma(x)}{dx} = \frac{\Gamma'(x)}{\Gamma(x)}$, and can be computed with a standard software, e.g., MATLAB®. Observe that Eq. (4.31) is the same as the Eq. (4.27) corresponding to the first moment fitting in the MM method.

Note that for the maximum likelihood estimator of u obtained from Eqs. 4.31 and 4.32, the expected deterioration at time t can be written as [39]:

$$\mathbb{E}[D(t)] = y_n \left(\frac{t}{t_n}\right)^b \tag{4.33}$$

Example 4.15 The objective of this example is to estimate the parameters of a gamma process using the two fitting methods described above (i.e., MM and ML). In this illustrative example, degradation data are obtained from simulation of a gamma process with shape parameter $v(t) = ct^2$ ($c = 0.005$), for $0 \leq t \leq 120$; and scale parameter $u = 1.5$. The results are used as if they were actual field data observations, for which the parameters of the gamma process will be obtained.

Thirty sets of data were obtained numerically; this information is assumed to correspond to field data for different artifacts. The thirty degradation data sets were divided in three groups of 10 artifacts each; in each group, data was collected at a specific and fix time interval; i.e., there were three different inspection strategies. The time intervals selected for each strategy are: $\Delta t = \{0.5, 1, 2.5\}$ years, thus obtaining $n = \{240, 120, 48\}$ measurements of an artifact condition in each set, respectively. The observed data of five artifacts of the set with $\Delta t = 2.5$, are shown in Fig. 4.6.

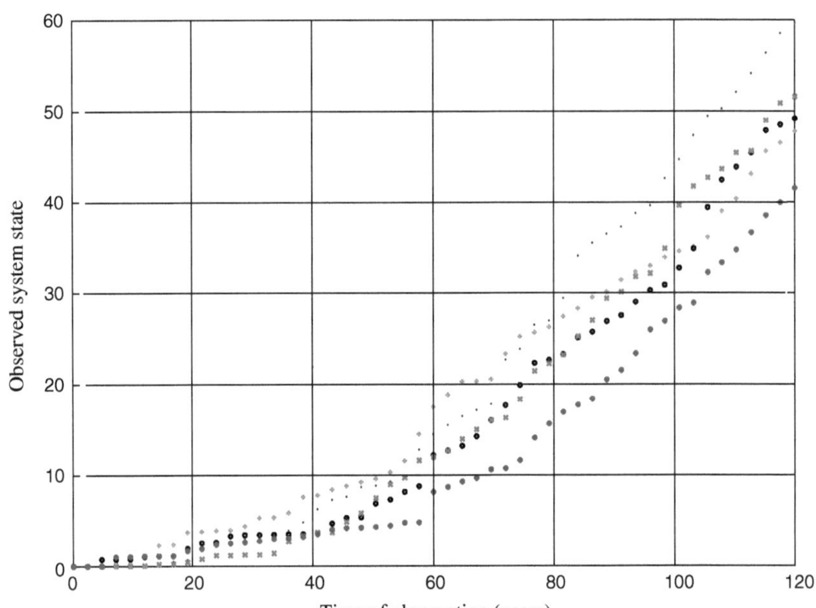

Fig. 4.6 Observations of the system state of various artifacts taken at times intervals of $\Delta t = 2.5$ years

Table 4.3 Mean relative error $\bar{\epsilon}$ (in %) for each data set

Method	Parameter	Set $j = 1$: $n = 48$ $\Delta t = 2.5$ (%)	Set $j = 2$: $n = 120$ $\Delta t = 1.0$ (%)	Set $j = 3$: $n = 240$ $\Delta t = 0.5$ (%)
MM:	$\hat{c}$	19	19	15
	$\hat{\beta}$	24	20	19
ML:	$\hat{c}$	17	11	5
	$\hat{\beta}$	22	14	11

Based on the previous discussion (Sects. 4.7.4 and 4.7.4.1), and given the form of the shape parameter (i.e., $v(t) = ct^2$), the value of $\hat{c}$ and $\hat{\beta}$ of the gamma process for each artifact data are calculated using both the MM and ML methods. Afterwards, the difference (i.e., error) of the estimative of the parameters for each artifact with respect to the parameters of the actual process, from which experimental data was generated, is calculated as: $\epsilon_i = (\hat{z}_i - z) \times 100/z$, where z can be either c or β. Then, the mean relative error was computed for each group, j, of ten artifacts (with observations at the same time interval) as: $\bar{\epsilon}_j = 0.1 \cdot \sum_i^{10} \epsilon_{i,j}$; with $j = 1, 2, 3$ and i the artifact number. The results are shown in Table 4.3.

Note first that, in this particular case, the ML method performs better than the MM method, for all data sets (i.e., smallest $\bar{\epsilon}$). Although for the first set the errors are quite similar (around 18 % for $\hat{c}$ and 23 % for $\hat{\beta}$), they become further apart as the number of data points increase. For instance, for the third data set, the error for $\hat{c}$ in the MM method is 15 % while in the ML method is 5 %, and the error for $\hat{\beta}$ is 19 % and 11 % for the MM and ML method, respectively. In summary, the error diminishes in both methods as more data points are available, but decreases faster for the ML method compared with the MM method. This is expected, as the ML method takes into account the entire density function.

In Figs. 4.7a, b we show various sample paths constructed with the parameters given by the estimators shown in Table 4.4; which correspond to specific artifacts. Besides, the mean deterioration $\mathbb{E}[D(t)]$ from the fitted gamma processes and the mean deterioration of the actual gamma process are plotted. Note that $\mathbb{E}[D(t)]$ of the fitted gamma processes are the same, for both algorithms. This is so, because $\mathbb{E}[D(t)]$ is proportional to the ratio $\hat{c}/\hat{\beta}$, which depends only on the last data point (t_n, y_n) for both algorithms, according to Eqs. (4.27) and (4.31). Note also that for this particular data set, the estimated mean deterioration is greater than the actual mean deterioration.

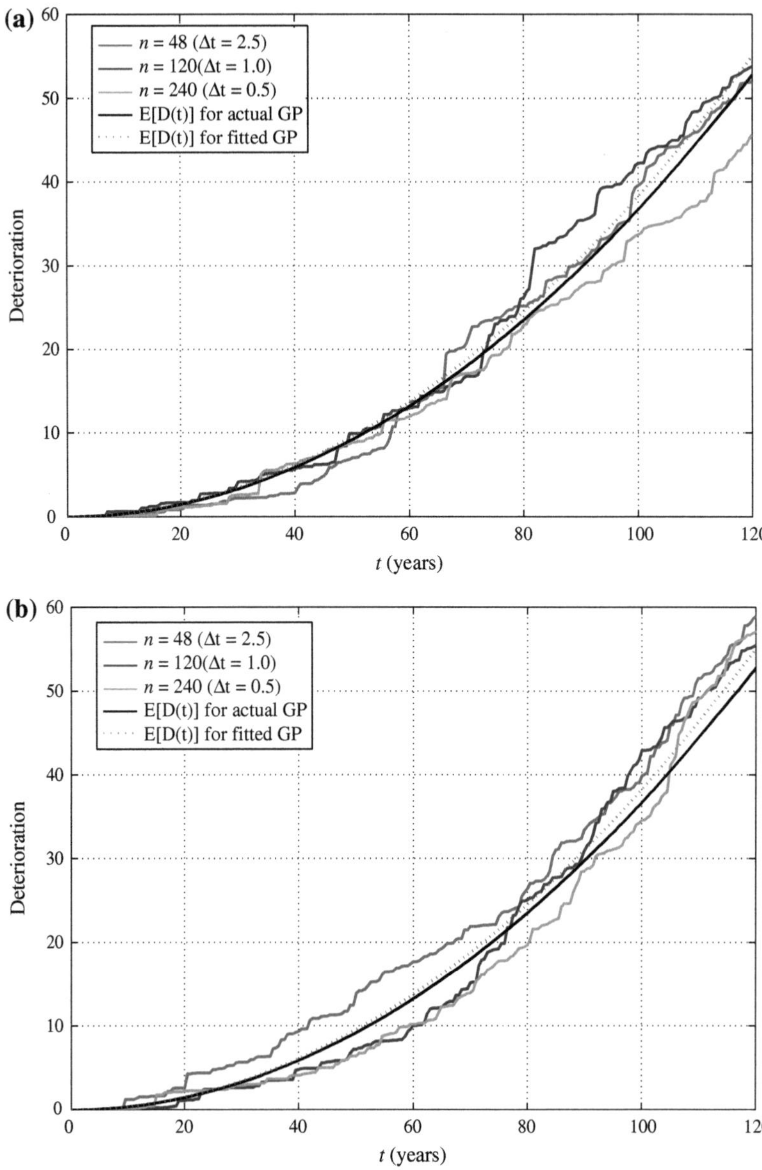

Fig. 4.7 Degradation sample paths evaluated using the parameters evaluated by (**a**) MM method; and (**b**) ML method

Table 4.4 Parameters of the gamma process used to build the sample paths shown in Figs. 4.7a, b

Method	Parameter	Set 1: $n = 48$ $\Delta t = 2.5$	Set 2: $n = 120$ $\Delta t = 1.0$	Set 3: $n = 240$ $\Delta t = 0.5$
MM:	$\hat{c}$	0.008	0.0074	0.0071
	$\hat{\beta}$	2.1011	1.9239	1.8484
ML:	$\hat{c}$	0.0078	0.0069	0.0065
	$\hat{\beta}$	2.034	1.804	1.7075

4.8 Analytical Degradation Models

In Sects. 4.4–4.7 we briefly discussed the importance of field data in modeling degradation and presented a first approximation using regression analysis. However, most of this book is concerned with analytical models. Then, in this and the following sections, we will provide a conceptual framework for characterizing system degradation over time and define appropriate random variables that will be used in the subsequent chapters.

4.8.1 A Brief Literature Review

Degradation modeling is challenging because it involves the interaction of environmental conditions with material and other physical properties of the system. There are many approaches available in the literature for modeling physical changes that can result in a reduction of system capacity. These approaches vary depending upon the problem at hand and the scope of the analysis. Physical changes such as crack initiation and growth, material corrosion, material removal, etc., and physical models of these phenomena may be quite detailed. However, it is not always an easy task to identify how these physical changes lead to a reduction in system capacity, which is how we define degradation. Then, in this book degradation models will focus not specifically on physical changes but rather on a more general model of reduction in capacity over time.

In the literature, many models assume that degradation is defined by a functional class with a set of parameters to be determined [13, 40, 41]. There are also models based mainly on the theory of stochastic processes; some examples can be found in [40, 42–44]. Markov processes have been used extensively, see for instance, [45–50]. Recently, a significant amount of research has been carried out based on models that use information obtained at different points in time to reevaluate the predictions about the system performance. Most of these methods include Bayesian probability; see, for example, [51–53]. A review of common probabilistic models for life-cycle performance of deteriorating structures can be found in [11]. Some additional references that may be of interest are [10, 11, 40, 51, 54–58].

To summarize, the literature on degradation modeling spans the spectrum from physical modeling of mechanical and chemical processes through life-cycle modeling of an idealized system state over time. What is clear is that degradation is a general response to the interaction of many different ongoing physical processes within the system. Each of these processes causes physical changes that lead to deterioration in performance. Moreover, some of these processes may be generally independent, while others may have complicated interactions. The reality is that actual physical changes in complex systems are often very difficult to observe and monitor in situ, leading us to embrace a more conceptual notion of degradation that allows modeling of a variety of physical mechanisms.

4.8.2 Basic Degradation Paradigms

Because of the challenges in modeling a variety of physical changes that cause system performance to degrade over time, most degradation modeling asserts two primary degradation classes, namely

- continuous (progressive or graceful) degradation; and
- degradation due to discrete occurrences (shocks).

Conceptually, it is convenient for a variety of reasons to classify degradation in this way. From an observational viewpoint, certain mechanisms, such as corrosion or continuous material removal due to friction or heat, fit naturally within the progressive deterioration category. These mechanisms generally involve very small changes in physical properties that occur continuously over a long timescale. Other changes, such as loss of material due to a sudden collision and disruptions due to failure of a component that may not cause immediate system failure, are more appropriately viewed as shock degradation. Mathematically, the stochastic models suitable for modeling continuous degradation are quite different from those suitable for modeling shock degradation. Because the drivers of progressive deterioration and shocks are typically different (and may be relatively independent), a general mathematical model of degradation can be constructed that consists of a *superposition* of models for each degradation class (see Chap. 7). In what follows, we provide practical examples and discuss models for both graceful and shock-based degradation separately before presenting a general model that incorporates both classes of degradation.

4.9 Progressive Degradation

4.9.1 Definition and Examples

Progressive degradation, also called *graceful* degradation, is the result of the system's capacity/ resistance (*life*) being *continuously* depleted at a rate that may change over time. As an example, three realizations of progressive degradation are shown in Fig. 4.8. Note that progressive deterioration may actually consist of a series of discrete damage occurrences, but if the actual damage at any point in time is very small, say

$$D(t) - D(t^{-}) < \epsilon, \tag{4.34}$$

for some arbitrarily small ϵ and the timescale is long, we model it as continuous degradation.

Progressive degradation is generally the result of a mechanical process that may be driven by internal or external system conditions. Some examples of well known, and widely studied, progressive mechanical degradation processes are:

- *Wearout of engineered devices* is observed in most mechanical devices that have been used for a time period close to their service life (e.g., tire treads or a piston continuously contacting a cylinder). This phenomenon is also observed in pavements of roadways and runways and bridge structures.
- *Material fatigue* is a degradation process that occurs in devices or structures subjected to repeated loading and unloading cycles. Fatigue leads to microscopic

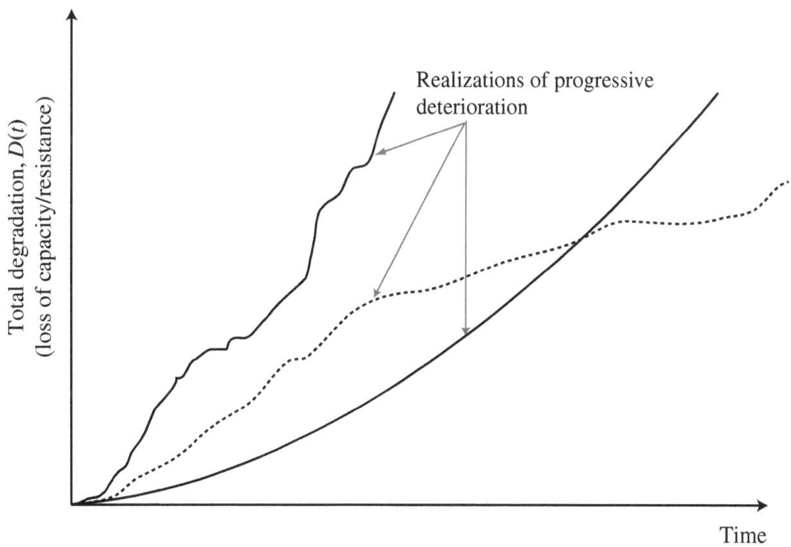

Fig. 4.8 Realizations of progressive (graceful) degradation of a system or component

cracks, which frequently form at the boundary (e.g., surface) of the element. Eventually a crack will reach a critical size, and the structure will fracture [59]. Fatigue problems have been widely studied in, for example, aeronautical engineering [60, 61]; and in pavement structures [62, 63].

- *Corrosion* is the gradual loss of material (primarily in metals) that reduces the component strength or deteriorate its appearance as a result of the chemical reaction with its environment, and it is frequently favored by the presence of chlorides or bacteria. Corrosion may concentrate on specific points forming "pits", which lead to crack initiation and propagation, or it can extend across a wide area corroding the surface uniformly. Deterioration models of steel structures have been widely discussed. Two cases in point are corrosion in marine environments (offshore structures); e.g., [64–66]; and corrosion in pipelines in [67].
- *Degradation of reinforced concrete structures* results from a reduction of the structural capacity caused mainly by chloride ingress, which leads to steel corrosion, loss of effective cross section of steel reinforcement, concrete cracking, loss of bond and spalling [68–70].
- *Concrete biodeterioration* is a consequence of the activity of bacteria that uses the sulfur found within the concrete microstructure, weakening it and increasing porosity; which, in turn, reduces the resistance and favors chloride ingress [71, 72].
- *Pavement deterioration* may be caused by three main processes: (1) fatigue cracking in asphaltic layers (or other stabilized layers), caused by the repetition of traffic loads, (2) permanent deformation or rutting in unbounded layers (mainly in the natural soil layer or subgrade), and (3) low temperature cracking in the asphalt course layer. Most pavement damage models are empirical and based on experimental data; however, some analytical models have been proposed recently. More information about these mechanisms can be found in [73, 74].
- *Moisture damage* refers to the effects that moisture causes on the structural integrity of any material. For example, it has been recognized as one of the main causes for early deterioration of adhesives and asphalt pavements. In the particular case of pavements, this phenomenon includes chemical, mechanical, thermodynamical and physical processes, each of them occurring at different magnitudes and rates [75, 76].

4.9.2 Models of Progressive Degradation

Progressive degradation is characterized by a continuous process; that is, loss of system capacity that has the form:

$$D(t) = \int_0^t \delta(\tau) d\tau, \tag{4.35}$$

where $\delta(t)$ is a degradation *rate* at time t, measured in capacity units per time unit; for example, the loss of material due to corrosion per year, or the annual increase of concrete porosity due to bacterial activity. The degradation rate over time $\{\delta(t), t \geq 0\}$ may itself be a stochastic process, or the parameters associated with an empirical deterioration law may be assumed to be unknown to reflect the variability observed in a sample of deterioration data [51].

In some cases it may be reasonable to assume a particular mathematical form for the degradation process based on experimental data or physical models, so that degradation may take the following general form:

$$D(t) = h(t - t_e) \text{ for } t > t_e, \tag{4.36}$$

where t_e is usually known as the time to deterioration initiation (e.g., time to corrosion initiation; see, for example, [69, 70]). The function h may take a linear, nonlinear, or any other form based on the problem at hand. It is important to note that the specific form chosen for the function h depends heavily on the physical properties of the specific system at hand (e.g., material characteristics, geometry, environmental conditions). Three examples of these type of models are presented in Fig. 4.9.

In many cases there are abundant data available to justify the form of Eq. 4.36 for specific deterioration processes. For example, [40] reports that many studies use degradation trends following a power form $h(t) = t^b$. For instance, for the expected degradation of concrete due to corrosion of reinforcement $b = 1$; for sulfate attack to concrete $b = 2$; for the diffusion-controlled aging $b = 0.5$ [9]; creep $b = 1/8$ [13]; and for scour-hole depth $b = 0.4$ [41].

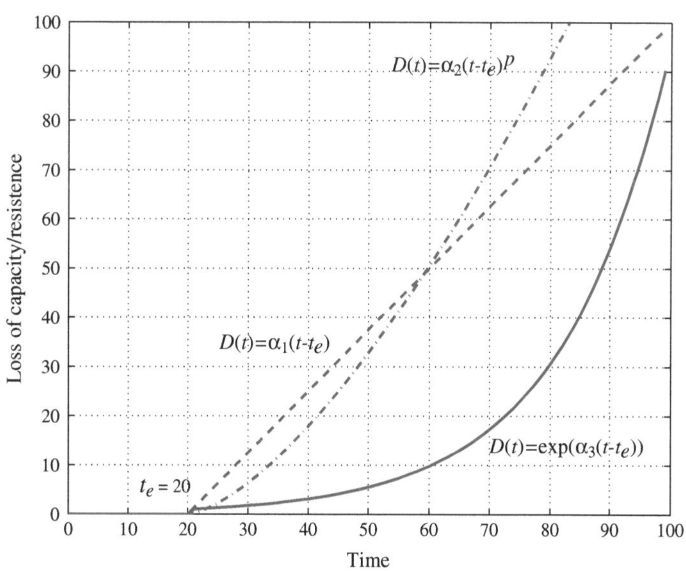

Fig. 4.9 Examples of progressive deterioration models; data: $u_0 = 100$, $\alpha_1 = 1.25$, $\alpha_2 = 0.2$, $\alpha_3 = 0.057$, and $p = 1.5$

4.9.3 Performance Evaluation

Let us assume that the system starts operating at time $t = 0$, and that the initial capacity has a known deterministic value $V(t = 0) = V_0 = v_0$. Then, the capacity of the system at time t can be expressed in terms of a deterioration rate as:

$$V(t) = v_0 - \int_0^t \delta(u)du \tag{4.37}$$

for $t \geq 0$. Note that the rate does not necessarily need to be constant over time. Some examples of degradation based on deterministic time-dependent rates are shown in Fig. 4.10.

An overview of random deterioration rate-based models can be found in [11]. If we assume that the minimum acceptable performance threshold is deterministic; i.e., k^*, the life of the system, i.e., L, or the time to failure, can be obtained as follows:

$$L = \inf\{t > 0 : \int_0^t \delta(u)du = v_0 - k^*\}. \tag{4.38}$$

Equation 4.38 basically states that the system fails once the capacity available, i.e., $v_0 - k^*$, is fully used.

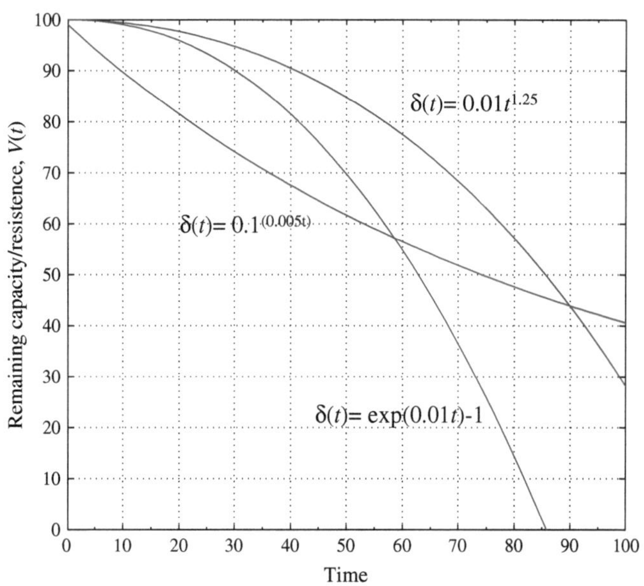

Fig. 4.10 Examples of rate-based deterioration models

4.10 Degradation Caused by Shocks

4.10.1 Definition and Examples

Shock-based degradation occurs when discrete amounts of the system's capacity are removed at distinct points in time. Shocks are events that cause a significant change in a system's performance indicator over a very small time interval. By *significant* we mean (Fig. 4.11)

$$D(t) - D(t - \Delta) > \xi, \tag{4.39}$$

where ξ is some arbitrary, positive, "large enough" value and Δ is some arbitrary, positive, "small enough" value, and we typically compress the time of occurrence of the damage to a single point. Generally, we use shock degradation when the damage that occurs at a particularly point in time is meaningful or *observable*. The size of the shock that occurs at time t is defined as the discontinuity in the degradation function $D(t) - D(t^-)$. Practically speaking, we may classify deterioration as shock degradation if significant damage occurs continuously but over a very short time interval (as shown in Fig. 4.11).

Shocks are assumed to occur randomly over time according to some physical mechanism, with each shock causing measurable damage to the system. We will denote the occurrence time of the ith shock as T_i and the size of the ith shock as Y_i; where,

$$Y_i = D(T_i) - D(T_i^-) \tag{4.40}$$

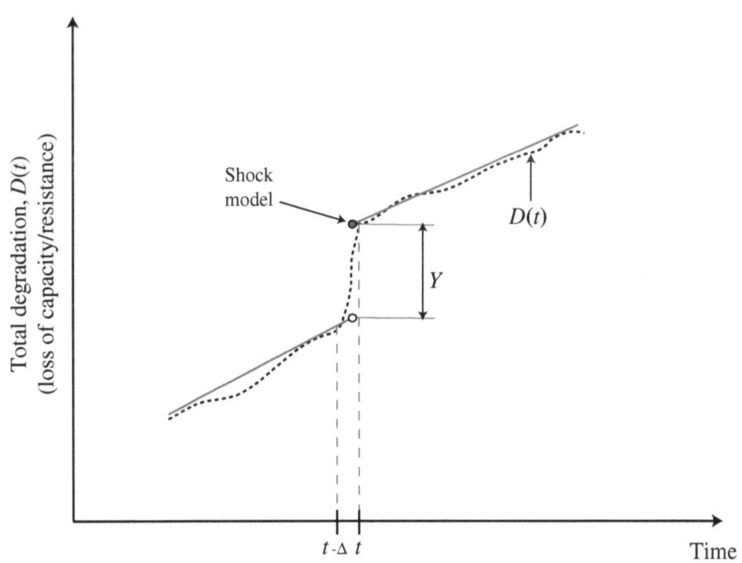

Fig. 4.11 Realization of a sudden event (i.e., shock)

Between the occurrence of shocks, the system state may or may not change continuously. For ease of exposition, in this section and in most of the book we will assume that the system degrades only at times where shocks occur.

Some examples of shock degradation include electrical, mechanical, or infrastructure systems subjected to, usually, unexpected extremely large demands; for example,

- *Overcurrent in electronic devices* occurs when a conductor experiences a spike in electric current, leading to excessive generation of heat. Possible causes for overcurrent include short circuits, excessive load, and incorrect design. In general overcurrent problems can be considered as shocks. However, in this case, if the failure does not occur (damage to equipment or electrical components of the circuit), the system remains in a condition "as good as new."
- *Earthquake damage* occurs when civil infrastructure (e.g., bridges, buildings) is subjected to a sudden acceleration which causes large inertial forces resulting in structural damage. This damage may result in the failure of one of various structural elements leading to the collapse of the structure. Mid-size earthquakes may not cause a collapse, but may cause damage (e.g., loss of stiffness) that accumulates with time reducing the structure's ability to withstand future events.

4.10.2 Models of Shock Degradation

Shock-based degradation has been used extensively in the literature (c.f. [77]), and several common assumptions are made that lead to different models.

The simplest models assume that the system will be unaffected by any disturbances below a specific threshold. Effectively, a system failure will occur only if the size of a shock exceeds a pre-specified threshold k^* (see Fig. 4.12) [78].

If damage does not accumulate, the system will be in one of two states: "as good as new," $V(t) = V_0$, or in a failed state, $V(t) \leq k^*$. Then, the system will fail at the ith shock if

$$Y_i > V_0 - k^*. \tag{4.41}$$

Furthermore, the life of the system L, which is the same as the time to first failure, is given by:

$$L = \inf\{t_n : Y_n > V_0 - k^*, n = 1, 2, \ldots\}, \tag{4.42}$$

This type of models have been used in modeling the fracture of brittle materials such as glass [79] and the failure of bridges due to overloads. Additional details can be found in [78], and a discussion on the applicability of this model will be presented in Chaps. 5–9.

The independent shock-based failure model given above is too simplistic to incorporate actual physical damage caused by successive shocks, therefore, models in which damage accumulates are generally more realistic. In cumulative damage models, the system is subjected to randomly occurring shocks, and each shock adds a

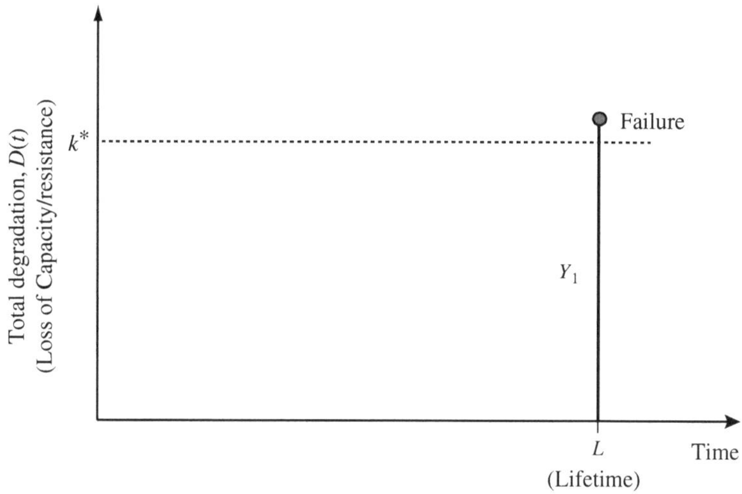

Fig. 4.12 Independent shock-based damage models

random amount of damage to the damage already accumulated. Here the total degradation $D(t)$ by time t is given by:

$$D(t) = \sum_{i=1}^{N(t)} Y_i \tag{4.43}$$

where $N(t)$ is the number of shocks that have occurred by time t. Note that in many practical applications the time between shocks is also random; therefore, $\{N(t), t \geq 0\}$ is a random process (a counting process as discussed in Chap. 3). A sample path of this type of process is given in Fig. 4.13 and described in [80, 81].

In this model, the remaining capacity of the system at time t is given by:

$$V(t) = V_0 - \sum_{i=1}^{N(t)} Y_i \tag{4.44}$$

and, as in Eq. 4.38, for a given failure or maintenance threshold k^*, the life, L, of the system is obtained by,

$$L = \inf\left\{t > 0 : \sum_{i=1}^{N(t)} Y_i \geq V_0 - k^*\right\} \tag{4.45}$$

Extensive research has been carried out on mathematical models for shock degradation; see for instance [77, 82–93].

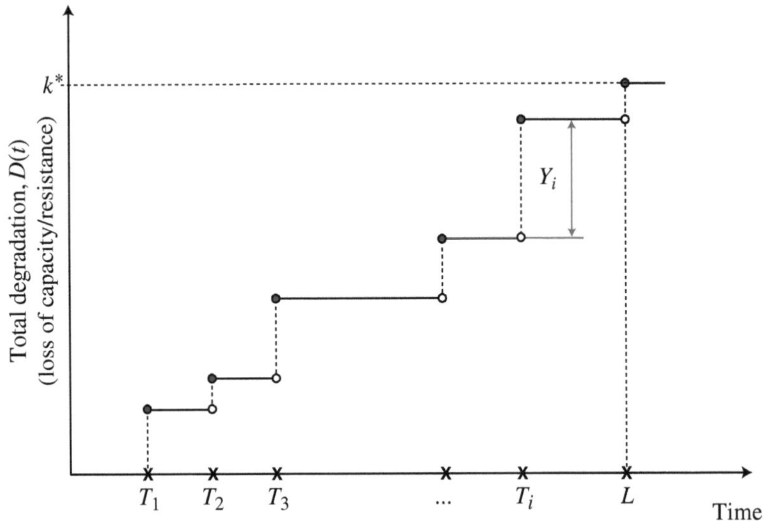

Fig. 4.13 Damage accumulation as a result of random shocks

4.10.3 Increasing Damage With Time

Increasing damage with time: in this type of model, shocks are independent but not necessarily identically distributed. Thus, the statistical properties of the shock size distribution may increase or decrease with time. This model is very convenient when dealing with the performance of systems where damage accumulates according to the previous state of the system. For instance, in the case of building structures located in seismic regions [95, 96]. Then, every earthquake causes some damage and the effect of the following event depends on the system state at the time of the event.

Two modeling alternatives are available for this type of problems. In the first, the shock size distribution parameters are not stationary; i.e., $Y_i \sim F(\mu(t), \eta(t), \ldots)$. The second option is that damage accumulates according to a function $g(Y, V)$, which should be continuous, nondecreasing in Y (shock size) and nonincreasing in V (system state). Then, if shock sizes, i.e., Y_i, are *iid* and occur at times $t_1, t_2, \ldots$. The degradation caused by shock Y_i is $g(Y_i, V(t_i-))$. Then, the accumulated damage at a given time t can be computed as:

$$D(t) = \sum_{i=1}^{N(t)} g(Y_i, V(t_i-)). \tag{4.46}$$

where, for instance,

$$g(y, v(t_i-)) = \beta \frac{y}{v(t_i-)} \quad (4.47)$$

Note that in this case, shocks are dependent on the system state [97].

4.11 Combined Degradation Models

Finally, in practice, there are problems that require some variations of progressive and shock models as described in previous sections. Here, we will describe some interesting cases.

4.11.1 Progressive and Shock Degradation

General life-cycle models describe the performance (i.e., degradation) of a system or a component throughout its lifetime. Then, once the system is put in service, damage starts accumulating as a result of progressive degradation or sudden events (i.e., shocks) until it fails. A sample path describing the performance of structural system throughout its lifetime is depicted in Fig. 4.14.

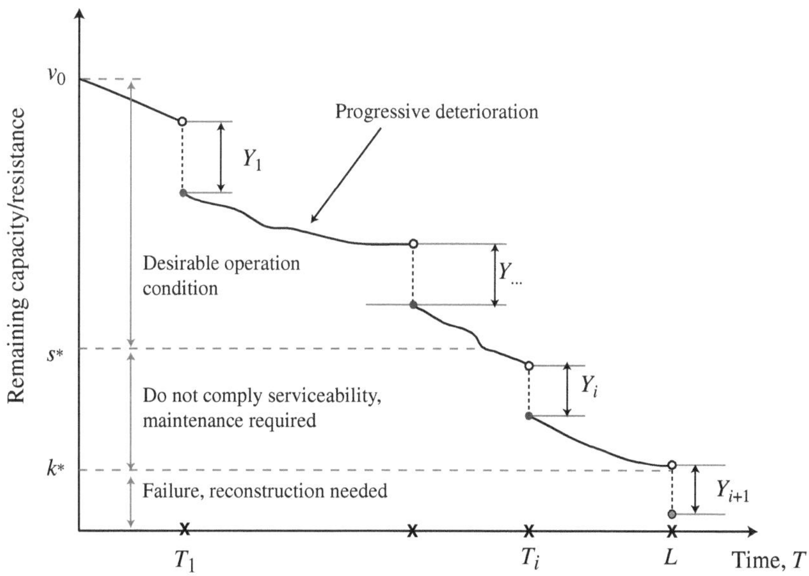

Fig. 4.14 Loss of remaining life as a result of both progressive degradation and random shocks

If the initial capacity of the system is v_0 and if $D(t)$ describes the degradation function, the capacity of the component by time t can be expressed as:

$$V(t) = v_0 - D(t) \tag{4.48}$$

Furthermore, based on the assumption that the structure is subjected to both continuous and sudden damaging events, and that they are independent, the degradation by time t can be computed as:

$$D(t) = \int_0^t \delta_p(u, \mathbf{p}(u))du + \sum_{i=1}^{N(t)} Y_i \tag{4.49}$$

where $N(t)$ is the number of shocks by time t, Y_i is the loss of capacity caused by shock i; $\delta_p(t, \mathbf{p}(t)) > 0$ describes the rate of some continuous progressive degradation process; and $\mathbf{p}(t)$ is a vector parameter that includes all random variables that influence the process. Then, combining Eqs. 4.48 and 4.49, the condition of the system by time t can be computed as:

$$V(t) = v_0 - \left[\int_0^t \delta_p(u, \mathbf{p}(u))du + \sum_{i=1}^{N(t)} Y_i\right] \tag{4.50}$$

and the life of the system requires solving,

$$\int_0^L \delta_p(u, \mathbf{p}(u))du + \sum_{i=1}^{N(L)} Y_i = v_0 - k^* \tag{4.51}$$

for L, if it exists.

4.11.2 Damage With Anealing

Damage with annealing. In some cases the system may recover a certain amount of capacity, ΔY, after the ith shock and before the shock $i + 1$ (see Fig. 4.15). Then, if the system recovers with a function $A(Y, t)$ after a shock of size Y, the accumulated damage (degradation) at any time t within the time interval between the ith and the $(i + 1)$th shock is:

$$Y_i - A(Y_i, t) \quad \text{for} \quad T_i \leq t \leq T_{i+1} \tag{4.52}$$

where Y_i is the shock size at time i. Therefore, the condition of the system at any time t would be

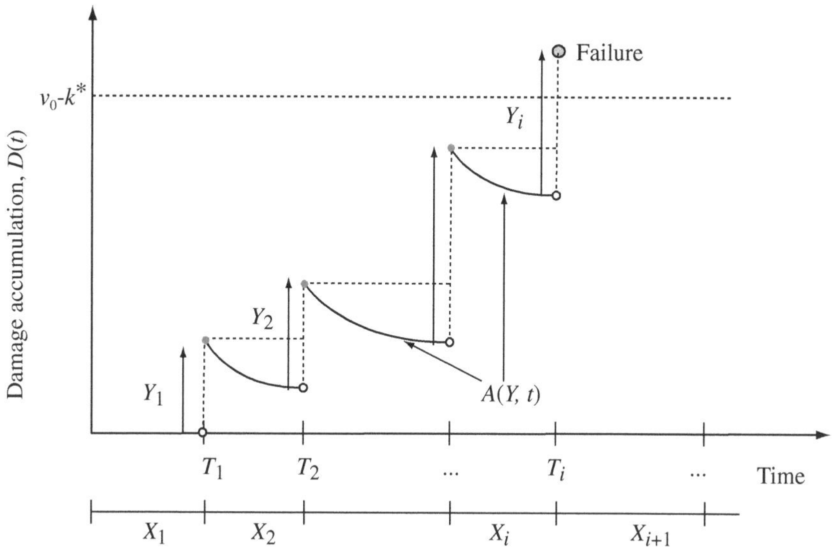

Fig. 4.15 Shock damage accumulation with annealing

$$D(t) = \left[\sum_{i=1}^{N(t)-1} Y_i - A(Y_i, (T_{i+1} - T_i))\right] + [Y_{N(t)} - A(Y_{N(t)}, (t - T_{N(t)}))] \quad (4.53)$$

where $T_{N(t)}$ is the time at which the $N(t)$ event occurs. Note that the time between shocks is a random variable and therefore $N(t)$ is also a random variable. In an application of this model, Takacs [94] considered the following recovery model: $A(Y_j, (t - T_j)) = Y_j \exp(-\alpha(t - T_j))$, where $0 < \alpha < \infty$. This type of behavior is common in some materials such as rubber, fiber reinforced plastics, asphalt, steel, and in general in most polymers [94]. Note that this type of behavior is a combined form of progressive and shock-based deterioration. The life of the system in this case can be computed similarly as in Eq. 4.45.

4.12 Summary and Conclusions

This chapter presents the fundamentals of degradation modeling. Thus, we first discuss important conceptual issues about the meaning of degradation and the way in which it affects the system's performance over time. Afterwards, we address the problem of data collection and analysis. It is argued that degradation models should be built based on actual data obtained from field observations of the physical performance of the system. This, however, is not an easy task, specially in the case of systems with expected long lifetimes such as civil infrastructure. Nevertheless, the most basic degradation model can be constructed using regression analysis. Although, this is a natural and common approximation, regression analysis by itself lacks completeness in the estimation of the physical nature of degradation and the uncertainties associated to the process.

We believe that understanding and modeling analytically the uncertain nature of the process is central to built useful degradation models. Then, in this chapter we have also presented the fundamentals of analytical degradation models. In particular we have focused on the formulation behind the two main degradation mechanisms: *progressive* and *shock-based*. In every case, we have briefly mentioned some examples of their manifestation in practice and outlined the mathematical formulation. In particular we have focused on explicitly define three aspects: (1) the degradation function, $D(t)$; (2) the condition state of the system at a given time t, $V(t)$; and (3) the life (time to failure) of the system, L. Also a general degradation model was outlined. In all cases various references were provided for the reader to find more detailed applications. The concepts treated in this chapter will be used extensively in the rest of the book.

References

1. W.Q. Meeker, L.A. Escobar, *Statistical Methods for Reliability Data* (Wiley, New York, 1998)
2. J.D. Kalbfleisch, R.L. Prentice, *The Statistical Analysis of Failure Time Data* (Wiley, New York, 1980)
3. M. Ben-Akiva, R. Ramaswamy, An approach for predicting latent infrastructure facility deterioration. Transp. Sci. **27**(2), 174–193 (1993)
4. S. Madanat, R. Mishalani, W.H.W. Ibrahim, Estimation of infrastructure transition probabilities from condition rating data. J. Infrastruct. Syst., ASCE **1**(2), 120–125 (1995)
5. B.S. Everitt, *An Introduction to Latent Variable Models* (Chapman and Hall, London, 1984)
6. M. Ben-Akiva, F. Humplick, S. Madanat, R. Ramaswamy, Latent performance approach to infrastructure management. Transp. Res. Rec. **1311**, 188–195 (1991)
7. M. Ben-Akiva, F. Humplick, S. Madanat, R. Ramaswamy, Infrastructure management under uncertainty: the latent performance approach. ASCE J. Transp. Eng. **119**, 43–58 (1993)
8. L. Nam, B.T. Adey, D.N. Fernando, Optimal intervention strategies for multiple objects affected by manifest and latent deterioration processes, in *Structure and Infrastructure Engineering*, 1–13 (2014)
9. B.R. Ellingwood, Y. Mori, Probabilistic methods for condition assessment, life prediction of concrete structures in nuclear power plants. Nucl. Eng. Des. **142**, 155–166 (1993)
10. Y. Mori, B. Ellingwood, Maintaining reliability of concrete structures. I: role of inspection/repair. J. Struct., ASCE, **120**(3), 824–835, (1994)
11. D.M. Frangopol, M.J. Kallen, M. van Noortwijk, Probabilistic models for life-cycle performance of deteriorating structures: review and future directions. Program. Struct. Eng. Mater. **6**(4), 197–212 (2004)
12. A. Petcherdchoo, J.S. Kong, D.M. Frangopol, L.C. Neves, NLCADS (New Life-Cycle Analysis of Deteriorating Structures) User's manual; a program to analyze the effects of multiple actions on reliability and condition profiles of groups of deteriorating structures. Engineering and Structural Mechanics Research Series No. CU/SR-04/3, Department of Civil, Environmental, and Architectural Engineering, University of Colorado, Boulder Co (2004)
13. E. Çinlar, Z.P. Bazant, E. Osman, Stochastic process for extrapolating concrete creep. J. Eng. Mech. Div. **103**(EM6), 1069–1088 (1977)

14. C. Karlsson, W.P. Anderson, B. Johansson, K. Kobayashi, *The Management and Measurement of Infrastructure: Performance, Efficiency and Innovation (New Horizons in Regional Science)* (Edward Elgar Publishing, Northampton, 2007)
15. C. Valdez-Flores, R.M. Feldman, A survey of preventive maintenance models for stochastically deteriorating single unit systems. Nav. Res. Logist. Q. **36**, 419–446 (1989)
16. D.-G. Chen, J. Sun, K.E. Peace, *Interval-Censored Time-to-Event Data: Methods and Applications* (Chapman & Hall/CRC Biostatistics Series, Boca Raton, 2012)
17. M.M. Desu, D. Raghavarao, *Nonparametric Statistical Methods For Complete and Censored Data* (Chapman & Hall/CRC Biostatistics Series, Boca Raton, 2003)
18. D.R. Helsel, *Non-detects and Data Analysis: Statistics for Censored Environmental Data* (Wiley, New Jersey, 2004)
19. W. Nelson, *Applied Life Data Analysis* (Wiley, New York, 1982)
20. K.B. Misra, *Reliability Analysis and Prediction: A Methodology Oriented Treatment* (Elsevier, Amsterdam, 1992)
21. P.A. Tobias, D.C. Trindade, *Applied Reliability*, 2nd edn. (Van Nostrand, Amsterdam, 1995)
22. M.S. Nikulin, N. Limnios, N. Balakrishnan, W. Kahle, C. Huber-Carol, *Advances in Degradation Modeling: Applications to Reliability, Survival Analysis and Finance*, Statistics for Industry Technology (Birkhauser, Boston, 2010)
23. B. Caicedo, J.A. Tristancho, L. Torel, Climatic chamber with centrifuge to simulate different weather conditions. Geotech. Test. J. **35**(1), 159–171 (2012)
24. J. Kastner, E. Arnold, When can a computer simulation act as substitute for an experiment: a case study from chemistry, in *Stuttgart Research Centre for Simulation Technology (SRC SimTech)*, pp. 1–18 (2011)
25. B. Anouk, S. Franceschelli, C. Imbert, Computer simulations as experiments. Synthese **169**, 557–574 (2009)
26. R. Frigg, J. Reiss, The philosophy of simulation: hot new issues or same old stew? Synthese **169**, 593–613 (2009)
27. M. Morrison, Models, measurement and computer simulation: the changing face of experimentation. Philos. Stud. **143**, 33–57 (2009)
28. E. Winsberg, *Science in the Age of Computer Simulation* (The University of Chicago Press, Chicago and London, 2010)
29. A. Haldar, *Recent Developments in Reliability-Based Civil Engineering* (World Scientific Press, New Jersey, 2006)
30. D.A. Ratkowsky, *Nonlinear Regression Modeling: A Unified Practical Approach* (Marcel Dekker, New York, 1983)
31. A.H.-S. Ang, W.H. Tang, *Probability Concepts in Engineering: Emphasis on Applications to Civil and Environmental Engineering.* (Wiley, New York, 2007)
32. C.J. Lu, W.Q. Meeker, Using degradation measures to estimate a time to failure distribution. Technometrics **34**, 161–174 (1993)
33. S. Caro, A. Diaz, D. Rojas, H. Nuez, A micro-mechanical model to evaluate the impact of air void content and connectivity in the oxidation of asphalt mixtures. Construct. Build. Mater. **61**, 181–190 (2014)
34. N.T. Kottegoda, R. Rosso, *Probability, Statistics and Reliability for Civil and Environmental Engineers* (McGraw Hill, New York, 1997)
35. B.M. Ayyub, R.H. McCuen, *Probability Statistics and Reliability for Engineering and Statistics*, 2nd edn. (Chapman & Hall/CRC Press, Boca Raton, 2003)
36. G.A.F. Seber, C.J. Wild, *Nonlinear Regression* (Wiley, New York, 1989)
37. D.M. Bates, D.G. Watts, *Nonlinear Regression Analysis and Its Applications* (Wiley, New York, 1988)
38. Technical committee AEN/CTN-41, *Bituminous mixtures. test methods for hot mix asphalt. Part 24: Resistance to fatigue.*AENOR—Asociacin Espaola de Normalizacin y certificacin, Madrid (2007)
39. J.M. Van Noortwijk, A survey of the application of gamma processes in maintenance. Reliab. Eng. Syst. Saf. **94**, 2–21 (2009)

40. J.M. van Noortwijk, A survey of the application of gamma processes in maintenance. Reliab. Eng. Syst. Saf. **94**, 2–21 (2009)
41. G.J.C.M. Hoffmans, K.W. Pilarczyk, Local scour downstream of hydraulic structures. Hydraul. Eng. **12**(14), 326–340 (1995)
42. T. Nakagawa, *Maintenance Theory of Reliability* (Springer, London, 2005)
43. H. Streicher, A. Joanni, R. Rackwitz, Cost-benefit optimization and risk acceptability for existing, aging but maintained structures. Struct. Saf. **30**, 375–393 (2008)
44. M. Sánchez-Silva, G.-A. Klutke, D. Rosowsky, Life-cycle performance of structures subject to multiple deterioration mechanisms. Struct. Saf. **33**(3), 206–217 (2011)
45. W. Harper, J. Lam, A. Al-Salloum, S. Al-Sayyari, S. Al-Theneyan, G. Ilves, K. Majidzadeh, Stochastic optimization subsystem of a network-level bridge management system. Transportation Research Record, page 1268 (1990)
46. S. Gopal, K. Majidzadeh, Application of Markov decision process to level-of service-based maintenance systems. Transp. Res. Rec. **1304**, 12–18 (1991)
47. Y. Kleiner, Scheduling inspection, renewal of large infrastructure assets. J. Infrastruct. Syst., ASCE **7**(4), 136–143 (2001)
48. R.G. Mishalani, S.M. Madanat, Computation of infrastructure transition probabilities using stochastic duration models. J. Infrastruct. Syst., ASCE **8**(4), 139–148 (2002)
49. V.M. Guillaumot, P.L. Durango, S. Madanat, Adaptive optimization of infrastructure maintenance and inspection decisions under performance model uncertainty. ASCE Infrastruct. Syst. **9**(4), 133–139 (2003)
50. O. Kubler, M.H. Faber, Optimal design of infrastructure facilities subject to deterioration, in *Proceedings of the ICASP'03 Der Kiureighian, Madanat & Pestana (Eds)*, 1031–1039 (2003)
51. M.D. Pandey, Probabilistic models for condition assessment of oil and gas pipelines. Int. J. Non-Destruct. Test. Eval. **31**(5), 349–358 (1998)
52. D. Straub, Stochastic modeling of deterioration processes through dynamic Bayesian networks. J. Eng. Mech., ASCE **135**(10), 1089–1098 (2009)
53. D. Straub, D. Kiureghian, Reliability acceptance criteria for deteriorating elements of structural systems. J. Struct. Eng., ASCE **137**(12), 1573–1582 (2011)
54. P. Thoft-Christensen, Reliability profiles for concrete bridges, in *Struct. Reliab. Bridge Eng.*, ed. by D.M. Frangopol, G. Hearn (McGraw-Hill, New York, 1996)
55. A.S. Nowak, C.H. Park, M.M. Szerszen, Lifetime reliability profiles for steel girder bridges, in *Optimal Perform. Civil Infrastruct. Syst.*, ed. by D.M. Frangopol (ASCE, Reston, Virginia, 1998), pp. 139–154
56. P. Thoft-Christensen, Assessment of the reliability profiles for concrete bridges. Eng. Struct. **20**(11), 1004–1009 (1998)
57. J.S. Kong, D.M. Frangopol, Life-cycle reliability-based maintenance cost optimization of deteriorating structures with emphasis on bridges. J. Struct. Eng. **129**(6), 818–828 (2003)
58. R.E. Melchers, C.Q. Li, W. Lawanwisut, Probabilistic modeling of structural deterioration of reinforced concrete beams under saline environment corrosion. Struct. Saf. **30**(5), 447–460 (2008)
59. S. Suresh, *Fatigue of Materials*, 2nd edn. (Cambridge University Press, Edimburgh, 1998)
60. V.V. Bolotin, *Mechanics of Fatigue*, Mechanical and Aerospace Engineering Series (CRC, Boca Raton, 1999)
61. A. Fatemi, *Metal Fatigue in Engineering* (Wiley, New York, 2000)
62. R. Lundstrom, J. Ekblad, U. Isacsson, R. Karlsson, Fatigue modeling as related to flexible pavement design, road materials and pavement design: state of the art. Road Mater. Pavement Des. **8**(2), 165–205 (2007)
63. E. Masad, V.T.F.C. Branco, N.L. Dallas, R.L. Lytton, A unified method for the analysis of controlled-strain and controlled-stress fatigue testing. Int. J. Pavement Eng. **9**(4), 233–243 (2007)
64. R.E. Melchers, Pitting corrosion of mild steel in marine immersion environment-1: maximum pit depth. Corrosion (NACE) **60**(9), 824–836 (2004)

65. R.E. Melchers, Pitting corrosion of mild steel in marine immersion environment-2: variability of maximum pit depth. Corrosion (NACE) **60**(10), 937–944 (2004)
66. R.E. Melchers, The effect of corrosion on the structural reliability of steel offshore structures. Corros. Sci. **47**, 2391–2410 (2005)
67. P.R. Roberge, W. Revie, *Corrosion Inspection and Monitoring* (Wiley, New York, 2007)
68. D. Val, M. Stewart, Decision analysis for deteriorating structures. Reliab. Eng. Syst. Saf. **87**, 377–385 (2005)
69. Y. Liu, R.E. Weyers, Modeling the time-to-corrosion cracking of the cover concrete in chloride contaminated reinforced concrete structures. ACI Mater. **95**, 675–681 (1988)
70. E. Bastidas, P. Bressolette, A. Chateauneuf, M. Sánchez-Silva, Probabilistic lifetime assessment of RC structures subject to corrosion-fatigue deterioration. Struct. Saf. **31**, 84–96 (2009)
71. E. Bastidas, M. Sánchez-Silva, A. Chateauneuf, M.R. Silva, Integrated reliability model of biodeterioration and chloride ingress for reinforced concrete structures. Struct. Saf. **20**(2), 110–129 (2007)
72. M. Sánchez-Silva, D.V. Rosowsky, Biodeterioration of construction materials: state of the art and future challenges. J. Mater. Civil Eng., ASCE **20**(5), 352–365 (2008)
73. Y.H. Huang, *Pavement Analysis and Design*, 2nd edn. (Pearson/Prentice Hall, New Jersey, 1998)
74. A.T. Papagiannakis, E. Masad, *Pavement Design and Materials* (Wiley, New Jersey, 2009)
75. S. Caro, E. Masad, A. Bhasin, D. Little, Moisture susceptibility of asphalt mixtures, part I: mechanisms. Int. J. Eng. Pavements **9**(2), 81–98 (2008)
76. R.G. Hicks, Moisture damage in asphalt concrete: synthesis of highway practice. Rep. No. NCHRP 175, National Cooperative Highway Research Program (1991)
77. T. Nakagawa, *Shock and Damage Models in Reliability* (Springer, London, 2007)
78. M.S. Finkelstein, V.I. Zarudnij, A shock process with a non-cumulative damage. Reliab. Eng. Syst. Saf. **71**, 103–107 (2001)
79. J.D. Esary, A.W. Marshall, F. Proschan, Shock models and wear processes. Ann. Prob. **1**, 627–649 (1973)
80. M. Abdel-Hameed, Life distribution properties of devices subject to a pure jump damage process. J. Appl. Prob. **21**, 816–825 (1984)
81. J. Grandell, *Doubly Stochastic Poisson Process Lecture Notes In Mathematics 529* (Springer, New York, 1976)
82. R.E. Barlow, F. Proschan, *Mathematical Theory of Reliability* (Wiley, New York, 1965)
83. Y.S. Sherif, M.L. Smith, Optimal maintenance models for systems subject to failure—a review. Nay. Res. Log. Q. **28**, 47–74 (1981)
84. T.J. Aven, U. Jensen, *Stochastic Models in Reliability*. Series in Applications of Mathematics: Stochastic Modeling and Applied Probability (41) (Springer, New York, 1999)
85. H.M. Taylor, Optimal replacement under additive damage and other failure models. Naval Res. Logist. Q. **22**, 1–18 (1975)
86. T. Nakagawa, On a replacement problem of a cumulative damage model: part 1. J. Oper. Res. Soc. **27**(4), 895–900 (1976)
87. T. Nakagawa, Continuous and discrete age replacement policies. J. Oper. Res. Soc. **36**(2), 147–154 (1985)
88. R.M. Feldman, Optimal replacement with semi-Markov shock models. J. Appl. Prob. **13**, 108–117 (1976)
89. R.M. Feldman, Optimal replacement for systems governed by Markov additive shock processes. Ann. Probab. **5**, 413–429 (1977)
90. R.M. Feldman, Optimal replacement with semi-Markov shock models using discounted costs. Math. Oper. Res. **2**, 78–90 (1977)
91. D. Zuckerman, Replacement models under additive damage. Naval Res. Logist. Q. **24**(1), 549–558 (1977)
92. M.A. Wortman, G.-A. Klutke, H. Ayhan, A maintenance strategy for systems subjected to deterioration governed by random shocks. IEEE Trans. Reliab. **43**(3), 439–445 (1994)

93. Y. Yang, G.-A. Klutke, Improved inspections schemes for deteriorating equipment. Probab. Eng. Inf. Sci. **14**, 445–460 (2000)
94. L. Takacs, *Stochastic Processes* (Wiley, New York, 1960)
95. J. Riascos-Ochoa, M. Sánchez-Silva, R. Akhavan-Tabatabaei, Reliability analysis of shock-based deterioration using phase-type distributions. Probab. Eng. Mech. **38**, 88–101 (2014)
96. J. Ghosh, J. Padgett, M. Sánchez-Silva, Seismic damage accumulation of highway bridges in earthquake prone regions. Earthquake Spectra **31**(1), 115–135 (2015)
97. M. Junca, M. Sánchez-Silva, Optimal maintenance policy for permanently monitored infrastructure subjected to extreme events. Probab. Eng. Mech. **33**(1), 1–8 (2013)

Chapter 5
Continuous State Degradation Models

5.1 Introduction

In this and the following chapters, the focus is on mathematical models for degradation that are based on stochastic processes. While very general deterioration models can be envisioned, we limit ourselves to models that are analytically tractable and which are widely used in practice. The models considered in this chapter describe the continuous evolution of system capacity over time. As discussed in Chap. 4, models of this type typically assume that loss of capacity occurs either due to discrete events (shocks), which occur randomly over time, or due to the effects of continuous (progressive) deterioration. In reality, of course, system capacity results from effects of both sources. In Chap. 7, we will present a general tractable paradigm for continuous-state degradation that incorporates both shocks and progressive degradation in a single mathematical model. For each model discussed, our main goals are to determine the distribution of time-dependent system capacity, $V(t)$, the distribution of system life (time to failure), L, and the instantaneous failure intensity. For simplicity, we consider the system only until first failure; maintained systems will be discussed in subsequent chapters (e.g., Chaps. 8–10).

The books of Nakagawa [1] and Nikulin et al. [2] provide an excellent discussion on the current status of mathematical degradation models. Also, there are many journal papers available that address this problem in different contexts, e.g., [3–10].

5.2 Elementary Damage Models

Perhaps the simplest model for system failure (often referred to in the literature as the "stress-strength" model [11]) proposes that failure occurs when the demand on a system exceeds the system capacity. Such model does not directly incorporate the *dynamics* of degradation, but it is useful as a starting point in considering more

M. Sánchez-Silva and G.-A. Klutke, *Reliability and Life-Cycle Analysis of Deteriorating Systems*, Springer Series in Reliability Engineering,
DOI 10.1007/978-3-319-20946-3_5

complex models. Suppose that a random variable V_0 represents initial capacity of a system, and an independent random variable D represents demand or load on the system. System failure occurs when the demand exceeds the capacity, so that the system fails with probability $P(V_0 \leq D)$ (see Chap. 2). Typically, initial system capacity V_0 is modeled based on the mechanical, electrical, and other physical properties of the engineered system, and incorporating randomness due to variability in materials, manufacturing processes, quality control, etc. Because our interest in this chapter is in modeling degradation, we will not be concerned with the evaluation of V_0 and henceforth we assume that the initial capacity of the system is a known quantity v_0. As this model does not explicitly incorporate a time component, it may be used to describe an initial failure (initial demand exceeds capacity) or one where the total demand over a *fixed* time horizon exceeds system capacity. Stress–strength models (Chap. 2) are used primarily in the design of systems that are intended for a fixed mission length and that are not maintained. Since time is not explicitly included in the model, the concept of the system lifetime does not have any meaning.

A similar, but slightly more complex model that incorporates time, can be constructed by assuming that the system starts operating at time $t = 0$ and that it remains in *as good as new* condition until a shock occurs, causing system failure (Fig. 5.1). If we define T_1 as the time of occurrence of the shock, then the lifetime of the system is simply $L = T_1$, and the lifetime distribution is the distribution of the time of the shock occurrence, F_1.

Now let us generalize this "first shoc" model further. Suppose that the system begins operating at time $t = 0$ and is subject to disturbances over time (we distinguish between shocks and disturbances here, in that disturbances do not necessarily cause damage to the system). Let the sequence of disturbances occur at (random) times $T_1, T_2, \ldots$, and let successive disturbances have magnitudes $Y_1, Y_2, \ldots$ Let us further assume that times between successive disturbances are independent,

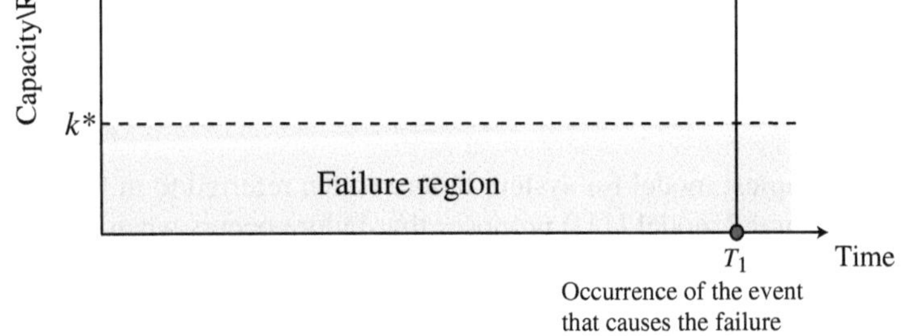

Fig. 5.1 System failure as a result of a single shock

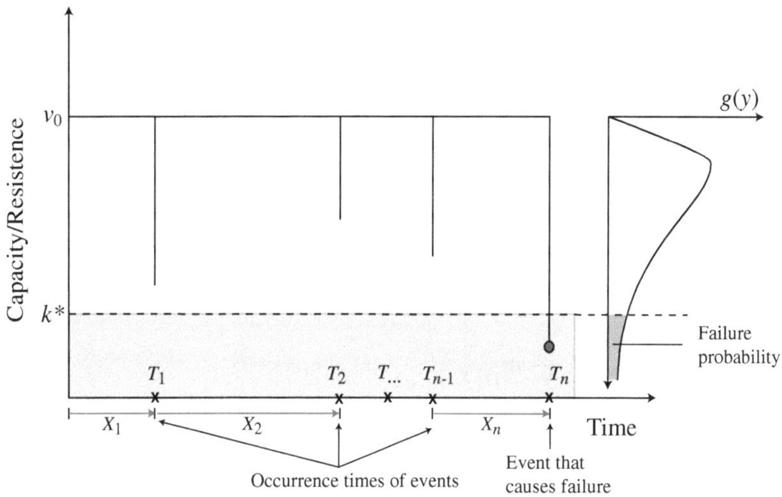

Fig. 5.2 System subject to multiple disturbances but failure observed as a result of a single event

identically distributed random variables with common distribution function F and mean $1/\lambda$. Besides, disturbance magnitudes are independent, identically distributed random variables with common distribution function G, independent of the times of disturbances. Suppose that a given disturbance causes the system to fail if the disturbance exceeds a threshold $q^* = v_0 - k^*$; otherwise, the system remains in "as good as new" condition. Note that the threshold value q^* is related to both the initial capacity and the limit state value, k^*. This model is known as the "Independent Damage Model" [1] and is illustrated in Fig. 5.2.

In this model, disturbance i causes the system to fail with probability π, where

$$\pi = P(Y_i > q^*) = 1 - G(q^*), \quad i = 1, 2, \ldots \tag{5.1}$$

The system will fail as a result of the Nth disturbance, where N is a geometrically distributed random variable with distribution function

$$P(N = n) = (1 - \pi)^{n-1}\pi, \quad n = 1, 2, \ldots \tag{5.2}$$

The distribution of system lifetime is then given by

$$\begin{aligned} P(L \le t) = P(T_n \le t) &= \sum_{n=1}^{\infty} P(T_n \le t | N = n) P(N = n) \\ &= \sum_{n=1}^{\infty} F_n(t)(1 - \pi)^{n-1}\pi \end{aligned}$$

$$= \sum_{n=1}^{\infty} F_n(t)[G(q^*)^{n-1} - G(q^*)^n]. \tag{5.3}$$

Here $F_n(t)$ denotes the n-fold convolution of F with itself, and represents the distribution of the time of the n-th shock.

The mean time to failure is [1]

$$\begin{aligned}\mathbb{E}[L] = \mathbb{E}[\mathbb{E}[L|N]] &= \sum_{n=1}^{\infty} \mathbb{E}[L|N = n]P(N = n) \\ &= \sum_{n=1}^{\infty} \frac{n}{\lambda} P(N = n) \\ &= \frac{1}{\lambda\pi} = \frac{1}{\lambda(1 - G(q^*))}.\end{aligned} \tag{5.4}$$

Example 5.16 Consider a structure with an initial capacity $v_0 = 100$ units that is subject to disturbances that occur randomly in time. Suppose the threshold that defines failure is $k^* = 25$ (in capacity units). Field data has shown that successive inter-arrival times of disturbances are independent exponentially distributed with mean $1/\lambda = 10$ years and that disturbance magnitudes are independent, identically distributed and follow a lognormal distribution G with parameters $\mu = 60$ and $\sigma = 18$. Compute the probability that the system fails by time $t = 5$, 10, and 30 years.

In this scenario, the system will fail if a disturbance exceeds $q^* = v_0 - k^* = 75$ units. Thus

$$\pi = 1 - G(75) = 0.182,$$

and the lifetime distribution is given by (Eq. 5.3)

$$P(L \le t) = \sum_{n=1}^{\infty} F_n(t)(1 - \pi)^{n-1}\pi = \sum_{n=0}^{\infty} F_{(n+1)}(t)(1 - \pi)^n\pi \tag{5.5}$$

Because the time between disturbances is exponentially distributed, the time of the nth disturbance follows an Erlang distribution with parameters n and $1/\lambda$, and therefore

$$F_n(t) = 1 - \sum_{i=0}^{n-1} \frac{(\lambda t)^i}{i!} e^{-\lambda t}, \quad t \ge 0 \tag{5.6}$$

Computing the required probabilities numerically, we have that $P(L \le 5) = 0.019$, $P(L \le 10) = 0.063$ and $P(L \le 30) = 0.3$.

In contrast, if the system fails at the occurrence of the first disturbance ($n = 1$), independent of the magnitude, we have

$$P(L \leq t) = P(T_1 \leq t) = 1 - e^{-\lambda t} = 1 - e^{-(0.1)t}, \quad (5.7)$$

and the corresponding probabilities are $P(L \leq 5) = 0.39$, $P(L \leq 10) = 0.63$, and $P(L \leq 30) = 0.95$.

5.3 Shock Models with Damage Accumulation

A somewhat more realistic model should include damage accumulation. Then, let us consider that shocks occur randomly over time, with each shock resulting in a random reduction in system capacity (damage), and that damage due to successive shocks is cumulative. Let us further assume that the system capacity is unchanged between occurrences of shocks. Thus, system capacity continues to be reduced after every shock until a shock occurs that drops capacity below the limit state; at that point in time, the system fails and is abandoned. Such damage models have been widely used in the literature; see for example, [1, 12, 13].

Shock-based degradation is typically modeled using a marked point process $\{(T_i, Y_i), \text{ with } i = 1, 2, \ldots\}$, where T_i represents the occurrence time of the ith shock and Y_i represents the amount of damage caused by the ith shock [14, 15]. This scenario is illustrated in Fig. 5.3.[1] Furthermore, denote the time between the ith and $i + 1$th shocks by X_i, i.e.,

$$X_i = T_{i+1} - T_i, \quad i = 1, 2, \ldots, \quad (5.8)$$

and let $\{N(t), t \geq 0\}$ denote the counting process for the number of shocks, that is, $N(t)$ gives the cumulative number of shocks by time t:

$$N(t) = \sum_{n=1}^{\infty} 1_{\{T_n \leq t\}}, \quad (5.9)$$

where 1_A is the indicator function for the event A.

Most models in the literature assume that successive times between shocks comprise an independent, identically distributed sequence (a *renewal* sequence, as clearly the times between shocks are nonnegative) and that $\{Y_i, i = 1, 2, \ldots\}$ is an independent, identically distributed sequence of nonnegative random variables, independent of $\{T_i, i = 1, 2, \ldots\}$. In the following sections, we will consider the case that

[1] Modeling the distribution of damage magnitudes is in general rather difficult, but data can be obtained, for example, from the so-called *fragility curves*, which describe the probability that the system reaches a certain damage level in terms of a specific demand parameter. Several approaches to compute these curves are available in the literature; see, for instance, [16].

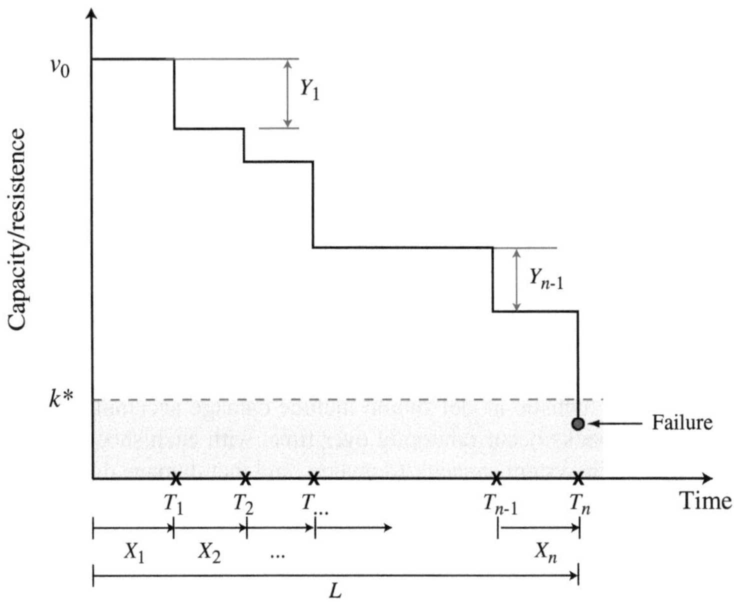

Fig. 5.3 Damage accumulation (loss of capacity/resistance) as a result of shocks

successive times between shocks form a (possibly nonhomogeneous) Poisson process as well as a more general case that they form an arbitrary renewal process. Our interest is in describing the capacity of the system at time t, $V(t)$, and the system lifetime L, with its associated distribution, mean time to failure, and failure intensity. In this model, the total damage by time $t \geq 0$ is given by

$$D(t) = \sum_{i=1}^{N(t)} Y_i, \tag{5.10}$$

and therefore the capacity of the system at time $t \geq 0$ is

$$V(t) = \max(v_0 - D(t), k^*). \tag{5.11}$$

The lifetime L can be analyzed as the first passage time of the process $\{V(t), t \geq 0\}$ to the limit state k^*. For our purposes, it is often easier to consider the lifetime in terms of the damage process $\{D(t), t \geq 0\}$ directly using the identity

$$\{V(t) \leq x\} \equiv \{D(t) \geq v_0 - x\}, \quad k^* < x < v_0, \tag{5.12}$$

so that the system fails when the damage $D(t)$ first exceeds the threshold $v_0 - k^*$.

5.3.1 Compound Poisson Process Shock Model and Generalizations

Perhaps the most widely employed cumulative damage shock model assumes that the process $\{(T_i, Y_i), i = 1, 2, \ldots\}$ forms a *compound Poisson process*, details of which were presented in Chap. 3. In this model, the times between shocks, $\{X_i; i = 1, 2, \ldots\}$ constitute a sequence of independent, exponentially distributed random variables with mean $1/\lambda$, and damage magnitudes $\{Y_i, i = 1, 2, \ldots\}$ are independent, identically distributed with common distribution function G with mean μ.

The compound Poisson process has stationary, independent increments, which makes it a particularly tractable model for accumulated shock damage. In particular, the number of shocks in the interval $[0, t]$ is given by

$$P(N(t) = n) = \frac{(\lambda t)^n}{n!} e^{-\lambda t}, \quad n = 0, 1, \ldots \tag{5.13}$$

and the damage accumulated by time t is then

$$D(t) = \begin{cases} 0 & \text{on } N(t) = 0 \\ \sum_{n=1}^{N(t)} Y_i & \text{on } N(t) > 0 \end{cases} \tag{5.14}$$

For ease of notation, we will denote the Poisson mass function with parameter a by $\{\Pi(n; a), n = 0, 1, \ldots\}$. Conditioning on the number of shocks in the interval $[0, t]$, the cumulative distribution function for $D(t)$ (i.e., total accumulated damage) is given by

$$\begin{aligned} P(D(t) \le d) &= \sum_{n=0}^{\infty} P(D(t) \le d | N(t) = n) P(N(t) = n) \\ &= \begin{cases} \Pi(0; \lambda t) = e^{-\lambda t} & d = 0 \\ \sum_{n=1}^{\infty} \Pi(n; \lambda t) G_n(d) & 0 < d < \infty, \end{cases} \end{aligned} \tag{5.15}$$

where G_n is the n-fold convolution of G with itself, and $G_0(\cdot) \equiv 1$. We note that the cdf of $D(t)$ has a discontinuity at zero that corresponds to the event that no shocks have occurred by time t and is absolutely continuous for $d > 0$.

Accordingly, we can compute the cumulative distribution function of *remaining* capacity as

$$\begin{aligned} P(V(t) \le x) &= P(D(t) > v_0 - x) \\ &= 1 - P(D(t) \le v_0 - x) \\ &= 1 - \sum_{n=0}^{\infty} \Pi(n; \lambda t) G_n(v_0 - x), \quad k^* < x < \infty, \end{aligned} \tag{5.16}$$

and the cumulative distribution function of the lifetime L as

$$P(L \leq t) = P(V(t) > k^*) = \sum_{n=0}^{\infty} \Pi(n; \lambda t) G_n(v_0 - k^*), \quad 0 < t < \infty. \tag{5.17}$$

The associated mean time to failure is given by

$$\mathbb{E}[L] = \lambda^{-1} \sum_{n=0}^{\infty} G_n(v_0 - k^*), \tag{5.18}$$

where $\sum_{n=0}^{\infty} G_n(v_0 - k^*)$ represents the expected number of shocks that cause the capacity to fall below $v_0 - k^*$.

Example 5.17 Consider a system whose initial condition is $v_0 = 100$ (capacity units) and that is subject to shocks that occur according to a Poisson process with rate $\lambda = 0.5$ events/year. If the ultimate limit state is defined by the the threshold $k^* = 25$, compute the probability that the system reaches the threshold before $t = 10$ years for the following cases if (1) shock sizes are deterministic $\mu = 6$ (capacity units); and (2) shocks sizes are exponentially distributed with parameter $\nu = 0.167$ (so mean shock size is again $\mu = 6$.)

In the first case where shocks have fix sizes, $\mu = 6$, the failure occurs if there are more than

$$n = \frac{(v_0 - k^*)}{b} = \frac{75}{6} = 12.5$$

shocks during the 10-year period. Therefore, the failure probability can be computed as

$$\begin{aligned} P(V(10) \leq 25) = P(N(10) > 12) &= \sum_{i=13}^{\infty} \frac{(0.5 \cdot 10)^i e^{-(0.5 \cdot 10)}}{i!} \\ &= 1 - \sum_{i=0}^{12} \frac{(0.5 \cdot 10)^i e^{-(0.5 \cdot 10)}}{i!} = 0.002 \end{aligned}$$

Let us now consider the case of exponentially distributed shock sizes with mean 6. Since G follows an exponential distribution, the nth convolution follows the Erlang density:

$$dG_n(y) = \frac{\nu^n y^{n-1}}{(n-1)!} e^{-\nu y} dy \tag{5.19}$$

where y is the amount of damage (i.e., loss of remaining capacity). Therefore, using Eq. 5.16, we have

$$
\begin{aligned}
P(V(10) \le 25) &= P(D(10) > 100 - 25) = P(D(10) > 75) \\
&= \sum_{n=1}^{\infty} (1 - G_n(v_0 - k^*))\Pi(n; \lambda t) \\
&= \sum_{n=1}^{\infty} \left(1 - \left[\int_0^{75} dG_n(y)\right]\right) \Pi(n; 5); \\
&= 0.025.
\end{aligned}
$$

here $\lambda t = (0.5)(10) = 5$. Note that in the second case the mean of shock sizes, i.e., $\mu_s = 1/0.167 = 6$, is the same as the shock sizes in the first case. However, the failure probability differs by approximately one order of magnitude, where clearly, the case of random shocks is larger than that of fixed deteriorating jumps.

The shock times form a stationary Poisson process may be generalized to allow for the times of shocks to form a nonhomogeneous Poisson process with intensity $\lambda(t)$; here, $\lambda(t)$ is a (nonnegative) deterministic function that controls the rate of shocks. The degradation process in this case (and hence, also the process tracking remaining capacity) still has independent increments, but the increments are no longer stationary (time homogeneous). For the non-homogeneous Poisson process, the increments have the distribution (see Chap. 3)

$$
P(N(t) - N(s) = n) = e^{-(m(t)-m(s))} \frac{(m(t) - m(s))^n}{n!}, \quad n = 0, 1, \ldots \tag{5.20}
$$

for $0 \le s < t < \infty$, where $m(t)$ is the cumulative intensity of the shock counting process, i.e.,

$$
m(t) = \int_0^t \lambda(u) du. \tag{5.21}
$$

Similar to expression 5.16, in this case the distribution of remaining capacity becomes

$$
P(V(t) \le v) = 1 - \sum_{n=0}^{\infty} \Pi(n, m(t)) G_n(v_0 - v), \quad k^* < v < \infty, \tag{5.22}
$$

and the lifetime distribution (see Eq. 5.17) is given by

$$
P(L \le t) = P(V(t) > k^*) = \sum_{n=i}^{\infty} \Pi(n; m(t)) G_n(v_0 - k^*), \quad 0 < t < \infty, \tag{5.23}
$$

where $\Pi(n; m(t)) = e^{-m(t)} m(t)^n / n!$.

The expected damage by time t is $m(t)/\mu$, where $1/\mu$ is the expected value of the shock size [1], and the mean time to failure (MTTF) can be computed as

$$\mathbb{E}[L] = \sum_{n=0}^{\infty}(1 - G_n(v_0 - k^*)) \int_0^{\infty} \frac{m(t)^n}{n!} e^{-m(t)} dt \tag{5.24}$$

Note that the central element of this model is the choice of the deterministic intensity function $\lambda(t)$ for the Poisson process, which, as mentioned before, is generally an increasing function with t indicating that as the system ages, degradation increases. A model for $\lambda(t)$ used commonly in practice is the Weibull model (also known as the *power law intensity* or Duane model [17]):

$$\lambda(t) = (\alpha t)^{\beta}, \quad \alpha > 0, \ \beta < \infty \tag{5.25}$$

For the case of a Weibull-type intensity function and exponentially distributed damage magnitudes, Zacks, [18], developed analytic expressions for the cumulative degradation by time t, as well as for the lifetime distribution. Also, Kahale and Wendt, [15, 19], discussed alternative intensity function models, including the log-linear and logistic intensity functions and provide additional details on the nonhomogeneous Poisson process shock model.

5.3.2 Renewal Process Shock Model

The (stationary) compound Poisson shock model can be generalized to allow for times between successive shocks to be independent, identically distributed, nonnegative random variables with common distribution function F, not necessarily exponential. In this case, $\{(T_i, Y_i), i = 1, 2, \ldots\}$ forms an (ordinary) *compound renewal process*. The increments in the counting process of the shocks are no longer independent, and these models are somewhat less tractable than their Poisson process counterparts, but are useful nonetheless. For the ordinary compound renewal process, the distribution of the number of shocks in $[0, t]$ is given by

$$P(N(t) = n) = F_n(t) - F_{n+1}(t); \quad n = 0, 1, 2, \ldots \tag{5.26}$$

where $F_0(t) \equiv 1$ and $F_n(t), n = 1, 2, \ldots$ is the n-fold Stieltjes convolution of $F(t)$ with itself. Similar to the compound Poisson process model, the accumulated damage by time t is

$$D(t) = \begin{cases} 0 & N(t) = 0 \\ \sum_{n=1}^{N(t)} Y_i & N(t) > 0 \end{cases} \tag{5.27}$$

The distribution of the accumulated damage in the interval $[0, t]$ for $d > 0$ can be computed as [1]

$$\begin{aligned} P(D(t) \le d) &= P\left(\sum_{i=0}^{N(t)} Y_i \le d\right) \\ &= \sum_{n=0}^{\infty} P\left(\sum_{i=0}^{N(t)} Y_i \le d | N(t) = n\right) P(N(t) = n) \\ &= \sum_{n=0}^{\infty} G_n(d)[F_n(t) - F_{n+1}(t)], \quad 0 < d < \infty, \end{aligned} \tag{5.28}$$

with $P(D(t) \le d) = 1 - F(t)$ for $d = 0$ and $G_n(d)$ the n-fold stieltjes convolution of $G(d)$ with itself. The expected damage by time t is

$$\begin{aligned} \mathbb{E}[D(t)] &= \int_0^{\infty} d \cdot dP(D(t) \le d) \\ &= \mathbb{E}[Y] \sum_{n=1}^{\infty} F_n(t) = \mathbb{E}[Y] M_F(t) \end{aligned} \tag{5.29}$$

where $M_F(t)$ is the renewal function of the distribution $F(t)$, i.e., the expected number of shocks in $[0, t]$. Note that if the expected value of the shocks is $\mathbb{E}[Y_1] = 1/\mu, \mathbb{E}[D(t)] = M_F(t)/\mu$, which is a result that was already presented and discussed in Chap. 3. In words, Eq. 5.29 states that the expected damage by time t is equal to the average damage caused by shocks multiplied by the expected number of shocks in the time interval $[0, t]$.

The distribution of remaining capacity at time t is given by

$$\begin{aligned} P(V(t) \le x) &= P(D(t) > v_0 - x) \\ &= 1 - \sum_{n=0}^{\infty} [F_n(t) - F_{n+1}(t)] G_n(v_0 - x) \\ &= \sum_{n=0}^{\infty} F_{n+1}(t)[G_n(v_0 - x) - G_{n+1}(v_0 - x)], \quad k^* < x < \infty \end{aligned} \tag{5.30}$$

where again v_0 is the initial state of the system and k^* is the minimum acceptable performance threshold.

For the case of renewal process shock-based damage accumulation, the distribution of time to failure can be computed as [1]

$$P(L \leq t) = P(D(t) > v_0 - k^*)$$
$$= \sum_{n=0}^{\infty} F_{n+1}(t)[G_n(v_0 - k^*) - G_{n+1}(v_0 - k^*)]. \tag{5.31}$$

and the mean time to failure (MTTF) is given by

$$\begin{aligned} \mathbb{E}[L] &= \int_0^{\infty} t \cdot dP(L \leq t) \\ &= \mathbb{E}[X] \sum_{n=0}^{\infty} G_n(v_0 - k^*) \\ &= \mathbb{E}[X][1 + M_G(v_0 - k^*)] \end{aligned} \tag{5.32}$$

where $M_G(v_0 - k^*)$ is the renewal function of the distribution $G(y)$ evaluated at $v_0 - k^*$, i.e., the expected number of shocks before the total damage exceeds the failure threshold $v_0 - k^*$. Two interesting results have been obtained to make an estimation of the mean time to failure. First, assume that the expected values of X_i and Y_i are described as $\mathbb{E}[X_i] = 1/\lambda$ and $\mathbb{E}[Y_i] = 1/\mu$, and that the variance of Y_i is σ_G^2. Then, it is possible to approximate $\mathbb{E}[L]$ as follows [1]:

$$\mathbb{E}[L] \approx \frac{1}{\lambda}\left(\mu(v_0 - k^*) + \frac{\mu^2\sigma_G^2 + 1}{2}\right). \tag{5.33}$$

Furthermore, if the distribution G has an increasing failure rate (IFR), it has been shown [1] that $\mu y - 1 < M_G(y) \leq \mu y$; and consequently,

$$\frac{\mu(v_0 - k^*)}{\lambda} < \mathbb{E}[L] \leq \frac{\mu(v_0 - k^*) + 1}{\lambda} \tag{5.34}$$

These bounds can be used to estimate the mean time to failure.

5.3.3 Solution Using Monte Carlo Simulation

In general, shock models may become very complex depending upon the distribution of inter-arrival times and shock sizes. In most cases, analytical expressions cannot be found. Therefore, simulation becomes a very good option to evaluate, among others, the main quantities of interest in degradation models; i.e., distribution of time to failure and probability distribution of the system condition at time t. The algorithm 1 presents the pseudocode to compute the distribution of time to failure and the mean time to failure for systems that deteriorate as a result of shocks only using Monte Carlo simulation.

Algorithm 1 Pseudocode for Monte Carlo simulations to compute the distribution of the time to failure and the MTTF of systems abandoned after first failure.

Require: T {Time window for the analysis}
F {Probability distribution of shock times}
G {Probability distribution of shock sizes}
v_0 {Performance condition at time $t = 0$}
k^* {Minimum performance condition}
N {number of simulations}
for $i = 1 : N$ **do**
  $t = 0$
  $s = 0$
  Generate a random value of the shock time, $\widehat{t_r}$, from F;
  $t = t + \widehat{t_r}$;
  while $t \leq T$ **do**
    Generate a random value of the shock size, $\widehat{s_r}$, from G;
    $s = s + \widehat{s_r}$;
    if $s \geq (v_0 - k^*)$ **then**
      $ET(i) = t$
      goto 16
    end if
    Generate a random value of the shock time, $\widehat{t_r}$, from F;
    $t = t + \widehat{t_r}$;
  end while
end for;
Compute Mean Time to Failure (MTTF) as: $\frac{1}{N} \sum_{i=1}^{N} \mathbb{E}T(i)$;
Fit a distribution of $\mathbb{E}T(1 : N)$ to find the probability distribution of the time to failure;

5.4 Models for Progressive Deterioration

Certain types of degradation, notably wear, erosion, and chloride ingress, tend to result in continuous reduction in system capacity over time. For instance, during normal use, a vehicle's tire tread declines continuously as a result of contact with the road surface. A number of different factors can help determine the rate at which the tread wears over time, such as driver behavior, tire inflation, and vehicle alignment. Thus the pattern of wear can appear nonconstant over time. As another example, in coastal areas with exposure to high humidity and salinity, metals, paint, concrete, and other materials can degrade continuously over time. This type of degradation is often referred to as *graceful* or *progressive* degradation. As mentioned in Chap. 4, in this case capacity is removed continuously over time rather than in discrete units such as with shock deterioration.

In this section, we discuss two types of models for continuous degradation, namely models based on an instantaneous degradation *rate*, either deterministic or stochastic, and those based on a continuous stochastic process, the Wiener process. Figure 5.4 shows several examples of sample paths for continuous degradation processes.

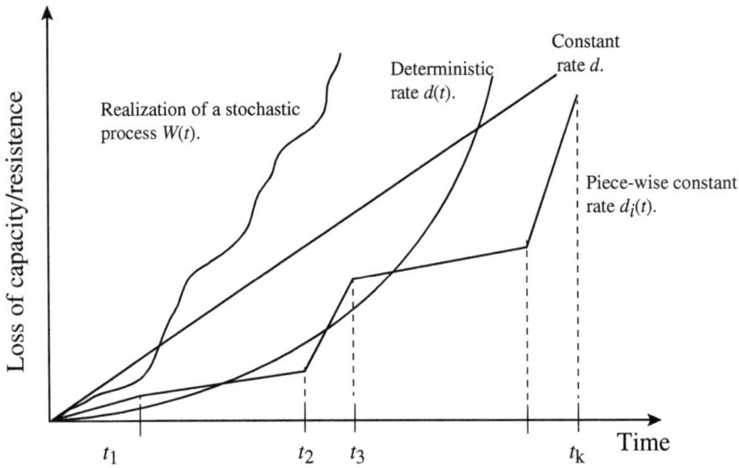

Fig. 5.4 Degradation rate-based models

5.4.1 Rate-Based Progressive Damage Accumulation Models

Rate-based models are among the most common models for *progressive deterioration* or *wear* [1, 20]. In rate-based models, damage is assumed to accumulate continuously over time driven by a (possibly random) instantaneous degradation rate $d(t)$. Then the accumulated damage at time t is given by

$$D(t) = \int_0^t d(\tau)d\tau, \tag{5.35}$$

and therefore the system lifetime is given by

$$T = \inf\{t \geq 0 : D(t) \geq v_0 - k^*\}. \tag{5.36}$$

If we assume that $\{d(t), t \geq 0\}$ is known with certainty, then the lifetime is also a deterministic quantity. In the simplest case, assume that deterioration rate is constant

$$d(t) \equiv d, t \geq 0. \tag{5.37}$$

In this case, capacity is removed from the system at rate d, and thus the lifetime is simply a linear function of the initial capacity and limit state value, i.e.

$$L = \frac{(v_0 - k^*)}{d}. \tag{5.38}$$

If the degradation rate is piecewise constant, namely

$$d(t) = d_i \qquad t_{i-1} \le t < t_i \qquad i = 1, 2, \ldots, n; \tag{5.39}$$

where $0 = t_0 < t_1 < t_2 < \cdots < t_n, n = 1, 2, \ldots,$ then the accumulated damage by time t is given by

$$D(t) = \sum_{i=0}^{n} d_{i+1}(t_{i+1} - t_i) \tag{5.40}$$

In general, if the deterioration rate is deterministic, the lifetime can be determined precisely (i.e., with certainty) using Eq. 5.36.

More complex models may be constructed under the assumption that the rate is the realization of a stochastic process $\{d(t); t \ge 0\}$ with independent increments. Suppose that the accumulated wear function takes the form $D(t) = A_t t + B_t$ with $A_t \ge 0$, and again suppose that the system fails when $D(t) \ge k^*$, where k^* is a prespecified performance threshold.

For these models, the complement of the lifetime distribution can be expressed as

$$P(L > t) = P(D(t) \le k^*) = P(A_t t + B_t \le k^*). \tag{5.41}$$

Nakagawa ([1]) considers several special cases of this model.

1. **Case 1**: $A_t \equiv a$, $B_t \equiv b$; and a, b and k^* constants. In this case, the problem is completely deterministic and the failure occurs at

$$t_f = \frac{k^* - b}{a} \tag{5.42}$$

2. **Case 2**: $A_t \equiv a$ (constant) and k^* also constant; if B_t is normally distributed with $\mu = 0$ and $Var = \sigma^2 t$, the reliability can be approximated as follows:

$$R(t) = P(at + B_t \le k^*) = P(B_t \le k^* - at) \approx \Phi\left(\frac{k^* - at}{\sigma\sqrt{t}}\right) \tag{5.43}$$

   where Φ is the standard normal distribution with mean 0 and standard deviation 1. Note that, for this particular case, the system may cross the threshold k^* at several points in time. The time to failure should then be computed as the time to the first passage.
3. **Case 3**: $B_t \equiv 0$, k^* constant and A_t normally distributed with mean a and $Var = \sigma^2 t$. Under this condition,

$$R(t) = P(A_t t \le k^*) = P(A_t \le k^*/t) = \Phi\left(\frac{k^* - at}{\sigma\sqrt{t}}\right) \tag{5.44}$$

Note that this equation is equal to Eq. 5.43. Besides, note that by making $\alpha = \sigma/\sqrt{ak^*}$ and $\beta = k^*/\alpha$ in Eqs. 5.43 and 5.44, the reliability can be rewritten as [21]

$$R(t) = \Phi\left[\frac{1}{\alpha}\sqrt{\frac{\beta}{t}} - \sqrt{\frac{t}{\beta}}\right] \tag{5.45}$$

which is called the Birnbaum–Saunders distribution [22] that is frequently used in fatigue-related problems [6, 23, 24].

4. **Case 4**: $A_t \equiv a$, $B_t = 0$ and the threshold is normally distributed with mean k^* and $Var = \sigma^2$.

$$R(t) = P(at \leq k^*) = \Phi\left(\frac{k^* - at}{\sigma}\right) \tag{5.46}$$

5.4.2 Wiener Process Models

Several authors, e.g., [25–30], have proposed the use of the Wiener process with drift to model degradation that accumulates continuously over time, for example, in modeling fatigue crack growth. The Wiener process (also referred to as *standard Brownian motion*) is a continuous-time process with stationary, independent increments and continuous sample paths, making it a potentially attractive stochastic process for modeling progressive deterioration. The Wiener process has been well studied for a wide variety of applications, including diffusion of small particles in a fluid medium and movement of stock prices in a market, and is often justified by assuming that increments in the degradation process are the result of a large number of very small effects, some of which may result in what we might term "anti-degradation." That is, although the significant trend may be toward increasing degradation (positive drift), the Wiener process does allow for degradation to decrease over time as well. We present an overview of the process here but also address several limitations that restrict its application in many practical situations.

In the simplest form, the degradation process $\{D(t), t \geq 0\}$ can be described by

$$D(t) = d_0 + \sigma W(t) + \mu(t), \quad 0 \leq t_0 \leq t, \tag{5.47}$$

where d_0 represents a constant initial degradation, $\{W(t), t \geq 0\}$ is a standard Brownian motion, and $\mu(t)$ and σ^2 are the mean drift and variance terms, respectively. As before, we assume that failure occurs when system capacity crosses a threshold (the limit state) k^*; we obtain the system lifetime as

$$L = \inf\{t \geq t_0 : D(t) \geq v_0 - k^*\}. \tag{5.48}$$

It is well known that the level crossings in a Wiener process follow an inverse Gaussian distribution. Then, by making $\mu(t) = \mu t$, the density of the system lifetime is given by

$$f_L(t) = \frac{v_0 - k^* - d_0}{\sqrt{2\pi\sigma^2 t^3}} \exp\Big(-\frac{(v_0 - k^* - d_0 - \mu t)^2}{2\sigma^2 t}\Big). \tag{5.49}$$

This model has not been used extensively in applications because it does not have monotonic sample paths. However, it has been used to model biomarker data [26, 28], situations where degradation data has been recorded, subject to measurement error [25], and for accelerated life testing [27, 30]. Waltraud and Lehmann [29] provide a thorough development of the parameter estimation associated with this model.

5.5 Approximations to Continuous Degradation Via Jump Processes

When modeling continuous deterioration, it is not always possible to evaluate explicitly the time-dependent nature of the degradation rate [31]. In this case, continuous degradation can be approximated by a sequence of small countable or uncountable discrete changes in the system condition. Several models have been proposed for this purpose, being the most common the gamma [3, 5] and the geometric [32] processes, which will be described in this section.

5.5.1 *Gamma Process*

Gamma processes have been used extensively to model degradation of materials [33–35], accumulation of flows into dams [36], and deterioration in many other engineering applications [3, 37, 38]. Like the compound Poisson process, the gamma process has independent increments, is right continuous, has left limits, is a.s. nondecreasing, and increases by discrete amounts (jumps). The increments in a gamma process follow a gamma distribution. The gamma process is defined as follows.

Definition 37 A (stationary) gamma process is a stochastic process $\{X(t), t \geq 0\}$ with $X(0) = 0$ a.s. and independent increments, whose distribution is given by

$$P(X(t) - X(s) \leq y) = \int_0^y \frac{u^{v(t-s)} x^{v(t-s)-1}}{\Gamma(v(t-s))} e^{-ux} 1_{(0,\infty)}(x) dx, \tag{5.50}$$

where $u > 0$ is known as the scale parameter and controls the rate of the jumps, and $v(t) > 0$ is known as the shape parameter and (inversely) controls the size of the jumps.

The gamma process has the property that jumps of size $[x, x+dx]$ ("small jumps") occur according to a Poisson process with rate dx. However, the gamma process is not a special case of the Poisson process except in the limit. Jump size follows a gamma distribution with constant scale parameter $u > 0$ and with a shape parameter that is a right continuous, nondecreasing, and real-valued function for $t \geq 0$, i.e., $v(t) > 0$ with $v(0) \equiv 0$ [3]. In the gamma process, the number of jumps in any time interval is countably infinite a.s.; however, "most" jumps are of small size so that the total jump size is finite over any finite interval. In this sense, the gamma process has been used to approximate continuous (progressive) degradation. Note that the gamma process is described directly by the distribution of its increments, while the compound Poisson process is usually described by the distribution of the jump sizes. Most applications that follow this approach use stationary gamma process, although nonstationary gamma process may be relevant in many cases. Some examples of nonstationary gamma processes can be found in [38–42].

A gamma process can be easily implemented using simulation. Then, a sample path can be constructed by simulating independent increments with respect to very small time intervals. Then, the procedure to construct one sample path can be summarized as follows [3]:

1. Define first a set of times at which the jumps occur, i.e., $\{t_1, t_2, \ldots, t_n\}$ with $\Delta t = (t_i - t_{i-1}) \to 0$ for $i = 1, 2, \ldots, (n-1)$.
2. Generate random independent increments $\{\delta_1, \delta_2, \ldots, \delta_n\}$ occurring at times $\{t_1, t_2, \ldots, t_n\}$; with $\delta_i = D(t_i) - D(t_{i-1})$, where $D(t_i)$ is the amount of degradation at time t_i. The increment, δ_i, is generated randomly from Eq. 5.52.
3. Construct the degradation sample path as

$$V(t_m) = v_0 - \sum_{i=1}^{m} \delta_i; \quad \text{with} \quad t_m = \sum_{i=1}^{m} t_i. \tag{5.51}$$

where v_0 is the system state at time $t = 0$.

In order to sample independent degradation increments δ_i, there are two simulation methods namely *increment sampling* and *bridge sampling* [43]. In the case of increment sampling, independent samples δ_i are obtained from the gamma density [3]:

$$f_{\Delta_i}(\delta_i | \Delta v_i, u) = \frac{u^{\Delta v_i} \delta_i^{\Delta v_i - 1}}{\Gamma(\Delta v_i)} e^{-u\delta_i} \tag{5.52}$$

where $\Delta v_i = v(t_i) - v(t_{i-1})$, i.e., the change in the shape parameter. Avramidis et al. [43] called this discrete-time simulation approach, *gamma sequential sampling* (GSS). An illustration of the use of gamma process for modeling progressive deterioration is presented in Fig. 5.5. The bridge sampling approach will not be presented inhere but the details can be found in [40, 43].

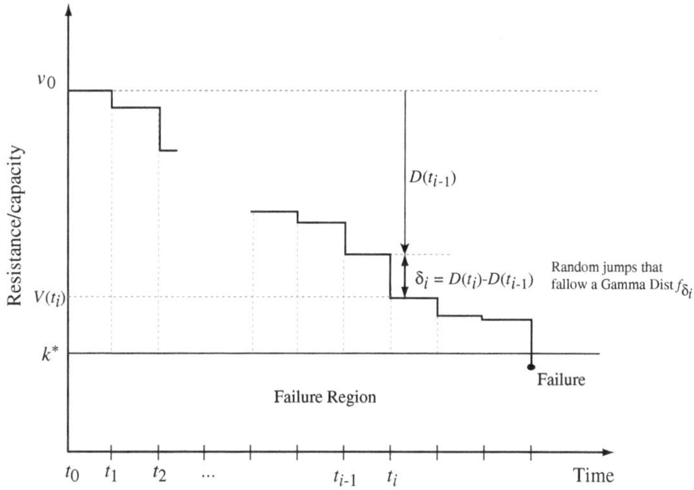

Fig. 5.5 Description of the generation of sample paths form a gamma process

The use of the gamma process requires estimating the parameters of the process (i.e., u and $v(t)$), which should be obtained from actual data observations. The problem of parameter estimation, for the specific case of the gamma processes, was discussed in Chap. 4 (Sect. 4.7.3). However, there is a significant amount of literature on the topic (e.g., see [44, 45]). Apart from the method of maximum likelihood (ML) and the method of moments, presented in Chap. 4, other methods available in the literature include the Bayesian estimation [46] and the use of expert judgement [39]. Noortwijk [3] describes in detail several approaches to find the parameters of the gamma process.

Example 5.18 Draw two realizations of two gamma process with *shape* parameters: $v(t) = 0.0055t^2$ and $v(t) = 5.5t^{0.5}$, and scale parameter $u = 1.5$. The time window selected for the analysis is $T = 120$. Finally, assume that the initial condition of the system is $v_0 = 100$ (capacity units).

In order to build the sample path of the degradation, the time domain was divided into 50 equally spaced intervals with $\Delta t = 2.4$ years. The sample paths of the degradation obtained by simulation using the gamma sequential sampling are presented in Fig. 5.6.

5.5.2 Geometric Process

A geometric process is a stochastic process $\{X_i,\ i = 1, 2, \ldots\}$ such that if there exists a real number $a > 0$, the sequence $\{a^{i-1}X_i,\ i = 1, 2, \ldots\}$ forms a renewal process [32]. The real number a is also called the ratio of the process. Then, for

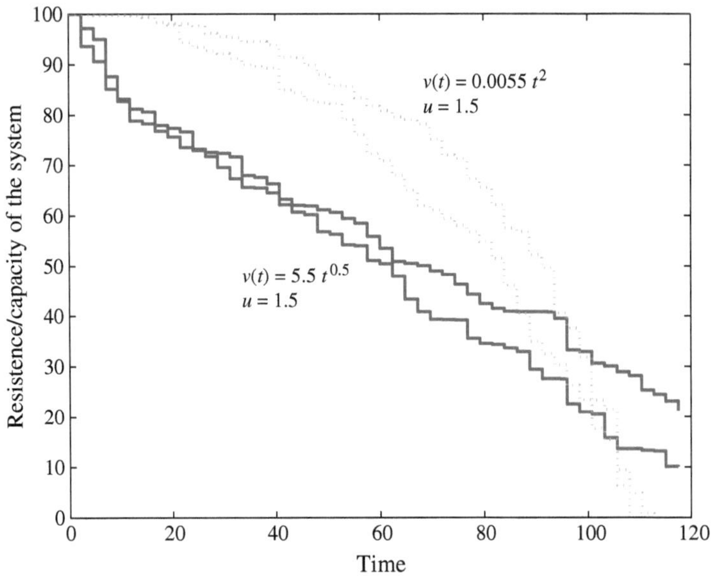

Fig. 5.6 Realizations of the degradation paths based on a Gamma process

$a > 1$ the process is stochastically decreasing, and for $0 \leq a < 1$ is increasing. For the particular case in which $a = 1$, it constitutes a renewal process; therefore, the geometric process is a *monotone* process and it is a generalization of the renewal process [32].

If the random variable X_1 has distribution $F(x)$ and density $f(x)$, then X_i has distribution $F(a^{i-1}x)$ with density $a^{i-1}f(a^{i-1}x)$. In practice, we will assume that $F(0) = P(X_1 = 0) < 1$. Furthermore, if for the initial distribution $\mathbb{E}[X_1] = \mu$ and $\text{Var}[X_1] = \sigma^2$, then

$$\mathbb{E}[X_i] = \frac{\mu}{a^{i-1}} \quad \text{and} \quad Var[X_i] = \frac{\sigma^2}{a^{2(i-1)}} \tag{5.53}$$

An important quantity for modeling degradation is

$$S_n = \sum_{i=1}^{n} X_i \tag{5.54}$$

where $S_0 = 0$. The first two moments of S_n are [32]

$$\mathbb{E}[S_n] = \mu \frac{1 - a^{-n}}{1 - a^{-1}} \qquad Var[S_n] = \sigma^2 \frac{1 - a^{-2n}}{1 - a^{-2}} \tag{5.55}$$

For $a > 1$ and $n \to \infty$,

$$\mathbb{E}[S_n] = \frac{a\mu}{a-1} \qquad Var[S_n] = \frac{a^2\sigma^2}{a^2-1} \tag{5.56}$$

where $S_0 = 0$. Note that for $a \leq 1$, $\mathbb{E}[S] \to \infty$ as $n \to \infty$. Clearly, the degradation process is not stationary and is highly defined by a nonlinear trend.

In some cases, there is a single monotone trend and the ratio a of the geometric process defines its direction and "intensity." However, sometimes real degradation data exhibit multiple trends (e.g., bathtub curve). In these cases, it may be convenient to use what is called a *threshold* geometric process. A stochastic process $\{Z_i,\ i = 1, 2, \ldots\}$ is called a *threshold* geometric process if there exist real numbers $\{a_m > 0, m = 1, 2, \ldots, k\}$ and integers $\{1 = M_1 < M_2 < \cdots < M_k < M_{k+1} = \infty\}$ such that for each m, $\{a_m^{i-M_m} Z_i, M_m \leq i < M_{m+1}\}$ forms a renewal process; for further details, see [32].

As in the gamma process, parameter estimation for a set of data is essential for modeling degradation. For the case of geometric process, it is required to find the best estimative of the mean μ, the variance σ^2, and the ratio of the process a. A description of existing approaches is presented in [32] where the authors describe data analysis methodologies considering specifically two models: the Cox-Lewis model and the Weibull process. Nonparametric models have been also discussed in [47–49]. In addition, some parametric estimations have been carried out under the assumption that X_1 has a lognormal distribution [50], and where X_1 has a gamma distribution [47]. Some other related work can be found in [51].

The geometric process, as a tool for modeling degradation, can be implemented using simulation. Thus, the procedure to construct one sample path can be summarized as follows:

1. Define first a set of times at which the jumps occur, i.e., $\{t_1, t_2, \ldots, t_n\}$ with small $\Delta t = (t_i - t_{i-1})$ for $i = 1, 2, \ldots, (n-1)$.
2. Generate random independent increments $\{\delta_1, \delta_2, \ldots, \delta_n\}$ occurring at times $\{t_1, t_2, \ldots, t_n\}$, with $\delta_i = D(t_i) - D(t_{i-1})$, where $D(t_i)$ is the amount of degradation at time t_i. The increment (jump), δ_i, is generated randomly from the distribution $F_{Y_i}(a^{i-1}y)$.
3. Construct the degradation sample path as

$$V(t_m) = v_0 - \sum_{i=1}^{m} \delta_i; \quad \text{with} \quad t_m = \sum_{i=1}^{m} t_i. \tag{5.57}$$

where v_0 is the system state at time $t = 0$.

Note that the "intensity" (speed) of degradation is defined by jumps, whose intensity is controlled by the ratio (i.e., the jump size probability distribution is $F_{Y_i}(a^{i-1}y)$), and that occur at specific (deterministic) and usually small time intervals. The selection of the ratio a defines the overall trend of the deterioration. In this model, special

Table 5.1 Distribution of Y_1 and the corresponding rates of the process for every case considered

Case	Distribution Y_1	μ_1	σ_1	Ratio a
1	Lognormal	0.05	0.01	0.75
2	Lognormal	0.05	0.01	0.95
3	Lognormal	25	5	1.5
4	Lognormal	25	5	2

care should be taken in tuning the relationship between the ratio a and the time interval between shocks, since shock size distributions depend on the number of shocks that have already occurred.

Finally, it is important to notice that when modeling progressive degradation shock sizes are expected to be small at the beginning and will grow (or decrease) in accordance with the ratio of the process. In particular, note that if $a > 1$, the expected total degradation will converge to $a\mu/(a-1)$ (Eq. 5.56), which means that failure will only occur if $a\mu/(a-1) < (v_0 - k^*)$ regardless of the number of time intervals considered. On the other hand, if $a < 1$, the task of estimating the number of jumps required for the system to fail is more difficult and requires some iterative approach.

Geometric processes can be used to model both progressive and shock-based degradation; in this section, we have focused on the former; its use for modeling shocks is presented in Sect. 5.6.2.

Example 5.19 Consider a system that degrades progressively and whose behavior will be modeled using a geometric process. Furthermore, assume that the initial state of the system is $v_0 = 100$ and that we want to model four possible degradation trends. In all cases, the initial jump sizes, i.e., Y_1, are lognormally distributed. The parameters of the distribution of Y_1 and the ratio of each process, a, are shown in Table 5.1.

Only one realization of each of the four models is presented in Fig. 5.7. Note first that in the cases considered, the ratio of the process defines whether the trend is concave or convex. Thus, for the case of $a > 1$, the shock size distribution will cause that the size of shocks decrease with time until they converge, implying that there is limit to damage (Fig. 5.7). This is observed in some physical phenomena such as fatigue through what is known as the fatigue or *endurance limit* [52]. Also, note that in these cases, as the ratio increases, more damage accumulates in the system.

For the particular case in which $a > 1$, we can use Eq. 5.56 to find the expected value of the total degradation:

$$\mathbb{E}[S_3] = \frac{a\mu}{a-1} = \frac{1.5 \cdot 25}{1.5 - 1} = 75 \qquad \mathbb{E}[S_4] = \frac{2 \cdot 25}{2-1} = 50 \tag{5.58}$$

which means that the expected minimum system condition will be $V_3(\infty) = 25$ and $V_4(\infty) = 50$, respectively. In the cases where $a < 1$, degradation starts slowly and increases with time. Smaller values of a lead to faster degradation, e.g., the decay

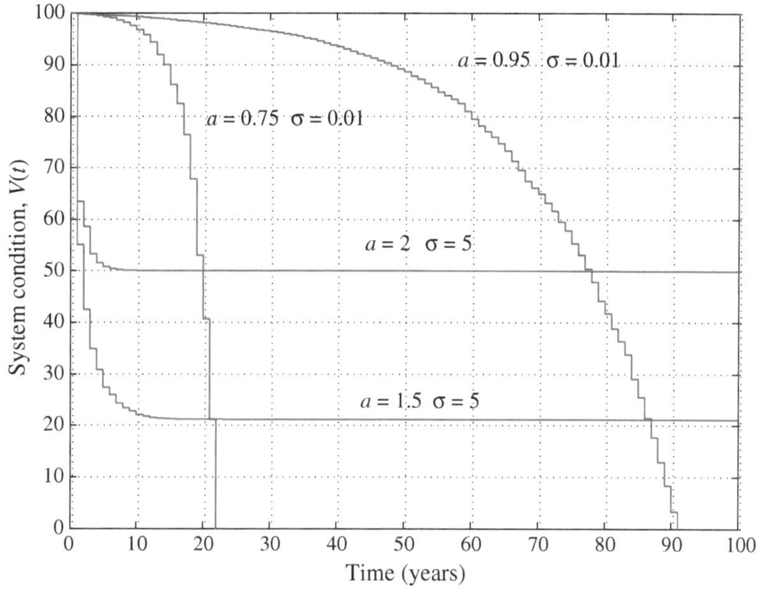

Fig. 5.7 Sample paths of the discrete representation of progressive deterioration based on a geometric process. Jump sizes are lognormally distributed

for $a = 0.75$ is much faster than for $a = 0.95$. Finally, note that the distribution probability of the initial distribution Y_1 when $a > 1$ has to be somewhat large compared with the case where $a < 1$.

5.6 Increasing Degradation Models

Frequently, the assumption that shock sizes are *iid* is too strong or not realistic. For instance, consider a bridge structure located in a seismic region, which is subjected to a series of earthquakes throughout its lifetime. Then, the damage caused by an earthquake is conditioned on the current state of the bridge structure at the time of its occurrence. This means that the probability distribution of a shock size (i.e., damage) depends on the current state of the system (i.e., level of damage at the time of the event). There are two basic approaches for modeling the increasing nature of damage accumulation with time; these are

- conditioning on the damage state; and
- defining a function of shock size distributions.

These two approaches will be discussed in the following subsections with emphasis on shock-based degradation.

5.6.1 Conditioning on the Damage State

Consider a system that starts operating with initial condition v_0, and it is damaged only as a result of *iid* shocks Y_i, which occur at times T_i, with $i = 1, 2, \ldots$. Then, the loss of capacity/resistance at time T_i, depends on the system state at time T_{i-1}. Assuming that there is no additional damage between any two shocks:

$$\Delta V_i = V(T_{i-1}) - V(T_i) = g(V(T_{i-1}), Y_i). \tag{5.59}$$

and therefore,

$$V(t) = V(T_{i-1}) - g(V(T_{i-1}), Y_i) \quad T_i \leq t < T_{i+1} \tag{5.60}$$

The state of the system at any time t can then be computed as

$$V(t) = v_0 - \sum_{i=1}^{N(t)} \Delta V_i = v_0 - \sum_{i=1}^{N(t)} g(V(T_{i-1}), Y_i) \tag{5.61}$$

where $V(T_0) = v_0$ (i.e., initial system state); and $N(t)$ is the number of shocks that have occurred by time t.

The central element of this model is to define the function g, which clearly is problem dependent. For example, functions of the form $g = \alpha Y_i / V(T_{i-1})$, with α a constant to be determined, can be used in many practical applications (Fig. 5.8).

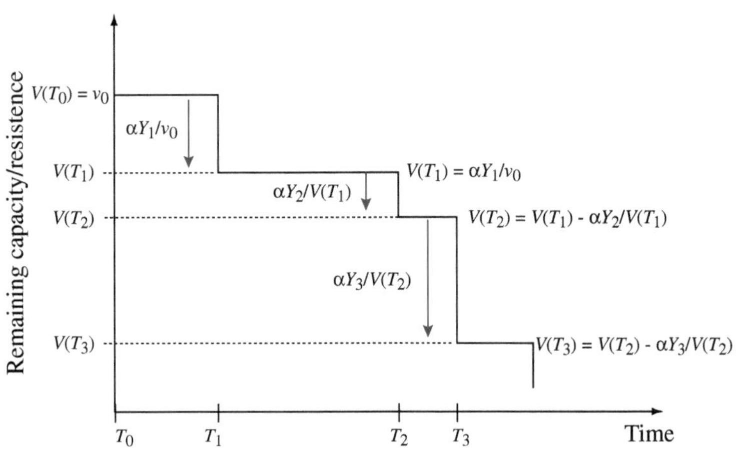

Fig. 5.8 Deterioration conditioned on damage state

For these types of problems, an analytical solution for the lifetime distribution and other important reliability quantities is clearly difficult to obtain. However, a reasonable solution can be found using Monte Carlo simulations. A simulation approach to compute the mean time to failure, i.e., $MTTF$, is shown in the algorithm 2. Note that by varying the value of k^*, it is possible to find the failure probability for a given performance level. Also, a modification of the algorithm can be made to compute the failure probability at a given point in time. In order to do this, an additional *While* should be included to control the evaluation time. Thus, the process stops when either the system fails before a reference time t or the time t is reached.

Algorithm 2 Monte Carlo simulation to compute MTTF for a deterioration conditioned on the system damage state for an arbitrary function g.

Require: T {Time window for the analysis}
  F {Probability distribution of shock times}
  G {Probability distribution of shock sizes}
  k^* {Minimum performance condition}
for $s = 1 : N$ **do**
  $V(t) = v_0$; {v_0 is the performance condition at time $t = 0$}
  $q = 0, T_q = 0, T_f = 0$;
  while $V(t_q) > k^*$ **do**
    $q = q + 1$;
    Generate a random value of the shock time $\widehat{T}_q$ from F;
    $T_f = T_f + \widehat{T}_q$;
    Generate a random value $\widehat{y}_q$ from G
    $V(\widehat{T}_q) = g(V(\widehat{T}_{q-1}), \widehat{y}_q)$ (e.g., $V(\widehat{T}_q) = \alpha\widehat{y}_q / V(\widehat{T}_{q-1})$);
  end while
  $T(s) = T_f$;
end for{N is the Number of simulations}
$MTTF = (1/N)\sum_{s=1}^{N} T(s)$;

Example 5.20 Let us consider a system where shocks are described by a Poisson process with $\lambda = 0.1$ and shock sizes Y are *iid* lognormally distributed with mean $\mu = 10$ and $\sigma = 2$. Evaluate the mean time to failure of the following state-dependent degradation models:

$$g_1(T_n) = \alpha \frac{Y_n}{V(T_{n-1})}$$

and

$$g_2(T_n) = \frac{Y_n}{(v_0 - V(T_{n-1}))^{\gamma(n-1)}}$$

Taken $\alpha = 1$ and $\gamma = 2$, and using simulation as described in algorithm 2, the results after 1000 simulations show the following mean times to failure: $MTTF_{g_1} = 58.03$ years and $MTTF_{g_2} = 24.28$ years.

5.6.2 Function of Shock Size Distributions

In this approach, we focus on evaluating damage accumulation not through the system state, as in previous section, but by evaluating the change in the shock size distribution. Consider that the sequence of shocks Y_i where $i = 1, 2, \ldots n$ indicates the order of the arrivals. Then, it is reasonably to assume that there exists a functional relationship between two successive shock distributions as follows:

$$F_{Y_{i+1}} = z(F_{Y_i}) \tag{5.62}$$

where z is a positive continuous increasing function. The selection of function z should be made carefully to keep some important stochastic properties of the process. A convenient way to manage this problem is through the so-called *Geometric* processes which was described in Sect. 5.5.2.

Example 5.21 Consider a system that deteriorates as a result of shocks. Shock sizes are lognormally distributed and shock arrivals are exponential with rate $\lambda = 0.5$.

Using Mote Carlo simulation, three sample paths of the process, with rate $a = 0.75$, are presented in Fig. 5.9. In addition, in Fig. 5.10, three sample paths of the same process, with varying rates $a = 0.25$, $a = 0.5$, and $a = 0.75$, are shown. It can be observed that as the process rate become smaller, the failure time becomes shorter. The mean times to failure for the three cases shown are $MTTF_{a=0.25} = 45.26$, $MTTF_{a=0.5} = 68.12$, and $MTTF_{a=0.75} = 102.34$.

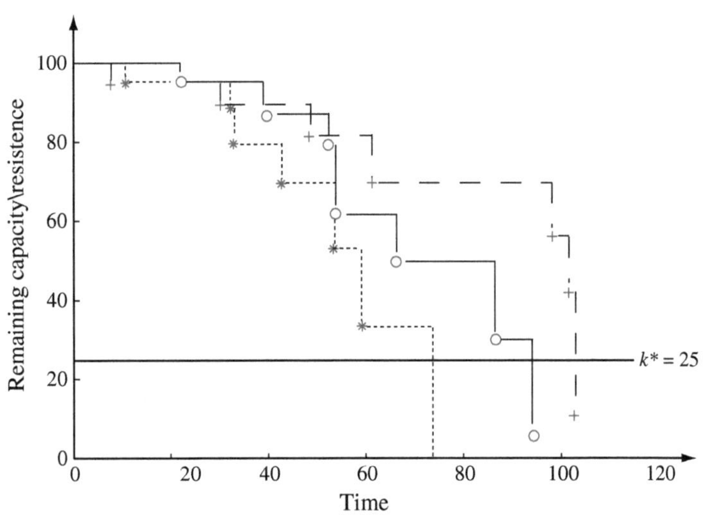

Fig. 5.9 Sample paths of a Geometric processes with the same ratio $a = 0.75$

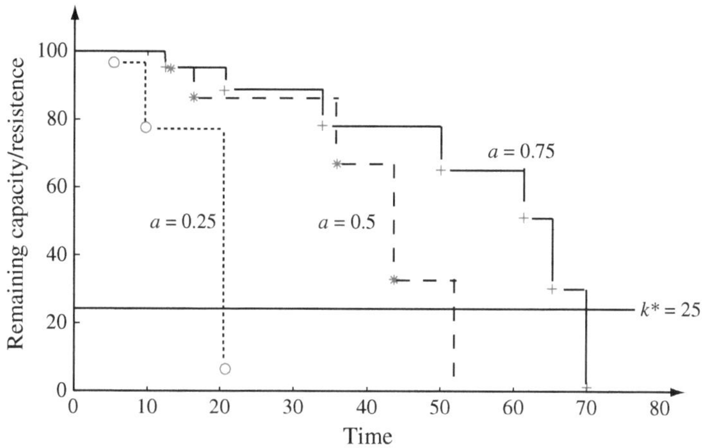

Fig. 5.10 Sample paths of a Geometric processes for various ratios, a

Let us expand the case of damage accumulation where the shock size distributions $\{Y_i, i = 1, 2, \ldots\}$ are described by a geometric process as described above. Thus, if shocks occur at random times, the total damage at time t can be computed as

$$S_{N(t)} = \sum_{i=1}^{N(t)} Y_i,$$

where $N(t)$ is a random variable that describes the number of shocks within the time window $[0, t]$. If $\mathbb{E}[Y_1] = \mu < \infty$, for $t > 0$ [32]; and recalling that $\mathbb{E}[Y_i] = \mu/a^{n-1}$ (Eq. 5.53), where a is the ratio of the process, then

$$\mathbb{E}[S_{N(t)+1}] = \mu \mathbb{E}\left[\sum_{i=1}^{N(t)+1} a^{-i+1}\right] \tag{5.63}$$

For $a \neq 1$, the Wald's equation for a geometric process [32] can be written as

$$\mathbb{E}[S_{N(t)+1}] = \frac{\mu}{1-a}(\mathbb{E}[a^{-N(t)}] - a) \tag{5.64}$$

and for which [32]

$$\mathbb{E}[a^{-N(t)}] = \begin{cases} > a + \frac{(1-a)t}{\mu} & 0 < a < 1 \\ = 1 & a = 1 \\ < a + \frac{(1-a)t}{\mu} & a > 1, t \leq \frac{a\mu}{a-1} \end{cases} \tag{5.65}$$

Note that the restriction on t is due to the convergence of the process. Equation 5.64 describes the expected damage caused by $N(t) + 1$ shocks.

Example 5.22 Evaluate the particular case of a geometric process for which X_1 follows an exponential distribution with parameter λ, i.e., $g_{Y_1}(y) = \lambda \exp(-\lambda y)$ and mean $\mu = 1/\lambda$.

According to Eq. 5.65, for $a < 1$ and $a > 1, t \leq \frac{a\mu}{a-1}$

$$\mathbb{E}[a^{-N(t)}] = a + \frac{(1-a)t}{\mu} \tag{5.66}$$

and therefore,

$$\mathbb{E}[S_{N(t)}] = \frac{\mu}{1-a}(\mathbb{E}[a^{-N(t)}] - a) \tag{5.67}$$

$$= \frac{\mu}{1-a}(a + \frac{(1-a)t}{\mu} - a) = t; \tag{5.68}$$

and more generally [32],

$$\mathbb{E}[S_{N(t)+n+1}] = \frac{a\mu}{a-1} + \frac{1}{a^{n+1}}\left(t - \frac{a\mu}{a-1}\right) \tag{5.69}$$

from which it can be obtained that $\mathbb{E}[S_{N(t)+1}] = \mu + t/a$.

5.7 Damage Accumulation with Annealing

In this model, which was described in Chap. 4, the system deteriorates as a result of shocks but between two consecutive shocks the system recovers part of the damage caused by the previous shock (see Fig. 4.14). If the recovery depends only on the previous shock size and not on the damage history, the total damage just before the $i+1$th shock is $D(t_{i+1}^-) = D(t_i) - Y_i h(t)$, where $D(t_i)$ is the total damage at time t_i and h is a decreasing function in t (system recovery) with $t_i \leq t \leq t_{i+1}$.

Then, the total damage at time t can be computed as

$$D(t) = \sum_{j=1}^{N(t)} Y_j h(t - S_j); \quad \text{with } S_j = \sum_{i=1}^{j} X_i \tag{5.70}$$

where $N(t) \equiv \max_j\{S_j \leq t\}$ and the random variable X_i represents the times between shocks. This model is usually refereed to as a *shot noise* model and it has been widely studied (e.g., see [53–55]). It has been used, for example, in river flow problems [55], dam behavior [56], and storage models [57].

A particular solution for this problem was proposed by Takcs [58] for a recovery function between shocks: $h(t) = e^{-\alpha t}$ with $0 < \alpha < \infty$. This means that if Y is the shock size at a given time and t the time that has passed after this last shock, the total damage accumulated will be $Yh(t) = Ye^{-\alpha t}$. Note that if $Y = 0$, there is no recovery and that the recovery of the system is larger as the size of Y increases. Also, for $t = 0$, there is no recovery at all, while for $t \to \infty$ the systems fully recovers.

Suppose that shocks occur according to a Poisson process with parameter λ. If we define $\Psi(t, y) = P(D(t) \leq y)$ and G is the probability distribution of shock sizes, after some mathematical manipulation, the Laplace transform of $P(D(t) \leq y)$ becomes [58]:

$$\Psi^*(t, s) = \exp\left[-\lambda \int_0^t [1 - G^*(se^{-\alpha u})]du\right] \tag{5.71}$$

where $G^*(s)$ is the Laplace transform of the shock size distributions, i.e., $G^*(s) = \int_0^\infty e^{-sx} dG(x)$. For $\mathbb{E}[Y] < \infty$ and $t \to \infty$ [1],

$$\Psi^*(\infty, s) = \exp\left[-\frac{\lambda}{\alpha} \int_0^1 \frac{[1 - G^*(su)]}{u} du\right] \tag{5.72}$$

Therefore, the distribution $P(D(t) \leq y)$ can be obtained by computing the Laplace inverse of Eqs. 5.71 or 5.72. Clearly, the solution of this problem may become very complex and the use of simulation techniques is necessary to find a solution.

Example 5.23 Find an expression for the time to failure of a system subject to shocks that occur according to a Poisson process with rate λ and where their sizes are distributed as $G(y) = 1 - e^{-\eta y}$ (adapted from [1]).

Replacing the Laplace transform of G in Eq. 5.71 and integrating,

$$\Psi^*(t, s) = \exp\left[-\lambda \int_0^t [1 - G^*(se^{-\alpha u})]du\right] \tag{5.73}$$

$$= \left(\frac{s + \eta e^{\alpha t}}{s + \eta}\right)^{\lambda/\alpha} e^{-\lambda t} \tag{5.74}$$

Then, computing the inverse Laplace transform [58],

$$P(D(t) \leq y) = e^{-\lambda t} \sum_{j=0}^{\infty} \binom{\frac{\lambda}{\alpha} + j - 1}{j} (1 - e^{-\alpha t})^j \sum_{i=j}^{\infty} \frac{(\eta y e^{\alpha t})^i}{i!} e^{-\eta y e^{\alpha t}} \tag{5.75}$$

For the case in which $t \to \infty$ (Eq. 5.72),

$$\Psi^*(\infty, s) = \exp\left[-\frac{\lambda}{\alpha} \int_0^1 \frac{[1 - G^*(se^{-\alpha u})]}{u} du\right] \tag{5.76}$$

$$= \left(\frac{\eta}{s + \eta}\right)^{\lambda/\alpha} \tag{5.77}$$

Then, by computing the inverse,

$$\lim_{t\to\infty} P(D(t) \leq y) = \int_0^y \frac{\eta(\eta u)^{(\lambda/\alpha)-1}}{\Gamma(\lambda/\alpha)} e^{-\eta u} du \tag{5.78}$$

which is the gamma distribution with mean $\lambda/(\alpha\eta)$.

5.8 Models with Correlated Shock Sizes and Shock Times

The last aspect that is important to characterize a shock model is the statistical dependence between the inter-arrival times X_i and the shock sizes Y_i. This requires to study the correlated pair of renewal sequences (X_n, Y_n) [59, 60]. Although there is little information available about this type of problems in the literature [60], two models that have been studied elsewhere are:

- Model I: this model assumes that the size of the kth shock, Y_k, is correlated only with the kth inter-arrival time, X_k.
- Model II: in this model, it is assumed that the kth shock, Y_k, affects the inter-arrival time of the subsequent $(k+1)$th shock, X_{k+1}.

The details of these models are beyond the scope of this book but can be found in [59, 60], where the properties of the associated renewal processes are provided and discussed.

5.9 Summary and Conclusions

In this chapter, we presented and discussed the main features of most common degradation models where the loss of capacity is defined within a continuous-state space. The chapter describes stochastic-based models that include both progressive and shock-based degradation. All models considered in this chapter focus on the cases of systems abandoned after first failure. The characteristics of models that are successively reconstructed are discussed toward the end of the book in the chapters that deal with maintenance and optimization. Although analytical solutions are provided for all cases presented, it has been highlighted the importance of using simulation as the complexity of models increases. In addition to the continuous-state degradation models presented in this chapter, models based on a discrete damage space can be found in Chap. 6. In Chap. 7, we will present a general approach to degradation that can accommodate the models presented in Chaps. 5 and 6.

References

1. T. Nakagawa, *Shock and Damage Models in Reliability* (Springer, London, 2007)
2. M.S. Nikulin, N. Limnios, N. Balakrishnan, W. Kahle, C. Huber-Carol, *Advances in Degradation Modeling: Applications to Reliability, Survival Analysis and Finance*, Statistics for Industry Technology (Birkhauser, Boston, 2010)
3. J.M. Van Noortwijk, A survey of the application of gamma processes in maintenance. Reliab. Eng. Syst. Saf. **94**, 2–21 (2009)
4. M.D. Pandey, Probabilistic models for condition assessment of oil and gas pipelines. Int. J. Non-Destr. Test. Eval. **31**(5), 349–358 (1998)
5. M.D. Pandey, X.X. Yuan, J.M. van Noortwijk, The influence of temporal uncertainty of deterioration on life-cycle management of structures. Struct. Infrastruct. Eng. **5**(2), 145–156 (2009)
6. C. Park, W.J. Padgett, New cumulative damage models for failure using stochastic processes as initial damage. IEEE Trans. Reliab. **54**, 530–540 (2005)
7. J. Ghosh, J. Padgett, M. Sánchez-Silva, Seismic damage accumulation of highway bridges in earthquake prone regions. Earthq. Spectra **31**(1), 115–135 (2015)
8. M. Sánchez-Silva, G.-A. Klutke, D. Rosowsky, Life-cycle performance of structures subject to multiple deterioration mechanisms. Struct. Saf. **33**(3), 206–217 (2011)
9. M. Junca, M. Sánchez-Silva, Optimal maintenance policy for permanently monitored infrastructure subjected to extreme events. Probab. Eng. Mech. **33**(1), 1–8 (2013)
10. I. Iervolino, M. Giorgio, E. Chioccarelli, Gamma degradation models for earthquake-resistant structures. Struct. Saf. **45**, 48–58 (2013)
11. K.C. Kapur, L.R. Lamberson, *Reliability in Engineering Design* (Wiley, New York, 1977)
12. J.L. Bogdanoff, F. Kozin, Probabilistic models of fatigue crack growth. Eng. Fract. Mech. **20**(2), 255–270 (1984)
13. F. Kozin, J.L. Bogdanoff, Probabilistic models of fatigue crack growth: results and specifications. Nucl. Eng. Des. **115**, 143–171 (1989)
14. T.J. Aven, U. Jensen, *Stochastic Models in Reliability*. Series in Applications of Mathematics: Stochastic Modeling and Applied Probability (41) (Springer, New York, 1999)
15. W. Kahle, H. Wendt, On accumulative damage process and resulting first passage times. Appl. Stoch. Models Bus. Ind. **20**, 17–26 (2004)
16. Federal Emergency Management Agency (FEMA), *Earthquake loss estimation methodology: technical manual*. National Institute of Building Sciences for the Federal Emergency Management Agency (FEMA), Washington (1997)
17. J.T. Duane, Learning curve approach to reliability monitoring. IEEE Trans. Aerosp. **2**, 563–566 (1964)
18. S. Zacks, Distributions of failure times associated with non-homogeneous compound poisson damage processes. Inst. Math. Stat.-Lect. Notes-Monogr. Ser. **45**, 396–407 (2004)
19. W. Kahle, H. Wendt, Parametric shock models, in *Advances in Degradation Modeling*, ed. by M.S. Nikulin, et al. (Birkhauser, Boston, 2010)
20. D.S. Reynolds, I.R. Savage, Random wear models in reliability theory. Adv. Appl. Probab. **3**, 229–248 (1971)
21. Z.W. Birnbaum, S.C. Saunders, A new family of life distributions. J. Appl. Probab. **6**, 319–327 (1969)
22. A. Desmond, Stochastic models of failure in random environments. Can. J. Stat. **13**, 171–183 (1985)
23. W.J. Owen, W.J. Padgett, Accelerated test models for system strength based on birnbaum-saunders distribution. Life Data Anal. **5**(2), 133–147 (1999)
24. D.B. Kececioglu, M.X. Jiang, A unified approach to random-fatigue reliability quantification under random loading, in *Proceedings of the Annals of Reliability Maintainability Symposium*, pp. 308–313 (1998)
25. G.A. Whitmore, Estimating degradation by a Wiener diffusion process subject to measurement error. Lifetime Data Anal. **1**, 307–319 (1995)

26. K. Doksum, S.L. Normand, Gaussian models for degradation processes-part I: methods for the analysis of biomarker data. Lifetime Data Anal. **1**(2), 131–144 (1995)
27. G.A. Whitmore, F. Schenkelberg, Modeling accelerated degradation data using wiener diffusion with a time scale transformation. Lifetime Data Anal. **3**, 27–45 (1997)
28. G.A. Whitmore, M.J. Crowder, J.F. Lawless, Failure inference from a marker process based on a bivariate Wiener model. Lifetime Data Anal. **4**, 229–251 (1998)
29. W. Kahle, A Lehmann, The Wiener process as a degradation model: modeling and parameter estimation, in *Advances in Degradation Modeling*, ed. by M.S. Nikulin et al. (eds.) (Birkhauser, Boston, 2010)
30. W.J. Padgett, M.A. Tomlinson, Inference from accelerated degradation and failure data based on Gaussian process models. Lifetime Data Anal. **10**, 191–206 (2004)
31. P. Kiessler, G.-A. Klutke, Y. Yang, Availability of periodically inspected systems subject to Markovian degradation. J. Appl. Probab. **39**, 700–711 (2002)
32. Y. Lam, *The Geometric Process and Its Applications* (World Scientific Press, New Jersey, 2007)
33. E. Çinlar, Z.P. Bazant, E. Osman, Stochastic process for extrapolating concrete creep. J. Eng. Mech. Div. **103**(EM6), 1069–1088 (1977)
34. E. Çinlar, On a generalization of gamma processes. J. Appl. Probab. **17**, 467–480 (1980)
35. N.D. Singpurwalla, Survival in dynamic environments. Stat. Sci. **1**, 86–103 (1995)
36. P.A.P. Moran, *The Theory of Storage* (Methuen, London, 1959)
37. J.D. Baker, H.J. van Der Graph, J.M. van Noortwijk, *Proceedings of the Eight International Conference on Structural Faults and Repair* (Edinburgh Engineering Technics Press, London, 1999)
38. M. Abdel-Hameed, A gamma wear process. IEEE Trans. Reliab. **24**(2), 152–153 (1975)
39. R.P. Nicolai, G. Budai, R. Dekker, M. Vreijling, A comparison of models for measurable deterioration: an application to coatings on steel structures. Reliab. Eng. Syst. Saf. **92**(12), 1635–1650 (2007)
40. N.D. Singpurwalla, S.P. Wilson, Failure models indexed by two scales. Adv. Appl. Probab. **30**(4), 1058–1072 (1998)
41. V. Bagdonavicius, M.S. Nikulin, Estimation in degradation models with explanatory variables. Lifetime Data Anal. **7**(1), 85–103 (2001)
42. W. Wang, P.A. Scarf, M.A.J. Smith, On the applications of a model of condition-based maintenance. J. Oper. Res. Soc. **51**(11), 1218–1227 (2000)
43. A.N. Avramidis, P. L'Ecuyer, P.A. Tremblay, Efficient simulation of gamma and variance gamma processes, in *Proceedings of the 2003 Winter Simulation Conference, IEEE*, Ed. by S. Chick, P.J. Sçnzhs, D. Ferrin, D.J. Morrice, pp. 319–323, Piscataway, August (2003)
44. N.T. Kottegoda, R. Rosso, *Probability, Statistics and Reliability for Civil and Environmental Engineers* (McGraw Hill, New York, 1997)
45. A.H-S. Ang, W.H. Tang, *Probability Concepts in Engineering: Emphasis on Applications to Civil and Environmental Engineering* (Wiley, New York, 2007)
46. F. Dufresne, H.U. Gerber, E.S.W. Shiu, Risk theory with gamma process. ASTIN Bul. **21**(2), 177–192 (1991)
47. J.S.K. Chang, Y. Lam, D.Y.P. Leung, Statistical inference for geometric processes with gamma distributions. Comput. Stat. Data Anal. **47**, 565–581 (2004)
48. Y. Lam, Non-parametric inference for geometric processes. Commun. Stat. Theory Methods **21**, 2083–2105 (1992)
49. Y. Lam, A shock model for the maintenance problem of reparable systems. Comput. Oper. Res. **31**, 1807–1820 (2004)
50. Y. Lam, S.K. Chang, Statistical inference for geometric processes with lognormal distributions. Comput. Stat. Data Anal. **27**, 99–112 (1998)
51. F.K.N. Leung, Statistical inferential analogies between arithmetic and geometric processes. Int. J. Reliab. Qual. Saf. Eng. **12**, 323–335 (2005)
52. S. Suresh, *Fatigue of Materials*, 2nd edn. (Cambridge University Press, Edimburgh, 1998)
53. J. Rice, On generalized shot noise. Adv. Appl. Probab. **9**, 553–565 (1977)

54. T.L. Hsing, J.L. Teugels, Extremal properties of shot noise processes. Adv. Appl. Probab. **21**, 513–525 (1989)
55. E. Waymire, V.K. Gupta, The mathematical structure of rainfall representations 1: a review of stochastic rainfall models. Water Res. Res. **17**, 1261–1272 (1981)
56. R.B. Lund, A dam with seasonal input. J. Appl. Probab. **31**, 526–541 (1994)
57. R.B. Lund, The stability of storage models with shot noise input. J. Appl. Probab. **33**, 830–839 (1996)
58. L. Takacs, *Stoch. Process.* (Wiley, New York, 1960)
59. U. Sumita, J. Shanthikumar, General shock models associated with correlated renewal sequences. J. Appl. Probab. **20**, 600–614 (1983)
60. U. Sumita, Z. Jinshui, *Analysis of correlated multivariate shock model generated from a renewal sequence*. Department of Social Systems and Management: discussion paper series No. 1194; University of Tsukuba, Tsukuba, Japan (2008)

Chapter 6
Discrete State Degradation Models

6.1 Introduction

This chapter presents and discusses models where the system state, as it degrades, takes values in a discrete state space. Furthermore, it is assumed that the change of the system state through time may occur at discrete or continuous points in time according to certain rules. These models assume that the system moves through a sequence of increasing damage states until failure or intervention. Under these assumptions, most models presented in this chapter are based on Markov processes and in particular on Markov chains, which may be discrete or continuous in time.

In the chapter, we present both the basic theory of Markov chains as well as extensions and generalizations of the Markov property to so-called semi-Markov processes. We also include several examples of each process and discuss estimation of model parameters. For further details on Markov and semi-Markov processes, the reader is referred to [1–4]. Finally, at the end of the chapter, we present some degradation models that take advantage of the characteristics and properties of phase-type distributions, originally inspired by Cox [5] and studied extensively by M.F. Neuts [6, 7].

6.2 Discrete Time Markov Chains

In this section, we introduce a discrete state stochastic processes whose future states are conditionally independent of their past states, provided that the present state is known. This condition is known as the *Markov property*, and these processes are called Markov chains. They are among the most widely studied and applied stochastic processes, particularly in engineering. While we limit ourselves to processes on a countable state space, we consider separately processes where time evolves by discrete epochs (discrete time Markov chains, or DTMC) and where time evolves continuously (continuous time Markov chains, of CTMC; see Sect. 6.3).

M. Sánchez-Silva and G.-A. Klutke, *Reliability and Life-Cycle Analysis of Deteriorating Systems*, Springer Series in Reliability Engineering,
DOI 10.1007/978-3-319-20946-3_6

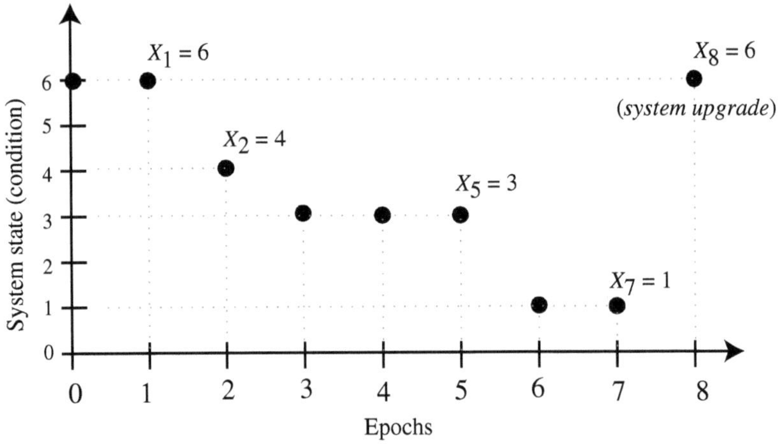

Fig. 6.1 Sample path of a discrete time markov chains (DTMC)

6.2.1 Definition

Consider, a stochastic process $\mathscr{X} = \{X_n, n = 0, 1, 2, \ldots\}$ that takes values in a countable state space S. The index set $\{n = 0, 1, 2 \ldots\}$ will be taken to represent *time* epochs, and we refer to X_n as the *state* of the process at time n. If $X_n = i \in S$, we say that the process is in state i at time n (Fig. 6.1).

The Markov property for a discrete time process can be stated as:

Definition 38 The stochastic process $\mathscr{X} = \{X_n, n \in \mathbb{N}\}$ with state space S satisfies the Markov property if

$$P(X_{n+1} = j | X_n = i, X_{n-1} = i_{n-1}, \ldots, X_1 = i_1, X_0 = i_0) = P(X_{n+1} = j | X_n = i) \quad (6.1)$$

holds for all i, i_n, and j in S and all $n \in \mathbb{N}$.

In words, the Markov property asserts that, for any reference time n, the "future" of the process (all states subsequent to n) is *conditionally* independent of the "past" (all states prior to n), given the "present" (the state at n). Such a process is called a discrete time Markov chain (DTMC). To simplify matters greatly, we will consider only *time homogeneous* Markov chains; i.e., those for which

$$P(X_{n+1} = j | X_n = i) = P_{ij}, \quad i, j \in S \quad (6.2)$$

is independent of n. The quantity P_{ij} is called the *one-step transition probability* from state i to state j. It represents the probability that, given the process is currently in state i, the process will be in state j at the next time epoch.

For a time homogeneous DTMC, we collect the one-step transition probabilities together in a matrix called the *one-step transition probability matrix* (or simply, the *transition probability matrix*) $\mathbf{P}$

$$\mathbf{P} = [P_{ij}], \quad i, j \in S \tag{6.3}$$

Note that $\mathbf{P}$ is a stochastic matrix; therefore, the elements in $\mathbf{P}$ are nonnegative and each row sums to 1. Note that the *2-step transition probabilities* ${P_{ij}}^{(2)} = P(X_2 = j|X_0 = i)$, are given by

$$\begin{aligned} {P_{ij}}^{(2)} &= \sum_{k \in S} P(X_2 = j|X_0 = i, X_1 = k) P(X_1 = k|X_0 = i) \\ &= \sum_{k \in S} P_{ik} P_{kj}, \end{aligned} \tag{6.4}$$

where the last equality follows by the Markov property. Thus, in matrix terms, the 2-step transition probability matrix $\mathbf{P}^{(2)}$ is given by

$$\mathbf{P}^{(2)} = \mathbf{P} \cdot \mathbf{P} = \mathbf{P}^2. \tag{6.5}$$

Determining the n-step transition probability matrix $\mathbf{P}^{(n)}$, whose elements are $P(X_n = j|X_0 = i), i, j \in S$, can be accomplished in a similar way. To this end, we introduce the *Chapman-Kolmogorov equations*

$$P_{ij}^{(n+m)} = \sum_{k \in S} P_{ik}^{(n)} P_{kj}^{(m)} \quad \text{for all } n, m \geq 0, i, j \in S \tag{6.6}$$

In matrix form, these equations are:

$$\mathbf{P}^{(n+m)} = \mathbf{P}^{(n)} \cdot \mathbf{P}^{(m)}, \tag{6.7}$$

and it therefore follows that

$$\mathbf{P}^{(n)} = \mathbf{P}^n, \quad n \geq 1. \tag{6.8}$$

Finally, we define the *state probability vector* at time n, $\mathbf{p}^n$, as the row vector whose elements are $\{P(X_n = i), i \in S\}$. The state probability vector provides the predictions on the state of the process at time n. Given, the initial state probability vector $\mathbf{p}^0$ and the one-step transition probability matrix $\mathbf{P}$, we can easily determine $\mathbf{p}^n$ for any $n \in \mathbb{N}$ by successive conditioning to obtain

$$\mathbf{p}^n = \mathbf{p}^{n-1}\mathbf{P} = \mathbf{p}^0\mathbf{P}^n \tag{6.9}$$

As the expression above indicates, the transient behavior of the DTMC is completely determined by its one-step transition probability matrix **P** and the initial state vector $\mathbf{p}^0$. For example, the *path probabilities*, namely the finite-dimensional joint distributions of $(X_1, X_2, \ldots, X_n)$ for any n, are given by

$$P(X_1 = i_1, X_2 = i_2, \ldots, X_n = i_n) = \sum_{i \in S} P(X_0 = i) P_{ii_1} P_{i_1 i_2} \cdot P_{i_{n-1} i_n} \tag{6.10}$$

Determining whether a limiting distribution of the DTMC exists and is independent of the initial state, that is, determining whether a probability distribution $\{\pi_j, j \in S\}$ exists, where

$$\lim_{n \to \infty} P(X_n = j | X_0 = i) = \pi_j, \tag{6.11}$$

with $\sum_{j \in s} \pi_j = 1$, involves classifying the states of the Markov chain into groups of states for which the first passage times between any two states in the group are finite with probability one. These groups of states comprise the *communicating classes* of the Markov chain, and one can determine from the matrix **P** whether a given communicating class is *recurrent* or *transient*. A communicating class of states is recurrent if it has the property that, once a state in the class is ever visited, it will be visited infinitely often; otherwise the class is transient. The limiting probability that the Markov chain is in a transient state is zero; the limiting probability that the Markov chain is in a recurrent state depends on the initial state as well as the transition probability matrix.

Markov chains may have *absorbing* states; these are recurrent states characterized by a 1 in the diagonal element corresponding to that state in the transition probability matrix. Absorbing states have the property that, once entered, the Markov chain remains in that state forever. For absorbing states, one can calculate the length of time to absorption, given the initial state.

For Markov chains whose states all communicate (so-called *irreducible* Markov chains) and are aperiodic[1] the limiting probabilities, if they exist, can be shown to satisfy the *balance equations*

$$\pi_j = \sum_{k \in S} \pi_k P_{kj}, \quad j \in S \tag{6.12}$$

and the *normalizing equation* $\sum_{j \in S} \pi_j = 1$.

Example 6.24 Consider a system of four identical components operating in parallel. Suppose that each component (independently of the others) survives for a geometrically distributed length of time with mean 2.5 units. The system will operate as long

[1] The term *periodic* means that the Markov chain can revisit a state only on steps that are a multiple of some integer $k > 1$.

as at least one of the components operates. Let X_n denote the number of failed components at the beginning of time period n, and suppose that initially all components are operational. The sequence $\{X_n, n = 0, 1, 2, \ldots\}$ comprises a Markov chain with state space $\{0, 1, 2, 3, 4\}$, where 0 means that all four components are working and 4 means that all four components have failed. Then, for example, $X_2 = 3$ means that there are three components that have failed at time $n = 2$.

Since the lifetimes of components are geometrically distributed, each component fails during a time period with probability $1/0.25 = 0.4$ and survives the time period with probability $1 - 0.4 = 0.6$. The transition probability matrix for this process is

$$\mathbf{P} = \begin{bmatrix} (0.6)^4 & 4(0.6)^3(0.4) & 6(0.6)^2(0.4)^2 & 4(0.6)(0.4)^3 & (0.4)^4 \\ 0 & (0.6)^3 & 3(0.6)^2(0.4) & 3(0.6)(0.4)^2 & (0.4)^3 \\ 0 & 0 & (0.6)^2 & 2(0.6)(0.4) & (0.4)^2 \\ 0 & 0 & 0 & 0.6 & 0.4 \\ 0 & 0 & 0 & 0 & 1 \end{bmatrix}$$
$$= \begin{bmatrix} 0.1296 & 0.3456 & 0.3456 & 0.1536 & 0.0256 \\ 0 & 0.216 & 0.432 & 0.288 & 0.064 \\ 0 & 0 & 0.36 & 0.48 & 0.16 \\ 0 & 0 & 0 & 0.6 & 0.4 \\ 0 & 0 & 0 & 0 & 1 \end{bmatrix}$$

where the value of $\mathbf{P}_{1,1}$ corresponds to the case in which all components are operating. To estimate the state probability vectors at time epochs 2, 5, 10, we use Eq. 6.9 with $\mathbf{p}^0 = [1, 0, 0, 0, 0]$ (i.e., all components are operating at time $t = 0$) to obtain

$$\begin{aligned} \mathbf{p}^2 &= [0.0168,\ 0.1194,\ 0.3185,\ 0.3775,\ 0.1678] \\ \mathbf{p}^5 &= [0,\ \ 0.0017,\ 0.0309,\ 0.2440,\ 0.7234] \\ \mathbf{p}^{10} &= [0,\ \ 0,\ \ 0.0002,\ 0.0238,\ 0.9760] \end{aligned}$$

For example, after five time intervals, the probability that the system does not operate (i.e., all components have failed) is 0.7234. Note that states 0,1, 2, and 3 are transient states and state 4 is an absorbing state, hence eventually the chain will end up in state 4 with probability 1 (e.g., $\mathbf{p}^{25} = [0, 0, 0, 0, 1]$).

Example 6.25 Now suppose we have a system whose functionality declines over time until the system fails. The system is inspected at periodic time epochs. At each inspection, if the system is within acceptable operating characteristics, it is classified into one of four states, with state 1 representing perfect operating condition and each higher state (2, 3, 4) representing decreased functionality. If an inspection determines that the system falls below acceptable operating performance, it is removed from service and classified as being in state 5, which represents system failure.

Suppose the system is abandoned at failure. If we let the discrete time index correspond to the sequence of inspections, we can define X_n to be the state of the system at (i.e., just after) the nth inspection. Inspections may or may not be equally spaced, but in order for us to model the process $\{X_n, n = 0, 1, \ldots\}$ as a DTMC, we must assume that the length of time the system spends in each state is memoryless. Under this assumption, suppose that data obtained from a large number of inspections yields the following estimates for transition probabilities:

$$\mathbf{P} = \begin{bmatrix} 0.312 & 0.156 & 0.375 & 0.063 & 0.094 \\ 0 & 0.414 & 0.069 & 0.276 & 0.241 \\ 0 & 0 & 0.359 & 0.256 & 0.385 \\ 0 & 0 & 0 & 0.8 & 0.2 \\ 0 & 0 & 0 & 0 & 1 \end{bmatrix}.$$

The objective of the analysis is to estimate the probability that the system is in a given state after n time steps.

This probability can be computed as:

$$\mathbf{p}^n = \mathbf{p}^0 \mathbf{P}^n$$

where $\mathbf{p}^0 = [1, 0, 0, 0, 0]$. Therefore, the state probabilities for $n = 1$, $n = 5$ and $n = 15$ are:

$$\begin{aligned} \mathbf{p}^1 &= [0.312,\ 0.156,\ 0.375,\ 0.063,\ 0.094] \\ \mathbf{p}^5 &= [0.003,\ 0.014,\ 0.029,\ 0.243,\ 0.711] \\ \mathbf{p}^{15} &= [0,\ \ 0,\ \ 0,\ \ 0.029,\ 0.971] \end{aligned}$$

And the evolution of the probability of failure as function of the number of transitions is shown in Fig. 6.2.

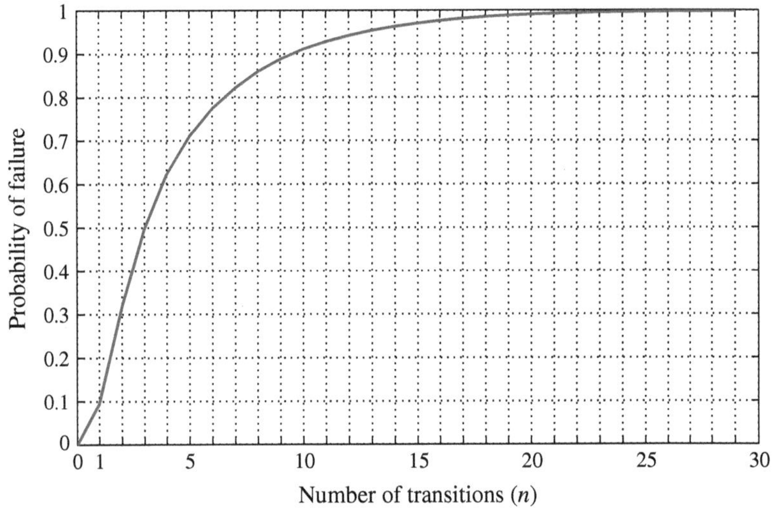

Fig. 6.2 Probability of failure as function of the number of transitions

Example 6.26 Consider the previous example, but suppose that when an inspection identifies that the system has degraded below acceptable operating conditions (state 5), it is taken out of service and replaced or refurbished to a "good as new" condition at the subsequent inspection. The transition probability matrix is then given by

$$\mathbf{P} = \begin{bmatrix} 0.312 & 0.156 & 0.375 & 0.063 & 0.094 \\ 0 & 0.414 & 0.069 & 0.276 & 0.241 \\ 0 & 0 & 0.359 & 0.256 & 0.385 \\ 0 & 0 & 0 & 0.8 & 0.2 \\ 1 & 0 & 0 & 0 & 0 \end{bmatrix}.$$

Note that, in this case, $[\mathbf{P}_{5,1} = 1$; which means that the system is taken to state "as good as new" once it reaches state 5. has The Markov chain in this example is irreducible; all states communicate with each other. Transient behavior may be determined as usual, but in this case the objective of the analysis is to estimate the steady-state probability that the system is in a given state.

$$\begin{aligned} \mathbf{p}^2 &= [0.191, \ 0.113, \ 0.262, \ 0.209, \ 0.224] \\ \mathbf{p}^5 &= [0.254, \ 0.075, \ 0.171, \ 0.328, \ 0.173] \\ \mathbf{p}^{10} &= [0.249, \ 0.067, \ 0.153, \ 0.361, \ 0.170] \\ \mathbf{p}^{20} &= [0.248, \ 0.066, \ 0.152, \ 0.364, \ 0.171] \end{aligned}$$

And for a large number of time steps e.g., $\mathbf{p}^{50} = [0.248, \ 0.066, \ 0.152, \ 0.364, \ 0.171]$.

6.2.2 *Estimating Transition Probabilities from Empirical Data*

The validity of the model results depends highly on the selection of the transition probability matrix, **P** (Eq. 6.3). However, generally, it is not easy to obtain it directly from field observations. Then, in many studies, its values are assigned arbitrarily or based on experience. In this section, we present a general approach to evaluate the matrix, **P**; in particular, we focus on the case in which the matrix **P** is constructed from system condition evaluations.

System Condition Evaluation

Engineering judgement has been widely used to describe the state of physical systems via condition ratings. Some examples of these ratings are the Pavement Condition Index (PCI) (scale 1 to 8) [8] and the bridge deck condition (scale 0 to 9) [9]. Rating data are discrete ordinal measurements with the purpose of ordering system states, and are not intended as a direct measure of the actual condition of the system [10]. Ratings are commonly described in linguistic terms and are associated with a discrete numerical scale; e.g., "excellent condition" = 5, "moderate condition" = 3, and "poor condition" = 1. In practice, the assessment and evaluation of these ratings are the bases for most maintenance and rehabilitation programs.

Since condition ratings provide a discrete assessment of the system at fixed points in time, Markov chains become a useful tool for estimating future system states. Thus, given some empirical data, the challenge is to obtain the transition probability matrices. Among many approaches available in the literature, the so-called *expected value* or *regression-based optimization* method have been widely used to obtain these probabilities [10–12]. In this method, transition probabilities are estimated by solving the nonlinear optimization problem that minimizes the sum of absolute differences between the regression curve that best fits the condition data and the conditions predicted using the Markov chain model.

Transition Probabilities from Experimental Data

Consider a system whose performance is defined on a discrete state space $\mathbf{S} = \{S_1, S_2, ..., S_k\}$. Suppose that observations of the system's state have been recorded for successive (time) intervals $n = 1, 2, ..., m$. Then, the stationary (i.e., time-independent) transition probabilities can be estimated by solving the following non-linear optimization problem [11]:

$$\begin{aligned} &\text{Minimize} && \textstyle\sum_{n=1}^{m} |Y(t) - \mathbb{E}[n, \mathbf{P}]| \\ &\text{Subject to:} && 0 \leq P_{ij} \leq 1 \text{ for } i, j = 1, 2, ..., k \\ & && \textstyle\sum_{i=1}^{m} P_{ij} = 1 \end{aligned} \tag{6.13}$$

where $Y(t)$ is the best regression model (Chap. 4); i.e., the average condition rating of the system at time t. $\mathbb{E}[n, \mathbf{P}]$ is the expected value of the system state predicted by using the Markov chain model; and $\mathbf{P}$ is the transition probability matrix, whose components P_{ij} are the decision variables. Note that when evaluating $Y(t) - \mathbb{E}[n, \mathbf{P}]$ the time t must correspond with the interval n of the assessments made using the Markov chain.

The expected value $\mathbb{E}[n, \mathbf{P}]$ is computed as follows:

$$\mathbb{E}[n, \mathbf{P}] = \mathbf{p}^n \cdot \mathbf{S} = [\mathbf{p}^0 \cdot \mathbf{P}^n] \cdot \mathbf{S} \tag{6.14}$$

where $\mathbf{p}^0$ is the vector of the condition state probabilities at age $n = 0$; the entries of $\mathbf{p}^0$ are obtained from a normalized histogram of frequencies of the system states at $n = 0$; and $\mathbf{P}^n$ is the n-step transition probability matrix. This matrix is determined by multiplying the transition matrix $\mathbf{P}$ by itself n times. Finally, the vector $\mathbf{S} = \{S_1, S_2, ..., S_k\}$ describes the system states and is usually a small value, e.g., $k \leq 10$ [10].

Some additional assumptions can be made to make the model more efficient computationally. First, if interventions are not allowed (e.g., maintenance), an additional restriction can be added so that $P_{ij} = 0$; for $i > j$. Also, in some cases it may be reasonable to assume that only changes from one state to the next are allowed; in other words, $P_{ij} = 0$ for $j > (i + 1)$. This restriction limits the search of the P_{ij} values [12].

This approach has received some criticism regarding difficulties in capturing the inherent nonstationary nature of the probabilities and its actual ability to describe the unobservable (see Chap. 4) deterioration mechanisms [10]. Other existing approaches to obtain transition probabilities from empirical data include ordered probit models [10, 12]; artificial intelligent techniques such as neural networks [13]; and the use of expert opinions [14]. These methods have been applied to many engineering fields, mostly related to infrastructure systems; for example, to the management of waste water systems [12], the prediction of bridge deck systems [15] and for pavement management [14, 16].

Example 6.27 The Federal Highway Administration keeps historical records about the condition of the transportation infrastructure throughout the US Among the many measurements they make, the National Bridge Inventory program [17] uses the Sufficiency Rating Index (SRI) to evaluate the condition of bridges. The SRI is an index that evaluates different structural and nonstructural properties of bridge performance and provides an overall assessment measured within the continuous range [0–100]. In this example, we consider the SRI data for the state of Florida, which reports assessments until 2011. All SRI data registered from bridge assessments over the last 100 years in Florida is shown graphically in Fig. 6.3. As it can be observed, and as expected, the dispersion of the data is quite large. Then, the purpose is to estimate the transition probability matrix and the probability of failure as function of time.

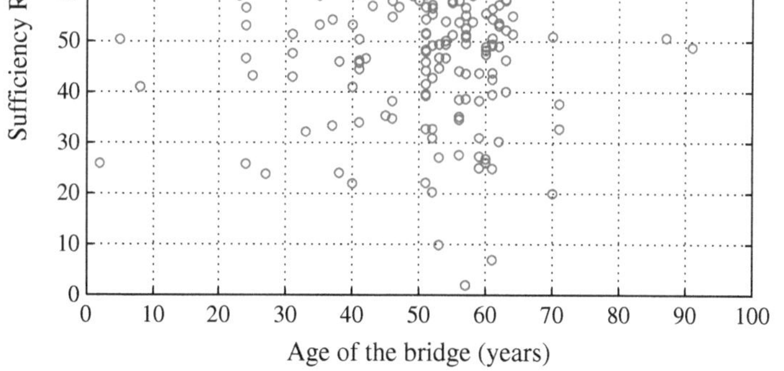

Fig. 6.3 Sufficiency rating versus age for bridges in florida

Table 6.1 Description of system estates

S	SRI range		Evaluation
1	0	15	Unacceptable
2	15	30	Deficient
3	30	50	Fair
4	50	65	Moderate
5	65	75	Good
6	75	90	Very good
7	90	100	Excellent

In order to develop a Markov model, the structural condition of bridges was grouped into the following states: $\mathbf{S} = \{1, 2, ..., 7\}$; these states were obtained after dividing the SRI values into the ranges shown in Table 6.1. In a Markov chain, the change between system states occurs at fixed time intervals. Therefore, for the purpose of this example, the 100 years observation time span was divided in 10 time steps of 10 years each. For example, all records between $t = 0$ and $t = 10$ were assigned as if they have occurred at $t = 10$. Clearly, the accuracy of the model depends on the length of time steps and the number of condition states.

Based on this classification, the next step consists of finding a good regression model for the system states. The model used in this case was:

$$Y(t) = 6.6291 - 0.0144t \tag{6.15}$$

where t is the age of the bridge and $Y(t)$ is the system state at time t. Clearly, the selection of this model requires some preprocessing of information. Then by solving the optimization problem formulated in Eq. 6.13, the following transition probability matrix is obtained:

$$\mathbf{P} = \begin{bmatrix} 0.99 & 0.01 & 0 & 0 & 0 & 0 & 0 \\ 0 & 0.69 & 0.31 & 0 & 0 & 0 & 0 \\ 0 & 0 & 0.52 & 0.39 & 0.09 & 0 & 0 \\ 0 & 0 & 0 & 0.47 & 0.37 & 0.16 & 0 \\ 0 & 0 & 0 & 0 & 0.51 & 0.42 & 0.07 \\ 0 & 0 & 0 & 0 & 0 & 0.62 & 0.38 \\ 0 & 0 & 0 & 0 & 0 & 0 & 1 \end{bmatrix}$$

Note that the use of a different regression model may of course lead to a different transition probability matrix. According to the Federal Highway Administration, the bridge is considered to require a mayor intervention if $k^* = \text{SRI} \leq 50$. Thus, it is said that the bridge is in a failed condition if it is in state 1, 2, or 3. Then, the failure probability at epochs (e.g., time intervals) $n = 1, 2, ...$ is computed by solving Eq. 6.9. The results show, for instance, the following failure probabilities: $P_f(10) = 0.017$,

$P_f(50) = 0.175$ and $P_f(100) = 0.322$. Note that failure probability grows slowly due to the values of the transition probability matrix derived from the regression selected (i.e., Eq. 6.15); but as expected, as n becomes larger, the failure probability approaches to 1.

6.3 Continuous Time Markov Chains

A continuous Time Markov chain (CTMC) is the continuous time analog of the DTMC, namely a continuous time process with a countable state space that satisfies the Markov property.

Definition 39 The stochastic process $\mathscr{X} = \{X(t), t \geq 0\}$ with countable state space S satisfies the Markov property if

$$P(X(t+s) = j | X(s) = i, X(u) = x(u), u < s) = P(X(t+s) = j | X(s) = i) \tag{6.16}$$

holds for all $i, j, x(u), u < s$ in S and all $s, t \geq 0$.

Again, for simplicity, we will consider only *time homogeneous* continuous time Markov chains; i.e., those for which

$$P(X(t+s) = j | X(s) = i) = P_{ij}(t), \tag{6.17}$$

is independent of s.

In the CTMC, the transitions from state to state occur in a structured manner. Then, suppose that the chain is in a particular state (call it state i) at time $t = 0$. By the Markov property, the length of time spent in state i during the initial sojourn must have the memoryless property; i.e., the length of time (sojourn time) spent in the state i before making a transition is an exponentially distributed random variable with parameter ν_i that depends only on state i. When the sojourn time in state i expires, the process instantaneously enters a different state. Just prior a state change epoch, the next state ("future") can depend only on the current state ("present") and neither on any previous states nor on the length of time spent in the current state ("past"). Thus, when the chain leaves state i, the next state is state $j \neq i$ with some probability P_{ij}. To summarize, state transitions occur as if according to a DTMC, with exponential sojourn times (with state dependent mean) in each state between transitions (Fig. 6.4).

We define the *transition probability functions* $P_{ij}(t)$ for each pair $i, j \in S$ and $t \geq 0$ as

$$P_{ij}(t) = P(X(t) = j | X(0) = i). \tag{6.18}$$

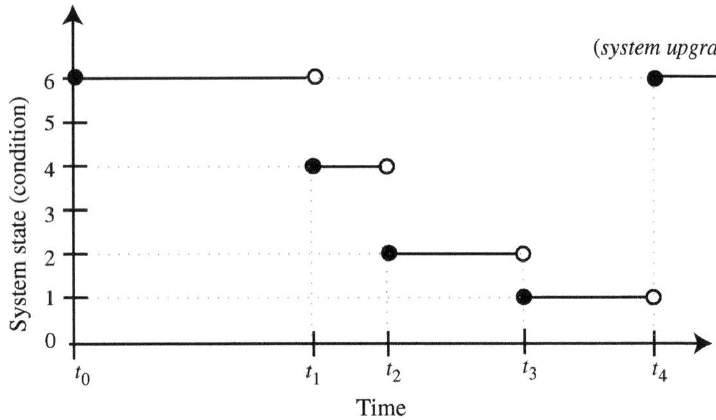

Fig. 6.4 Sample path of a continuous time Markov chain

These functions satisfy the continuous time Chapman-Kolmogorov equations

$$P_{ij}(t+s) = \sum_{k \in S} P_{ik}(t)P_{kj}(s), \quad i, j \in S \quad \text{and} \quad t, s \geq 0, \tag{6.19}$$

which follow directly from the Markov property.

The transition probability functions of the CTMC play a role analogous to the n-step transition probabilities of the DTMC in determining the transient behavior of the process. The transition probability functions arise as the solution to a system of differential equations, known as the *Kolmogorov differential equations.* To develop these equations, we first state the following lemma (for a proof see [2]), which defines the fundamental parameters of the CTMC.

Lemma 40

$$\lim_{h \to 0} \frac{1 - P_{ii}(h)}{h} = \nu_i \tag{6.20}$$

$$\lim_{h \to 0} \frac{P_{ij}(h)}{h} = q_{ij} \quad i \neq j, \tag{6.21}$$

The parameters $\{\nu_i, \quad i \in S\}$ and $\{q_{ij}, \quad i, j \in S, \quad i \neq j\}$ are the fundamental parameters of the CTMC. In fact, with respect to the informal description of transitions of the CTMC given above, ν_i is the parameter of the exponential sojourn time of each visit to state i, and q_{ij} has the representation

$$q_{ij} = \nu_i \cdot P_{ij}, \tag{6.22}$$

where P_{ij} is the probability that the next state is j at a transition epoch from state i. For this reason, we refer to the $q_{ij}, i, j \in S$ as the transition *rates* of the CTMC, and the probabilities P_{ij}, $i, j \in S$ as the transition probabilities of the *embedded* Markov chain; i.e., the DTMC viewed strictly at transition epochs. Note that

$$\sum_{j \in S} P_{ij}(h) = P_{ii}(h) + \sum_{j \neq i} P_{ij}(h) = 1 \Rightarrow \lim_{h \to 0} \frac{P_{ii}(h) - 1 + \sum_{j \neq i} P_{ij}(h)}{h} = 0,$$

and therefore the lemma above implies that

$$-\nu_i + \sum_{j \neq i} q_{ij} = 0, \quad i \in S, \tag{6.23}$$

therefore,

Definition 41 The infinitesimal generator matrix (or simply, the generator) of the CTMC is the matrix comprised of the parameters above, arranged as follows (here we list the states as $\{1, 2, 3, \ldots\}$):

$$\mathbf{Q} = \begin{bmatrix} -\nu_1 & q_{12} & q_{13} & q_{14} & \cdots \\ q_{21} & -\nu_2 & q_{23} & q_{24} & \cdots \\ q_{31} & q_{32} & -\nu_3 & q_{34} & \cdots \\ \vdots & \vdots & \vdots & \vdots & \vdots \end{bmatrix} \tag{6.24}$$

The generator matrix $\mathbf{Q}$ is somewhat analogous to the one-step transition probability matrix of the DTMC; both transient and steady-state behavior can be characterized in terms of $\mathbf{Q}$. Two sets of differential equations (collectively known as the *Kolmogorov differential equations*) can be used to determine the transient behavior of the CTMC. These equations follow directly from the continuous time Chapman-Kolmogorov equations (6.19) and the lemma above, and we state them here without proof (see [2]):

Theorem 42 (Kolmogorov Backward equations) *For all* $i, j \in S$ *and* $t \geq 0$,

$$P'_{ij}(t) = \sum_{k \neq i} q_{ik} P_{kj}(t) - \nu_i P_{ij}(t). \tag{6.25}$$

Theorem 43 (Kolmogorov Forward equations) *Under suitable regularity conditions, for all* $i, j \in S$ *and* $t \geq 0$,

$$P'_{ij}(t) = \sum_{k \neq j} q_{kj} P_{ik}(t) - \nu_j P_{ij}(t). \tag{6.26}$$

In a few limited cases, the Kolmogorov differential equations can be explicitly solved, but in the vast majority of cases, we must rely on numerical solutions to obtain the transient behavior of the CTMC. To that end, consider the backward Kolmogorov differential equations in matrix form

$$\mathbf{P}'(t) = \mathbf{Q}\mathbf{P}(t), \tag{6.27}$$

where $\mathbf{P}(t)$ is the matrix of transition probability functions at time t. Written in this form, the unknown matrix $\mathbf{P}(t)$ would appear to have a solution of "exponential nature," namely

$$\mathbf{P}(t) = e^{t\mathbf{Q}}. \tag{6.28}$$

In fact, numerically we may consider a solution approach that exploits this property by evaluating $e^{t\mathbf{Q}}$ as [1, 2]:

$$e^{t\mathbf{Q}} = \sum_{i=0}^{\infty} \frac{t^i}{i!}\mathbf{Q}^i, \tag{6.29}$$

with $\mathbf{P}(0) \equiv \mathbf{I}$, the identity matrix.

Determining the limiting behavior of the CTMC as $t \to \infty$ again involves classifying the states into sets of communicating classes, determining the recurrence property of the class, and evaluating the disposition of the process based on the initial state. Then, we say that the CTMC is irreducible, aperiodic, positive recurrent if its underlying Markov chain has those properties. In this case, the limiting behavior is again determined by balance equations.

Let $\alpha_j = \lim_{t\to\infty} P(X(t) = j|X(0) = i)$ be the limiting probability that the CTMC is in state j (independent of the initial state); these probabilities are given by

$$\alpha_j = \frac{\pi_j/\nu_j}{\sum_{i\in S}\pi_i/\nu_i}, \tag{6.30}$$

where the π_i are the solution to the balance Eq. 6.12 of the embedded DTMC with $\sum_i \pi_i = 1$. Note that in terms of the parameters of the CTMC, Eq. 6.30 and the normalizing equation are equivalent to

$$\nu_j\alpha_j = \sum_{i\in S}\alpha_i q_{ij}, \tag{6.31}$$

with

$$\sum_{j\in S}\alpha_j = 1. \tag{6.32}$$

Example 6.28 Consider a system that alternates between operating and failed states. The system operates for an exponentially distributed length of time with mean $1/\mu =$ 25 days. When the system fails, it is sent immediately for repair. Each repair lasts an exponentially distributed length of time with mean $1/\lambda = 4$ days and returns the system to a "good as new" state, and it recommences operation. Let $X(t)$ describe the operating status of the system, with $X(t) = 0$ if the system is being repaired at time t, and $X(t) = 1$ if the system is operating at time t. Then $\{X(t), t \geq 0\}$ comprises a two-state CTMC with generator

$$\mathbf{Q} = \begin{bmatrix} -\lambda & \lambda \\ \mu & -\mu \end{bmatrix} = \begin{bmatrix} -0.25 & 0.25 \\ 0.04 & -0.04 \end{bmatrix}$$

For the two-state CTMC, we can explicitly solve the Kolmogorov differential equations to find $\mathbf{P}(t)$. Then, considering the backward Kolmogorov differential equations (Eq. 6.27),

$$\mathbf{P}'(t) = \mathbf{QP}(t) = \begin{bmatrix} \lambda(P_{10}(t) - P_{00}(t)) & \lambda(P_{11}(t) - P_{01}(t)) \\ \mu(P_{00}(t) - P_{10}(t)) & \mu(P_{01}(t) - P_{11}(t)) \end{bmatrix}$$

and, similarly, the forward Kolmogorov differential equations lead to

$$\mathbf{P}'(t) = \mathbf{P}(t)\mathbf{Q} = \begin{bmatrix} -\lambda P_{00}(t) + \mu P_{01}(t) & \lambda P_{00}(t) - \mu P_{01}(t) \\ -\lambda P_{10}(t) + \mu P_{11}(t) & \lambda P_{10}(t) - \mu P_{11}(t) \end{bmatrix}$$

Then, solving for $P'_{00}(t)$ and $P'_{10}(t)$ we get (see derivation in e.g., [3]):

$$\begin{aligned} P'_{00}(t) &= \frac{\mu}{\mu+\lambda} + \frac{\lambda}{\mu+\lambda} e^{-(\mu+\lambda)t} \\ &= \frac{0.04}{0.04+0.25} + \frac{0.25}{0.04+0.25} e^{-(0.04+0.25)t} \\ P'_{10}(t) &= \frac{\mu}{\mu+\lambda} + \frac{\mu}{\mu+\lambda} e^{-(\mu+\lambda)t} \\ &= \frac{0.04}{0.04+0.25} + \frac{0.04}{0.04+0.25} e^{-(0.04+0.25)t} \end{aligned}$$

Then, since $P_{00}(t) + P_{01}(t) = P_{10}(t) + P_{11}(t) = 1$,

$$P_{01}(t) = 1 - P_{00}(t) = 1 - \left(\frac{0.04}{0.04+0.25} + \frac{0.25}{0.04+0.25} e^{-(0.04+0.25)t} \right)$$

$$P_{11}(t) = 1 - P_{10}(t) = 1 - \left(\frac{0.04}{0.04+0.25} + \frac{0.04}{0.04+0.25} e^{-(0.04+0.25)t} \right)$$

Then, for $t = 5$ years,

$$\mathbf{P}(5) = \begin{bmatrix} 0.3401 & 0.6599 \\ 0.1703 & 0.8297 \end{bmatrix}$$

and the limiting probabilities (i.e., $t \to \infty$) for every state are [3]:

$$\lim_{t\to\infty} \mathbf{P}(t) = \frac{1}{\mu+\lambda} \begin{bmatrix} \mu & \lambda \\ \mu & \lambda \end{bmatrix} = \begin{bmatrix} 0.1379 & 0.8621 \\ 0.1379 & 0.8621 \end{bmatrix}$$

which means that

$$\alpha_0 = 0.1379 \quad \text{and} \quad \alpha_1 = 0.8621$$

Note that these values can be computed directly by taking the limits on t above, or by solving the balance Eq. 6.31 with the normalizing Eq. 6.32.

Example 6.29 Consider a system that can take five possible states describing its condition; i.e., $S = \{1, 2, 3, 4, 5\}$, where state 1 indicates that the system operates in as good as new condition, states 2, 3, 4 indicates that the system functions but in an increasingly degraded condition, and state 5 that it is not operating at all (i.e., the system has failed).

The time between changes in the system states is assumed to be exponentially distributed with vector rate $\nu = \{0.1, 0.2, 0.3, 0.4, 0\}$. Note that for this example, the mean length of time spent in a particular state decreases as the index of the state increases. If the system is brand new (state 1) at time $t = 0$, compute the probability that the system has failed, i.e., $P(X(t) = 5)$, by times $t = 10, 20, 50$ years, and draw the failure and survival probability functions.

The transition probability matrix of the underlying Markov chain is:

$$\mathbf{P} = \begin{bmatrix} 0 & 1 & 0 & 0 & 0 \\ 0 & 0 & 1 & 0 & 0 \\ 0 & 0 & 0 & 1 & 0 \\ 0 & 0 & 0 & 0 & 1 \\ 0 & 0 & 0 & 0 & 1 \end{bmatrix}$$

Note that the form of matrix $\mathbf{P}$ implies that the system cannot jump between states without passing through all intermediate states. According to Eq. 6.22, the infinitesimal generator matrix $\mathbf{Q}$ has terms $q_{ij} = v_i \cdot P_{ij}, i \neq j$ and $q_{ii} = -\nu_i$. Thus,

$$\mathbf{Q} = \begin{bmatrix} -0.1 & 0.1 & 0 & 0 & 0 \\ 0 & -0.2 & 0.2 & 0 & 0 \\ 0 & 0 & -0.3 & 0.3 & 0 \\ 0 & 0 & 0 & -0.4 & 0.4 \\ 0 & 0 & 0 & 0 & 0 \end{bmatrix}$$

Note that in matrix $\mathbf{Q}$, the position $\mathbf{Q}_{5,5} = 0$ indicates that state 5 is an absorbing state; in other words, once the system enters this state it never leaves. The transition probability functions evaluated at time $t = 10$ years can be obtained by using Eq. 6.29:

$$\mathbf{P}(10) = \begin{bmatrix} 0.3679 & 0.2325 & 0.1470 & 0.0929 & 0.1597 \\ 0 & 0.1353 & 0.1711 & 0.1622 & 0.5313 \\ 0 & 0 & 0.0498 & 0.0944 & 0.8558 \\ 0 & 0 & 0 & 0.0183 & 0.9817 \\ 0 & 0 & 0 & 0 & 1.0000 \end{bmatrix}.$$

If the system is put in operation (i.e., "as good as new" condition) at $t = 0$, then the probabilities of being in each state at time 10 is given by the first row of the matrix $\mathbf{P}(10)$ above. In particular, the probability that the system has failed by time 10 is $P_{1,5}(10) = 0.1597$. Computing in a similar fashion, the first rows of the matrices $\mathbf{P}(20)$ and $\mathbf{P}(50)$ are given by

$$P_{1,\cdot}(20) = [0.1353, \ \ 0.1170, \ \ 0.1012, \ \ 0.0875, \ \ 0.5590]$$
$$P_{1,\cdot}(50) = [0.0067, \ \ 0.0067, \ \ 0.0066, \ \ 0.0066, \ \ 0.9733]$$

which means that the probabilities that the system has failed by times 20 and 50 are 0.5590 and 0.9733, respectively. The change of the failure probability (i.e., the probability that the system is in state 5) and the probability of survival over time is presented in Fig. 6.5.

Example 6.30 Consider the previous example again, but suppose that when the system reaches state 5, it is reconstructed and taken back to its original "good as new" condition (state 1). We assume that the time required for reconstruction is an exponential random variable with $\nu_5 = 0.7$. Note that ν_5 is larger than the other values since we are assuming the mean repair time is shorter. In this case, the transition probability matrix is:

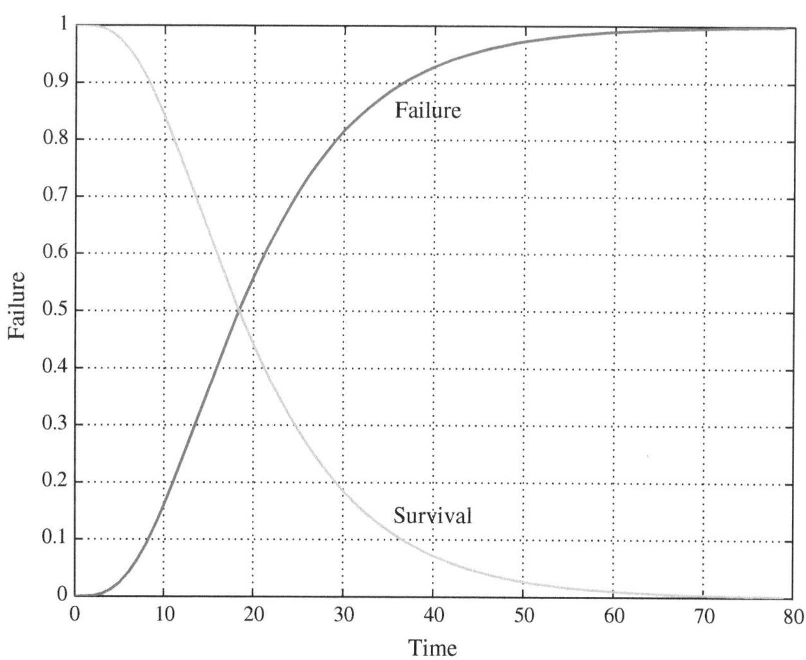

Fig. 6.5 Probability of failure as function of time

$$\mathbf{P} = \begin{bmatrix} 0 & 1 & 0 & 0 & 0 \\ 0 & 0 & 1 & 0 & 0 \\ 0 & 0 & 0 & 1 & 0 \\ 0 & 0 & 0 & 0 & 1 \\ 1 & 0 & 0 & 0 & 0 \end{bmatrix}$$

and the infinitesimal generator matrix $\mathbf{Q}$ becomes:

$$\mathbf{Q} = \begin{bmatrix} -0.1 & 0.1 & 0 & 0 & 0 \\ 0 & -0.2 & 0.2 & 0 & 0 \\ 0 & 0 & -0.3 & 0.3 & 0 \\ 0 & 0 & 0 & -0.4 & 0.4 \\ 0.7 & 0 & 0 & 0 & -0.7 \end{bmatrix}$$

The transition probability functions evaluated at time $t = 10$ years (again obtained by using Eq. 6.29) are now given by:

$$\mathbf{P}(10) = \begin{bmatrix} 0.4593 & 0.2491 & 0.1514 & 0.0944 & 0.0459 \\ 0.3212 & 0.2102 & 0.1952 & 0.1712 & 0.1021 \\ 0.5178 & 0.1606 & 0.1126 & 0.1216 & 0.0873 \\ 0.5490 & 0.2262 & 0.1071 & 0.0721 & 0.0457 \\ 0.4917 & 0.2503 & 0.1398 & 0.0803 & 0.0378 \end{bmatrix}$$

If the system begins in state 1 at time 0, then the probabilities that the system is in a given state for $t = 10, 20, 50$ years are:

$$P_{1,\cdot}(10) = [0.4593, \;\; 0.2491, \;\; 0.1514, \;\; 0.0944, \;\; 0.0459]$$
$$P_{1,\cdot}(20) = [0.4437, \;\; 0.2239, \;\; 0.1517, \;\; 0.1149, \;\; 0.0658]$$
$$P_{1,\cdot}(50) = [0.4498, \;\; 0.2241, \;\; 0.1498, \;\; 0.1124, \;\; 0.0648]$$

Note that in this case, the system is irreducible, and therefore, the probabilities $P_{1,\cdot}(n)$ are approaching the limiting probabilities of the CTMC given by (6.30) or (6.31) and (6.32), which are independent of the starting state of the process.

6.4 Markov Renewal Processes and Semi-Markov Processes

In some cases, the assumption that the times between system state changes are exponentially distributed does not reflect the actual behavior of the system. If the distribution of the time between changes of state of the system has an arbitrary distribution, then, the memoryless property required of a Markov process does not hold. In this section, we discuss a process termed a *semi-Markov process* that generalizes the continuous time Markov chain to allow for non-exponential sojourn times between state changes. Such a process will make transitions between states according to a

Markov chain, but the amount of time (the *sojourn time*) that the process spends in a given state i before making a transition into a different state j will have a distribution that depends on both states i and j. In order to develop this more general process, we use the approach of [1] and first define the so-called *Markov renewal process*, which describes the evolution of state changes and holding times in each state.

Consider, a sequence of random variables $\{X_n, n = 0, 1, 2, \ldots\}$ taking values in a countable state space S, and a sequence of random variables $\{T_n, n = 0, 1, 2, \ldots\}$, taking values in $[0, \infty)$, with $0 = T_0 \leq T_1 \leq T_2 \leq \cdots$. Here, the random variable X_n represents the nth system state and the random variable T_n represents the time of the nth transition, $n = 0, 1, 2, \ldots$.

Definition 44 The stochastic process $(\mathscr{X}, \mathscr{T}) = \{X_n, T_n, n \in \mathbb{N}\}$ is called a Markov renewal process (MRP) if

$$\begin{aligned} P(X_{n+1} = j, T_{n+1} - T_n \leq t | X_0 = i_0, \ldots, X_{n-1} = i_{n-1}, X_n = i, T_0, \ldots, T_n) \\ = P(X_{n+1} = j, T_{n+1} - T_n \leq t | X_n = i) \end{aligned} \tag{6.33}$$

holds for all $i, j, i_m, m = 0, \ldots, n-1 \in S$, all $n \in \mathbb{N}$, and all $t \in [0, \infty)$.

As usual, we will assume that the process $(\mathscr{X}, \mathscr{T})$ is time homogeneous, so that for any $i, j \in S$, and $t \geq 0$,

$$P(X_{n+1} = j, T_{n+1} - T_n \leq t | X_n = i) = Q_{ij}(t) \tag{6.34}$$

independent of n. The functions $\{Q_{ij}(t), \quad i, j \in S, \quad t \geq 0\}$ comprise the *semi-Markov kernel* of the MRP.

Definition 45 Let $(\mathscr{X}, \mathscr{T})$ be a Markov renewal process. The process $\mathscr{Y} = \{Y(t), t \geq 0\}$, where $Y(t) = X_n$ for $T_n \leq t < T_{n+1}$, is called the *semi-Markov process* (SMP) associated with $(\mathscr{X}, \mathscr{T})$.

The Markov renewal process $(\mathscr{X}, \mathscr{T})$ describes the evolution of the process explicitly in terms of the discrete sequence of states visited and successive sojourn times spent in each state, while the semi-Markov process $\mathscr{Y}$ tracks the state of the process continuously over time. It can be shown (see [1]) that $\mathscr{X} = \{X_0, X_1, \ldots\}$ forms a Markov chain (the *embedded Markov chain*) with transition probabilities

$$P_{ij} = \lim_{t \to \infty} Q_{ij}(t).$$

We say that the Markov renewal process (and the associated semi-Markov process) is irreducible if the embedded Markov chain is irreducible. We now define

$$G_{ij}(t) = \frac{Q_{ij}(t)}{P_{ij}}, \tag{6.35}$$

with the convention that $G_{ij}(t) \equiv 1$ if $P_{ij} = 0$. Then, as a function of t, each $G_{ij}(t)$ is a (conditional) distribution function with the following interpretation

$$G_{ij}(t) = P(T_{n+1} - T_n \le t | X_n = i, X_{n+1} = j). \tag{6.36}$$

That is, $G_{ij}(t)$ is the distribution function of the sojourn time in state i, given that the next state visited is state j. We generally assume that the distributions $G_{ij}(t)$ are continuous with density functions $g_{ij}(t)$. Note that the CTMC can be viewed as a Markov renewal process where

$$G_{ij}(t) = P(T_{n+1} - T_n \le t | X_n = i, X_{n+1} = j) = 1 - e^{-\nu_i t}, t \ge 0, \tag{6.37}$$

independent of j. Moreover, we have that for any integer $n \ge 1$, states $i_0, \ldots, i_n \in S$, and any $t_1, \ldots, t_n \in [0, \infty)$,

$$P(T_1 - T_0 \le t_1, \ldots, T_n - T_{n-1} \le t_n | X_0 = i_0, \ldots, X_n = i_n) \tag{6.38}$$
$$= G_{i_0 i_1}(t_1) \cdots G_{i_{n-1} i_n}(t_n). \tag{6.39}$$

so that the sojourn times in successive states are conditionally independent, given the sequence of states visited by the Markov chain. For each fixed state $i \in S$, the epochs T_n for which $X_n = i$, i.e., the successive visits of the process to state i, form a (possibly delayed) renewal process.

In terms of the semi-Markov process $\mathscr{Y}$, each time the process enters state i, it spends a random length of time in that state with distribution $H_i(t)$, where

$$H_i(t) = \sum_j P_{ij} G_{ij}(t). \tag{6.40}$$

Let μ_i denote the mean sojourn time in state i. Assuming $G_{ij}(t)$ is continuous, it follows that $H_i(t)$ has a density $h_i(t)$ and a hazard rate function $\lambda_i(t)$, given by

$$\lambda_i(t) = \frac{h_i(t)}{\overline{H}_i(t)}, \quad i \in S \tag{6.41}$$

The semi-Markov process can be analyzed as a Markov process, where we define the "state" of the process at any time as the pair (i, x) [2]. Here i is the current state, and x is the amount of time the process has spent in state i on the current visit. This method of analysis is known as the method of *supplementary variables*, as each state is supplemented with the length of time spent in that state before transition to a new state. In this way, instantaneous transitions from a given supplemented state are independent of past states, and the supplemented state process possesses the Markov property. From state (i, x), the process moves instantaneously to state $(j, 0)$ with probability intensity $\lambda_i(x) P_{ij}(x)$. Note that this two-dimensional Markov process has a continuous state space, and hence, the techniques required to analyze it are somewhat more complicated than those required for analysis of a CTMC. However, the general approach is the same and involves developing a set of differential equations involving the state probabilities and the hazard rate functions.

If the semi-Markov process is irreducible and positive recurrent, and under appropriate conditions on functions $H_i(t)$ (non-lattice with finite mean), a limiting density $p_i(x)$ exists, such that

$$p_i(x) = \lim_{t\to\infty} P(Y(t) = i, \quad \text{time spent in state } i \text{ on current visit} = x), \tag{6.42}$$

and is given by

$$p_i(x) = \frac{\pi_i \mu_i}{\sum_{j\in S} \pi_j \mu_j} \frac{\overline{H}_i(x)}{\mu_i}. \tag{6.43}$$

with limiting probabilities π_i, $i \geq 0$ and μ_j is the mean sojourn time in state j. Furthermore, from (6.43), the limiting probabilities of the states of the semi-Markov process $\mathscr{Y}$ are

$$P_i = \lim_{t\to\infty} P(Y(t) = i | Y(0) = j) = \frac{\pi_i \mu_i}{\sum_{j\in S} \pi_j \mu_j} \tag{6.44}$$

independent of the initial state j, and the limiting probability for the length of time spent in the current state, given the state is i is the *equilibrium* distribution of H_i, namely

$$H_i^e(y) = P(\text{time in state} \leq y \mid \text{state is } i) = \int_0^y \frac{\overline{H}_i(x)}{\mu_i} dx \tag{6.45}$$

The time-dependent behavior of the semi-Markov process is quite difficult to obtain and is generally approached via Laplace transforms ([4]). However, we can exploit the fact that successive visits to a given state form a renewal process to justify using Monte Carlo simulation as an efficient method for estimating the time-dependent state probabilities [18]. Let $i = 1, 2, ...N_{Tot}$ be the number of simulations; and let $S(t) = j$ be the state of the system at time t, where the possible system states are $j = 1, 2, ...m$. Then, simulation can be implemented as shown in Algorithm 3.

Algorithm 3 Semi-Markov Processes: Monte Carlo for computing the probability of being in a given state at time T; i.e., $S(T)$.

1: $i = 1$
2: **repeat**
3: generate a random trajectory (realization) of the system by sampling sequential transitions (state and times) of the semi-Markov model; i.e., $Tr_i = \{(t_k, S(t_k)), 0 < t \leq T, k = 1, 2, ...\}$.
4: $Sim_{i,S(T)} = 1$
5: $i = i + 1$
6: **until** $i > N_{Tot}$ $\{N_{Tot}$ is the number of simulations$\}$
7: $P(S(T) = j) = 1/N_{Tot} \cdot \sum_i Sim_{i,j}$ {probability of being in a given state}

In this algorithm, a trajectory Tr is defined as a set of both times at which the system changes its state (i.e., t_k) and the new system state (i.e., $S(t_k)$). The term $Sim_{i,S(T)}$ corresponds to the state of the system (i.e., $j = 1, 2, ...$) at time T in the ith simulation, which is obtained from the randomly generated trajectory; and N_{Tot} is the total number of simulations. We demonstrate this approach with an example.

Example 6.31 Consider a sewer system whose condition may be evaluated as: "*Good*", "*Acceptable*," "*Poor*," and "*Unacceptable*," which is represented by the state space $S = \{1, 2, 3, 4\}$. For this system, the transition probability matrix is:

$$P = \begin{bmatrix} 0.6 & 0.25 & 0.10 & 0.05 \\ 0 & 0.52 & 0.23 & 0.25 \\ 0 & 0 & 0.65 & 0.35 \\ 0 & 0 & 0 & 1 \end{bmatrix}$$

Let us also assume that the holding time distributions are lognormally distributed, i.e., $F_{ij} \sim LN(M_{ij}, S_{ij}))$ with the following means and variances:

$$\mathbf{M} = \begin{bmatrix} 5 & 7 & 9 & 11 \\ 0 & 3 & 5 & 7 \\ 0 & 0 & 2 & 4 \\ 0 & 0 & 0 & 1 \end{bmatrix} \qquad \mathbf{S}^2 = \begin{bmatrix} 1.56 & 1.1 & 3.24 & 14.8 \\ 0 & 0.56 & 0.56 & 1.96 \\ 0 & 0 & 0.49 & 1 \\ 0 & 0 & 0 & 0.02 \end{bmatrix}$$

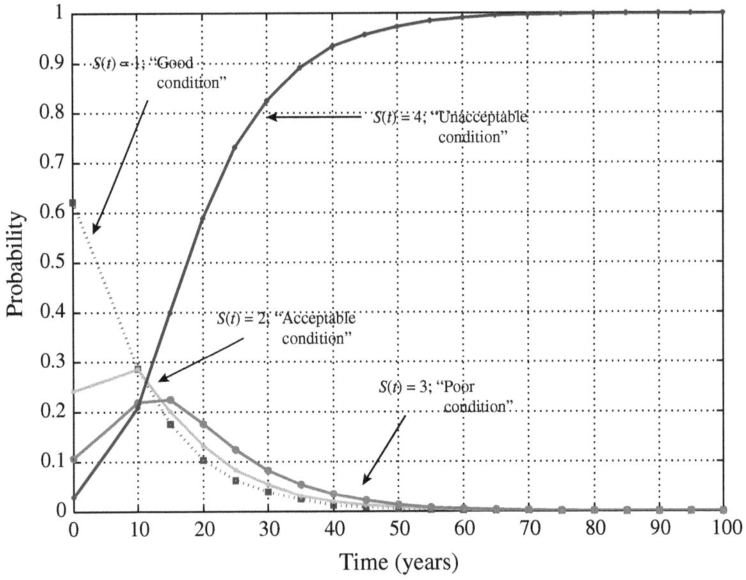

Fig. 6.6 State of the system at different time windows; solution obtained using Monte Carlo simulation (20,000 sample paths)

The objective of the study is to compute the probability of being in a given state at time t.

The state of the system, obtained using simulation, for various time windows is presented in Fig. 6.6. It is important to keep in mind that the accuracy of the prediction depends on the number of simulations; thus, as this number increases, the estimative of the probability improves.

6.5 Phase-Type Distributions

6.5.1 Overview of PH Distributions

The previous section highlights the computational difficulty of relaxing the requirement of exponential sojourn times in the CTMC; if sojourn times are allowed to follow arbitrary distributions in each state, the method of supplementary variables results in a Markov process defined on a continuous state space. An alternative approach to modeling non-exponential sojourn times is to approximate the sojourn times via a family of sojourn times known as *phase-type* or *PH*-distributions. Phase type distributions retain a Markovian structure on a discrete (although more complex) state space. One of the simplest members of this family, first studied by A.K. Erlang around 1910 and known as the Erlang distribution, is the distribution of the sum of k independent, identically distributed exponential random variables. Such a distribution can be thought of as the length of time required to pass through a sequence of stages (or phases), each consisting of an exponential holding time. For the Erlang distribution, the "memory" of the sojourn time is embedded in the current stage, and therefore, a Markov process can be constructed where the state is the stage, sojourn times in states are exponential, and transitions between states are described simply by the number of stages.

This simple idea led to the development of the class of phase-type distributions that generalize the concept of convolution/mixture of exponential stages. As in the Erlang case, the "memory" of the sojourn time in a given state is encoded in a discrete phase, so that knowledge of the current phase (and the transition structure) are sufficient to invoke the Markov property. Originally inspired by Cox ([5]), phase-type distributions were studied extensively by Neuts [6, 7] and others [19–21], who developed the so-called "matrix-geometric method" for their analysis. These distributions, which include, among others, the Erlang, hyperexponential, hypoexponential, and Coxian distributions, have a number of appealing properties as sojourn time models for Markovian systems.

Phase-type distributions have been used extensively in many engineering and computer science applications, such as telecommunications and queueing [6, 22, 23], reliability [24], and finance [25]. In this section, we summarize the formulation, properties, and solution techniques of this class of distribution functions.

6.5.2 Formulation of Continuous Phase-Type Distributions

In its most general form, a PH distribution is formulated as the distribution of the time to absorption in a finite Markov chain with a single absorbing state. The Markov chain can be either a DTMC, which results in a discrete distribution, or a CTMC, which results in a continuous distribution. For simplicity, we describe PH distributions based on CTMCs, but those based on DTMCs follow similarly. Let $\mathcal{X}$ be a CTMC on the state space $\{1, 2, \ldots, m, m+1\}$, $m \geq 1$, where state $m+1$ is an absorbing state and states $\{1, 2, \ldots, m\}$ are transient states. Let us assume that the infinitesimal generator of the CTMC is given by

$$\mathbf{Q} = \begin{bmatrix} \mathbf{T} & \mathbf{t} \\ \mathbf{0} & 0 \end{bmatrix}, \tag{6.46}$$

Here $\mathbf{T}$ is an $m \times m$ subgenerator matrix whose (off-diagonal) (i, j)th element is the transition rate between transient state i and transient state j, and whose diagonal element in row i is the negative of the mean holding time in state i, and $\mathbf{t}$ is a $m \times 1$ column vector consisting of the transition rates from each of the transient states to state $m+1$. Note that $\mathbf{t}$ is determined by $\mathbf{T}$ and the fact that row sums of $\mathbf{Q}$ must be zero. Let the initial probability vector for the Markov chain be given by $[\boldsymbol{\alpha}, a_{m+1}]$, where $\boldsymbol{\alpha}$ is a $1 \times m$ row vector. Here $\alpha_i, i = 1, \ldots, m$ represent the probabilities that the chain starts out in each of the transient states $\{1, 2, \ldots, m\}$, and α_{m+1} is the probability that the chain begins in the absorbing state.

Definition 46 The time until absorption in the CTMC $\mathcal{X}$ given above,

$$\tau = \inf\{t \geq 0 | X(t) = m+1\}$$

is said to have a phase-type distribution with parameters $\mathbf{T}$ and $\boldsymbol{\alpha}$. We denote the distribution by writing

$$\tau \sim PH(\boldsymbol{\alpha}, \mathbf{T})$$

and we say that the PH distribution has order m.

The cumulative distribution function F_τ, density f_τ, and moments $\mathbb{E}[X^n]$ of τ are given by [19]:

$$\begin{aligned} F_\tau(x) &= 1 - \boldsymbol{\alpha} \exp(\mathbf{T}x)\mathbf{1}, \\ f_\tau(x) &= \boldsymbol{\alpha} \exp(\mathbf{T}x)\mathbf{t} \\ \mathbb{E}[X^n] &= (-1)^n n! \boldsymbol{\alpha} \mathbf{T}^{-n} \mathbf{1} \end{aligned}$$

where $\mathbf{1}$ is an $m \times 1$ column vector whose elements are all 1, and $\exp(\cdot)$ is the matrix exponential operator, defined as $\exp(\mathbf{T}x) = \sum_{k=1}^{\infty} (\mathbf{T}x)^k / k!$ (as in 6.29). Note that if $a_{m+1} > 0$, then F_τ has a jump of size a_{m+1} at the origin.

Example 6.32 As mentioned previously, the family of Erlang distributions are examples of PH distributions. The 2-stage Erlang distribution (E_2) with mean $1/\lambda$ can be modeled as a PH distribution with transient states $\{1, 2\}$ and

$$\mathbf{T} = \begin{bmatrix} -2\lambda & 2\lambda \\ 0 & -2\lambda \end{bmatrix}, \quad \mathbf{t} = \begin{bmatrix} 0 \\ 2\lambda \end{bmatrix} \quad \text{and} \quad \boldsymbol{\alpha} = \begin{bmatrix} 1 & 0 \end{bmatrix}$$

then, according to Eq. 6.46

$$\mathbf{Q} = \begin{bmatrix} \mathbf{T} & \mathbf{t} \\ \mathbf{0} & 0 \end{bmatrix} = \left[\begin{array}{cc|c} -2\lambda & 2\lambda & 0 \\ 0 & -2\lambda & 2\lambda \\ \hline 0 & 0 & 0 \end{array}\right]. \tag{6.47}$$

As mentioned previously, the Erlang distribution E_2 models the time spent in passing through two consecutive, independent, and identical exponentially distributed stages, each with mean sojourn time $1/2\lambda$. The Markovian transition rate diagram for this distribution is shown in Fig. 6.7. The k-stage Erlang distribution E_k follows analogously.

If the mean sojourn times in the exponential stages are different, we obtain the family of hypoexponential distributions, which are also PH distributions. The name hypoexponential refers to the fact that the variance of these distributions is smaller than that of the exponential.

Example 6.33 Hyperexponential distributions arise as probabilistic (i.e., convex) mixtures of exponential distributions and are also basic PH distributions. A 2-stage Hyperexponential distribution (H_2) can be modeled as a PH distribution with transient states $\{1, 2\}$ and

$$\mathbf{T} = \begin{bmatrix} -\lambda_1 & 0 \\ 0 & -\lambda_2 \end{bmatrix}, \quad \mathbf{t} = \begin{bmatrix} \lambda_1 \\ \lambda_2 \end{bmatrix}, \quad \text{and} \quad \boldsymbol{\alpha} = \begin{bmatrix} p_1 & p_2 \end{bmatrix},$$

with $p_2 = 1 - p_1$. The hyperexponential distribution H_2 models the time spent when the sojourn time is selected to be exponential with mean $1/\lambda_i$ with probability $p_i, i = 1, 2, p_1 + p_2 = 1$. In reliability theory, hyperexponential distributions (and their generalizations) are frequently used in modeling the time to failure in systems with competing failure modes. The name hyperexponential refers to the fact that the variance of this distribution exceeds that of the exponential, and consequently, these distributions are useful in approximating heavy-tailed sojourn times. Figure 6.8 shows the Markovian transition rate diagram of the distribution H_2.

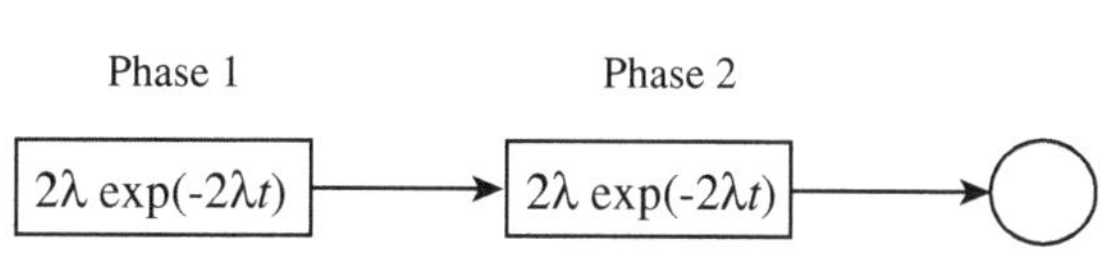

Fig. 6.7 Phase holding time diagram for the E_2 phase-type distribution

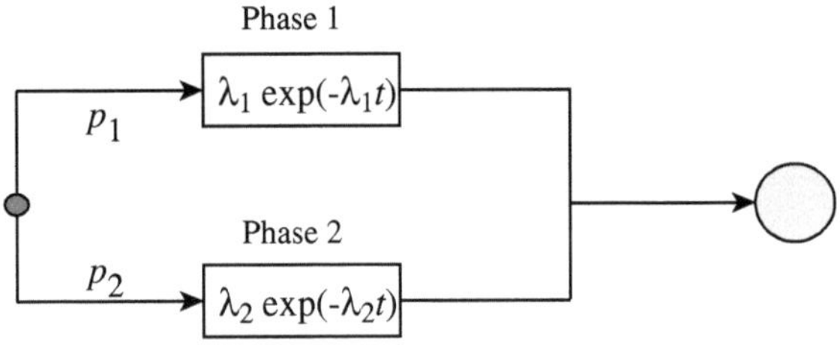

Fig. 6.8 Phase holding time diagram for a two-phase hyperexponential distribution

Many PH distributions can be constructed using the building blocks of the hypoexponential and hyperexponential distributions; i.e., as probabilistic mixtures of convolutions of exponential distributions. Others, such as coxian distributions, are constructed similarly to the hypoexponential, but may allow transition to the absorbing state from any of the transient states.

6.5.3 Properties of PH Distributions and Fitting Methods

As mentioned previously, PH distributions can be modeled using a discrete state Markov chain; thus the well-developed algorithmic machinery for analyzing Markov chains can be applied to a large class of non-exponential sojourn times. Two further properties of PH distributions are of particular importance and justify their use as *approximations* of general distributions.

1. *Denseness property*: PH distributions are *dense* in the set of continuous density functions with support on $[0, \infty)$ (Latouche and Ramaswami, Theorem 2.6.5 [19]). The term dense refers to the complete coverage of the continuous density functions (in the sense of weak convergence of distribution), and means that there is any continuous distribution can be approximated arbitrarily closely by a member of the PH distribution family. A number of efficient algorithms have been proposed in the literature to fit a PH distribution to arbitrary (positive) datasets (numerically generated from any continuous distribution or from field measurements) [26–29].
2. *Closure under convolutions*: Latouche and Ramaswami, Theorem 2.6.1 [19] show that if X and Y are two independent random variables with distributions $PH(\boldsymbol{\alpha}, \mathbf{T})$ of order m and $PH(\boldsymbol{\beta}, \mathbf{S})$ of order n, respectively, then the sum $X+Y$ is distributed $PH(\boldsymbol{\gamma}, \mathbf{U})$ of order $m+n$ with:

$$\boldsymbol{\gamma} = \begin{bmatrix} \boldsymbol{\alpha} & a_{m+1}\boldsymbol{\beta} \end{bmatrix} \quad \text{and} \quad \mathbf{U} = \begin{bmatrix} \mathbf{T} & \mathbf{t}\cdot\boldsymbol{\beta} \\ 0 & \mathbf{S} \end{bmatrix}, \tag{6.48}$$

where $\mathbf{T}\cdot\mathbf{1}+\mathbf{t} = \mathbf{0}$ and $(\mathbf{t}\cdot\boldsymbol{\beta})_{ij} = t_i\beta_j$. This result is easily seen if we imagine the total holding time as consisting of passage through the transient phases associated with X (label these 1 through m) followed by passage through the transient phases

associated with Y (label these $m+1$ through $m+n$). The terms $t_i\beta_j$ in the matrix $\mathbf{U}$ represent the transition rates out of transient phases of X and into transient phases of Y.
The term a_{m+1} corresponds to the probability that the holding times in the transient phases associated with X are 0. Then, $a_{m+1}\boldsymbol{\beta}$ is the probability that the Markov chain associated with $X+Y$ starts in the transient states associated to Y.

Property 2 above shows that the PH representation of a sum of k independent PH distributed random variables can be obtained by successive application of (6.48). In Sect. 6.7 we use this property to determine the PH representation of the cumulated damage D_k when successive damage magnitudes are independent, PH distributed random variables.

6.6 Numerical Considerations for PH Distributions

PH distributions are used to fit (positive) datasets that may come from field or experimental measurements, or might be generated numerically from any continuous distribution. While now quite common in many engineering applications, the drawback to the use of PH distributions lies in the dimensionality of the Markov chain required to adequately approximate a particular distribution. Complex distributions, particularly, those with relatively large tails, may require dozens or even hundreds of parameters for a satisfactory approximation. Once an acceptable approximation is obtained, then efficient Markov chain algorithms are required to evaluate system performance.

There are two main statistical techniques used to fit PH distributions to data; these are moment matching techniques (MM), and techniques based on maximum likelihood estimators using an expectation-maximization procedure (EM) (see also Chap. 4). In the MM approach, a PH distribution is sought that matches the mean, variance, and possibly higher moments of the dataset. MM techniques for PH distribution fitting were first described in [30–32]. These methods are usually employed to fit 2 to 3 moments of a dataset and have the advantage of resulting in a PH distribution with a relatively small number of phases.

When the dataset is influenced by the behavior of many higher moments, for example, heavy-tailed behavior, moment-based approaches cannot appropriately capture the features of the dataset in PH form. In these cases, maximum likelihood-based methods are superior to those based on moments. The EM algorithm first developed in [33] has become the standard for estimating parameters for PH distributions. Although the EM methods are generally slower, may be numerically unstable, and result in higher order PH distributions than the MM approach, they are generally seen as preferable, and much recent effort has been devoted to algorithmic improvements in the EM algorithm. Recent work has employed variance reduction techniques, such as data set partitioning, segmentation, and cluster-based approaches to improve the fitting procedure (cf. [28, 34, 35]).

The selection of a particular fitting algorithm is a matter of experience and depends on the problem at hand. As a general rule, three important aspects should be taken into account: 1) the availability of information (e.g., the number of data points or moments); 2) the level of accuracy needed for the PH fitting; and 3) the computational effort for the fitting and later for the evaluation of reliability.

6.7 Phase-Type Distributions for Modeling Degradation: Examples

Because PH distributions are closed under convolutions, they are appealing as models for accumulated shock degradation. If we assume that successive shocks sizes are independent and follow a PH distribution with known parameters, the accumulated damage after n shocks also has a PH distribution whose parameters are easily determined. In this section, we present examples that illustrate the applicability and convenience of using PH distributions for modeling degradation.

Example 6.34 Consider, a structural system that deteriorates as a result of earthquakes; with inter-arrival times X_i distributed exponential with mean $\mu_X = 10$ years and shock sizes Y_i, lognormally distributed with mean $\mu_Y = 20$ (in appropriate capacity units) and coefficient of variation COV_Y. If it is assumed that the initial structural capacity is $v_0 = 100$ and the failure threshold $k^* = 0$; the purpose of the example is to evaluate the reliability function. For comparative purposes, several COV_Y will be evaluated (i.e., $\text{COV}_Y = 0.2, 0.5, 0.8, 1.0$).

Note first that X_i is already PH distributed since the exponential distribution is the simplest form of PH. However, the size of shocks are lognormally distributed, which does not comply with the PH structure. Therefore, the shock size distribution was adjusted to a PH distribution using both the MM and EM fitting algorithms. The process consisted in generating randomly a large data set of Y values, in this case $N = 10^5$, and then fitting the data to a phase-type distribution (see [36] for more details).

In order to validate the results, the density $f(t)$ of the system's lifetime was evaluated and it is shown in Fig. 6.9 for the various COV_Y considered. The results show clearly that the PH models fit very well the data obtained by Monte Carlo Simulation. Note that in the case where $\text{COV}_Y = 2.0$, the difference in the approximation can be easily resolved by adjusting better the PH parameters.

A significant advantage of using PH distributions is its computational efficiency. Then, the results of the analysis are summarized in Table 6.2. They show very close fits of both PH approximations with Monte Carlo simulations. In particular, the EM fitting shows relative errors of around 1 % in the $MTTF$ and the COV_L for all of the values of COV_Y considered. Besides, MM has relative errors for COV_L above 3 %, while the relative errors of the EM are around 1 %. These relative errors in both fittings might be considered small for the most practical applications. The differences between EM and MM for values of $\text{COV}_Y > 0.5$ are due to the fact that EM uses

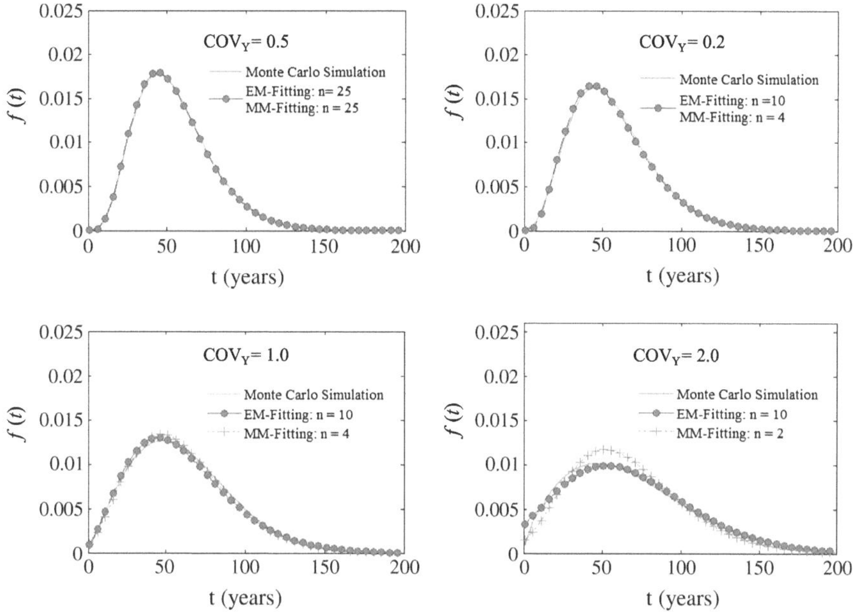

Fig. 6.9 Density of the system's lifetime, $f(t)$, computed using Monte Carlo simulation and the PH shock model (with the MM and EM algorithms for the fitting)

$n = 10$ PH phases for the fitting, while MM uses 2 or 3. In contrast, the results for $\mathrm{COV}_Y < 0.5$ are similar in both cases because both MM and EM have a good fit of the variable Y.

Table 6.2 also shows the execution times (ET) for Monte Carlo and the PH shock model. The time performance of the PH shock model estimation for both fitting approaches is $\approx 10^{-1}$ s, which is better than Monte Carlo simulations ($\approx$1 s). Clearly, the ET depends on the number of shocks to failure, which in this example take values from 8 to 22. However, even with a greater number of shocks ($K \approx 100$) the computation with the PH shock model is less expensive than with Monte Carlo simulations.

Several studies (empirical and from physical principles) have derived expressions for the *deterioration trends* (i.e., the expected value $\overline{D}(t) = \mathbb{E}[D(t)]$ of the deterioration over time) of components and materials of structures under different degradation mechanisms [37–39]. The proposed PH shock model can be applied to reproduce deterioration trends for several of such mechanisms and to compute the reliability quantities in a straightforward manner; we will illustrate this in the following example.

Example 6.35 In concrete and steel components, general deterioration due to chemical, physical, or environmental factors can be modeled as [40] (see also Chap. 4):

$$\mathbb{E}[D(t)] = ct^b, \tag{6.49}$$

Table 6.2 Reliability estimation of a structure subject to earthquakes

	$COV_Y = 0.2$	$COV_Y = 0.5$	$COV_Y = 0.8$	$COV_Y = 1.0$
Monte Carlo				
Simulation				
ET: Execution time (s)	3.2	3.0	2.6	2.8
MTTF	54.8	56.2	58.0	60.1
COV_L	0.44	0.47	0.50	0.52
PH shock model				
MM algorithm				
n: Number of PH states	25	4	2	2
K: Number of shocks	8	12	17	20
ET: Execution time (s)	0.1	0.11	0.11	0.14
MTTF (% Error)	55.2 (**0.7 %**)	56.2 (**0.1 %**)	58.1 (**0.2 %**)	60.0 (**0.2 %**)
COV_L (% Error)	0.44 (**0.1 %**)	0.47 (**0.2 %**)	0.52 (**3.4 %**)	0.55 (**5.9 %**)
PH shock model				
EM algorithm				
n: Number of PH states	25	10	10	10
K: Number of shocks	8	11	14	18
ET: Execution time (s)	0.1	0.12	0.21	0.22
MTTF (% Error)	55.2 (**0.7 %**)	56.2 (**0.1 %**)	58.8 (**1.3 %**)	59.9 (**0.2 %**)
COV_L (% Error)	0.44 (**0.1 %**)	0.47 (**0.3 %**)	0.50 (**0.2 %**)	0.53 (**1.2 %**)

Results from Monte Carlo simulation and with the PH shock model by using the PH representations of Y_i with the MM and EM algorithms

for constants $c > 0$ and $b > 0$. As mentioned in Sect. 4.9.2, for the case of diffusion-controlled degradation $b = 0.5$, which gives a square root relationship; if degradation is caused by sulfate attack on concrete, $b \geq 1$ (usually $b = 2$ which defines a quadratic law); corrosion of reinforcement follows a linear law ($b = 1$); and for creep in concrete, $b = 1/8$ (see more details in [37, 40]). Another example is the case of fatigue in materials subjected to cyclic loading, which could be modeled as a cumulative deterioration shock model [38]). Finally, an interesting application is the case of aftershocks after a major earthquake. In this case, the rate of their arrival decreases over time following the wel- known Omori's Law [41, 42]: $n(t) = K(t+c)^{-1}$, where K and c are constants. Then, the total number of aftershocks $N(t)$ in the time interval between 0 and t is given by: $N(t) = \int_0^t n(s)ds = K\ln(t/c+1)$. If each aftershock produces a mean damage μ_Y, the total deterioration until time t is given by [42]:

$$D(t) \approx \mu_Y N(t) = K\mu_Y \ln\left(\frac{t}{c+1}\right). \tag{6.50}$$

If we consider nonidentically distributed inter-arrival times or shock sizes, different functional forms of $\overline{D}(t)$ can be obtained. The approach followed in this analysis consists of two steps (see [36] for more details):

1. Define PH distributions for the first inter-arrival time X_1 and shock size Y_1 as:

$$X_1 \sim PH(\boldsymbol{\tau}_1, \mathbf{T}_1) \quad \text{and} \quad Y_1 \sim PH(\boldsymbol{\gamma}_1, \mathbf{Y}_1). \tag{6.51}$$

2. For the next shocks ($k \geq 2$), define X_k equally distributed as $g(k)X_1$ and Y_k as $h(k)Y_1$, i.e. (see Chap. 5):

$$X_k \stackrel{d}{=} g(k)X_1 \quad \text{and} \quad Y_k \stackrel{d}{=} h(k)Y_1, \tag{6.52}$$

where $g(k)$ and $h(k)$ are functions of the shock number k. Hence, the PH representations, distributions and means of X_k and Y_k are given by:

$$\begin{aligned} X_k &\sim PH(\boldsymbol{\tau}_1, \mathbf{T}_1/g(k)), \quad F_{X_k}(t) = F_{X_1}(t/g(k)), \quad \mu_{X_k} = \mu_{X_1} g(k), \\ Y_k &\sim PH(\boldsymbol{\gamma}_1, \mathbf{Y}_1/h(k)), \quad F_{Y_k}(y) = F_{Y_1}(y/h(k)), \quad \mu_{Y_k} = \mu_{Y_1} h(k). \end{aligned} \tag{6.53}$$

Note that while the PH-matrices $\mathbf{T}_k$ and $\mathbf{Y}_k$ change for each k, but keep the sizes n_X and n_Y for the first shock $k = 1$, the initial probability vectors $\boldsymbol{\tau}_k$ and $\boldsymbol{\gamma}_k$ remain equal to $\boldsymbol{\tau}_1$ and $\boldsymbol{\gamma}_1$, respectively.

As an example, suppose that $X_k \stackrel{d}{=} X_1$ and $Y_k \stackrel{d}{=} kY_1$ for all $k \geq 1$ (i.e. $g(k) = 1$ and $h(k) = k$ in Eq. (6.52). Hence, $X_k \sim PH(\boldsymbol{\tau}_1, \mathbf{T}_1)$ with mean $\mu_{X_k} = \mu_{X_1}$ and $Y_k \sim PH(\boldsymbol{\gamma}_1$ for $k\mathbf{Y}_1)$ with mean $\mu_{Y_k} = k\mu_{Y_1}$, $k \geq 1$. The results for different PH representations of X_1 and Y_1 show that for large ratios (t/μ_{X_1}) the asymptotic behavior of $\overline{D}(t)$ is quadratic. More precisely, the empirical result from the simulations show that: $\overline{D}(t) \to \frac{1}{2}\mu_{Y_1}(t/\mu_{X_1})^2$ when $(t/\mu_{X_1}) \to \infty$. Note that this particular

Table 6.3 Cases considered for the distributions of inter-arrival times X_k and shock sizes Y_k ($k \geq 2$)

Case	$X_k(\stackrel{d}{=})$	$Y_k(\stackrel{d}{=})$	PH-matrix $\mathbf{T}_k$	PH-matrix $\mathbf{Y}_k$	Mean of X_k: μ_{X_k}	Mean of Y_k: μ_{Y_k}
1	X_1	Y_1	$\mathbf{T}_1$	$\mathbf{Y}_1$	μ_{X_1}	μ_{Y_1}
2	X_1	kY_1	$\mathbf{T}_1$	$\frac{1}{k}\mathbf{Y}_1$	μ_{X_1}	$k\,\mu_{Y_1}$
3	X_1	k^2Y_1	$\mathbf{T}_1$	$\frac{1}{k^2}\mathbf{Y}_1$	μ_{X_1}	$k^2\mu_{Y_1}$
4	X_1	$b^{k-1}Y_1$	$\mathbf{T}_1$	$\frac{1}{b^{k-1}}\mathbf{Y}_1$	μ_{X_1}	$b^{k-1}\mu_{Y_1}$
5	kX_1	Y_1	$\frac{1}{k}\mathbf{T}_1$	$\mathbf{Y}_1$	$k\mu_{X_1}$	μ_{Y_1}
6	k^7X_1	Y_1	$\frac{1}{k^7}\mathbf{T}_1$	$\mathbf{Y}_1$	$k^7\mu_{X_1}$	μ_{Y_1}
7	$a^{k-1}X_1$	Y_1	$\frac{1}{a^{k-1}}\mathbf{T}_1$	$\mathbf{Y}_1$	$a^{k-1}\mu_{X_1}$	μ_{Y_1}

Table 6.4 Deterioration trends $\overline{D}(t)$ (asymptotic, i.e., when $(t/\mu_{X_1}) \rightarrow \infty$) and degradation mechanisms obtained from different definitions of the distributions of inter-arrival times X_k and shock sizes Y_k ($k \geq 2$)

Case	$X_k(\overset{d}{=})$	$Y_k(\overset{d}{=})$	$\overline{D}(t) \rightarrow$	Trend	Degradation mechanism
1	X_1	Y_1	$\mu_{Y_1}(t/\mu_{X_1})$	Linear	Corrosion of reinforcement
2	X_1	kY_1	$\frac{1}{2}\mu_{Y_1}(t/\mu_{X_1})^2$	Quadratic	Sulfate attack on concrete
3	X_1	k^2Y_1	$\frac{1}{3}\mu_{Y_1}(t/\mu_{X_1})^3$	Cubic	———
4	X_1	$b^{k-1}Y_1$	$\frac{\mu_{Y_1}}{1-b}$, $0 < b < 1$	Constant	———
			$\frac{\mu_{Y_1}}{b-1}b^{(t/\mu_{X_1})}$, $b > 1$	Exponential	Growth of cracks in metals
5	kX_1	Y_1	$\mu_{Y_1}\sqrt{2(t/\mu_{X_1})}$	Square root	Diffusion-controlled aging
6	k^7X_1	Y_1	$\mu_{Y_1}\sqrt[8]{8(t/\mu_{X_1})}$	Eighth root	Creep in concrete
7	$a^{k-1}X_1$	Y_1	$\frac{\ln[(a-1)t/\mu_{X_1}+1]}{\ln a}$,	Logarithmic	Aftershock arrivals
			$a > 1$		(Omori's Law, Eq. (6.50))

case may describe, for example, the deterioration trend of concrete when subjected to sulfate attack, presented in Eq. (6.49).

Another special an interested case is where either $h(k)$ or $g(k)$ are equal to a^k. This conditions defines a geometric process for X_k or Y_k. The geometric process was discussed in Chap. 5. In Tables 6.3 and 6.4 we present some other relationships between X_k and Y_k (i.e., varying $g(k)$ and $h(k)$), their corresponding PH representations (matrices $\mathbf{T}_k$ and $\mathbf{Y}_k$), the (asymptotic) deterioration trends, and the specific degradation mechanisms that can be modeled [36].

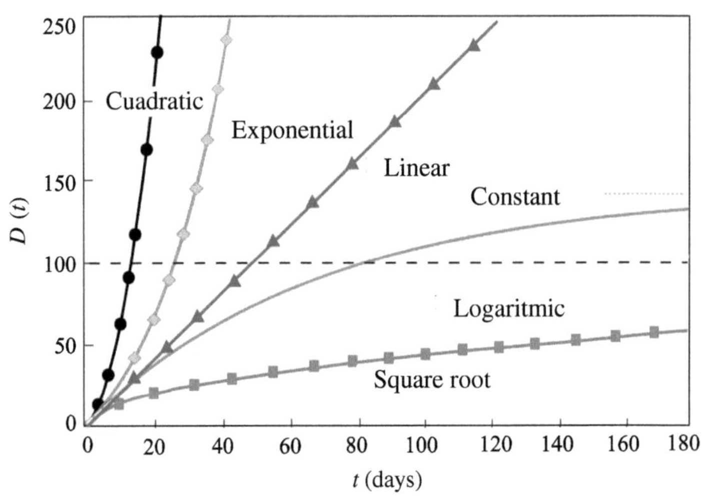

Fig. 6.10 Trends of $D(t)$ for different definitions of X_k and Y_k ($k \geq 2$) obtained from de distributions of X_1 and Y_1 (Tables 6.3 and 6.4). The distributions of X_1 and Y_1 were obtained using the MM algorithm and assuming $\mu_{X_1} = 2.5$ days, $\text{COV}_{X_1} = \text{COV}_{Y_1} = 0.5$, and $\mu_{Y_1} = 5$

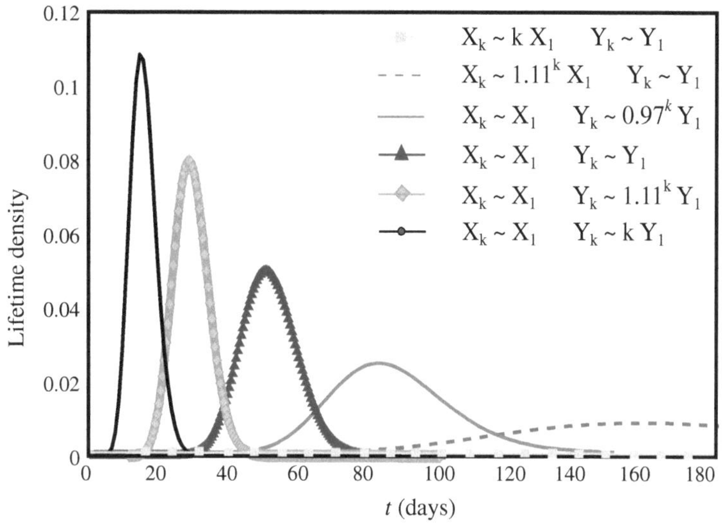

Fig. 6.11 Density of the lifetime of a system with the degradation models defined in Tables 6.3 and 6.4)

Figure 6.10 shows the plots of $\overline{D}(t)$ for particular examples shown in Tables 6.3 and 6.4. For all the cases the mean of X_1 was $\mu_{X_1} = 2.5$ days with coefficient of variation $\text{COV}_{X_1} = 0.5$, and shock size Y_1 with mean $\mu_{Y_1} = 5$ and $\text{COV}_{Y_1} = 0.5$. The PH representation of these variables was obtained by the MM algorithm, which requires 4 states for the fitting. Also, Fig. 6.11 shows the density of the lifetime for an initial performance $z = 100$ (in appropriate units depending on each application case) and threshold $k^* = 0$.

These results show that PH shock-based deterioration can be used to model and estimate the reliability of a wide range of degradation mechanisms with different deterioration trends and rate of shocks. This is done by relaxing the identical distribution assumption and by assuming that the random variables X_k and Y_k are distributed proportional to X_1 and Y_1, respectively, with proportional factor depending on k (see Chap. 5).

6.8 Summary and Conclusions

Markov processes exhibit very useful properties for modeling deterioration of systems whose state (condition) can be defined as a discrete space. *Markov chains* models focus on the transition between states at fixed time intervals and hold the Markov property, which implies that the next state of the system depends only on its current state and not on the history. On the other hand, *semi-Markov* processes allow for the time between transitions to be random with arbitrary distribution.

Semi-Markov processes can be discrete or continuous depending upon the distribution of the time between system state changes. A special case of Semi-Markov processes are *continuous time Markov* processes in which the distribution of the time between system state changes follows an exponential distribution; and therefore, the Markov property holds. In addition to traditional Markovian models, in this chapter, we have also discussed the so-called phase-type distributions, which have a number of useful properties as sojourn time models for Markovian and non-Markovian systems. Provided that exists information to construct transition probability matrices, and then, the system performance restrictions can be satisfied, Markovian models can be of great value in modeling degradation. In particular, phase-type distributions can be use with advantage to handle problems such as computing convolution for shock-based degradation.

References

1. E. Çinlar, *Introduction to Stochastic Processes* (Prentice Hall, New Jersey, 1975)
2. S.M. Ross, *Introduction to Stochastic Dynamic Programming* (Academic Press, New York, 1983)
3. S.M. Ross, *Stochastic Processes*, 2nd edn. (Wiley, New York, 1996)
4. R.A. Howard. *Dynamic probabilistic systems, volume II: semi-Markov and decision processes*, 2nd edn. (Wiley, New York, 2007)
5. D.R. Cox, A use of complex probabilities in the theory of stochastic processes. Math. Proc. Camb. Philos. Soc. **51**, 313–319 (1955)
6. M.F. Neuts, K.S. Meier, On the use of phase type distributions in reliability modelling of systems with two components. OR Spektrum **2**, 227–234 (1981)
7. M.F. Neuts, Structured stochastic matrices of M/G/1 type and their applications. Math. Proc. Camb. Philos. Soc. New York (1985)
8. J.V. Carnahan, W.J. Davis, M.Y. Shahin, Optimal maintenance decisions for pavement management. J. Trans. Eng. ASCE **113**(5), 554–572 (1987)
9. Federal Highway Administration (FHA). *Recording and coding guide for structure inventory and appraisal of the nation's bridges*. U.S. Department of Transportation, Washington D.C. (1979)
10. S. Madanat, R. Mishalani, W.H.W. Ibrahim, Estimation of infrastructure transition probabilities from condition rating data. J. Infrastruct. Syst. ASCE **1**(2), 120–125 (1995)
11. A.A. Butt, M.Y. Shahin, K.J. Feighan, S.H. Carpenter, Pavement performance prediction model using the markov process. Trans. Res. Rec. **1123**, 12–19 (1987)
12. H.-S. Baik, H.S. Jeong, D.M. Abraham, Estimating transition probabilities in markov chain-based deterioration models for management of wastewater systems. J. Water Resour. Plan. Manag. ASCE **132**(15), 15–24 (2006)
13. D.H. Tran, B.J.C. Perera, A.W.M. Ng, Hydraulic deterioration models for storm-water drainage pipes: ordered probit versus probabilistic neural network. J. Comput. Civil Eng. ASCE **24**, 140–150 (2010)
14. S.B. Ortiz-Garc, J.J. Costello, M.S. Snaith, Derivation of transition probability matrices for pavement deterioration modeling. J. Trans. Eng. ASCE **132**(2), 141–161 (2006)
15. G. Morcous, Performance prediction of bridge deck systems using markov chains. J. Perform. Constr. Facil. ASCE **20**(2), 146–155 (2006)
16. M. Ben-Akiva, R. Ramaswamy, An approach for predicting latent infrastructure facility deterioration. Trans. Sci. **27**(2), 174–193 (1993)

17. Federal Highway Administration (FHA). *National Bridge Inventory (NBI)*, Washington D.C. (2011). http://www.fhwa.dot.gov/bridge/nbi.htm
18. M. Hauskrecht, *Monte Carlo approximations to continuous-time semi-Markov processes*. Technical Report: CS-03-02, Department of Computer Science, University of Pittsburgh (2002)
19. G. Latouche, V. Ramaswami, *Introduction to matrix analytic methods in stochastic modeling* (Society for Industrial and Applied Mathematics, Philadelphia, 1999)
20. E.P.C. Kao, *An Introduction to Stochastic Processes* (Duxbury Press, Belmont, 1997)
21. C. O'Cinneide, Characterization of the phase-type distribution. Commun. Stat. Stoch. Models **6**, 1–57 (1990)
22. M.F. Neuts, R. Pérez-Ocón, I. Torres-Castro, Repairable models with operating and repair times governed by phase type distributions. Adv. Appl. Probab. **32**, 468–479 (2000)
23. R. Akhavan-Tabatabaei, F. Yahya, J.G. Shanthikumar, Framework for cycle time approximation of toolsets. IEEE Trans. Semicond. Manuf. **25**(4), 589–597 (2012)
24. O.O. Aalen, Phase type distributions in survival analysis. Scand. J. Stat. **22**, 447–463 (1995)
25. S. Asmussen, F. Avram, M.R. Pistorius, Russian and american put options under exponential phase-type lévy models. Stoch. Process. Appl. **109**, 79–111 (2004)
26. A. Bobbio, A. Horváth, M. Telek, Matching three moments with minimal acyclic phase type distributions. Stoch. Models **21**, 303–326 (2005)
27. T. Osogami and M. Harchol-Balter. A closed-form solution for mapping general distributions to minimal PH distributions. *Computer Performance Evaluation. Modelling Techniques and Tools.*, 63(6):200–217, 2003
28. A. Thümmler, P. Buchholz, M. Telek, A novel approach for phase-type fitting with the em algorithm. IEEE Trans. Dependable Secur. Comput. **3**(3), 245–258 (2006)
29. J.P. Kharoufeh, C.J. Solo, M.Y. Ulukus, Semi-markov models for degradation based reliability. IIE Trans. **42**(8), 599–612 (2010)
30. M.A. Johnson, M.R. Taaffe, Matching moments to phase distributions: mixtures of erlang distributions of common order. Stoch. Models **5**, 711–743 (1989)
31. M.A. Johnson, M.R. Taaffe, An investigation of phase-distribution moment-matching algorithms for use in queueing models. Queueing Syst. **8**, 129–148 (1991)
32. M.A. Johnson, M.R. Taaffe, A graphical investigation of error bounds for moment-based queueing approximations. Queueing Syst. **8**, 295–312 (1991)
33. S. Asmussen, O. Nerman, M. Olsson, Fitting phase type distributions via the em algorithm. Scand. J. Stat. **23**, 419–441 (1996)
34. A. Riska, V. Diev, E. Smimi, Efficient fitting of long-tailed data sets into phase-type distributions. SIGMETRICS Perform. Eval. Rev. **30**, 6–8 (2002)
35. P. Reinecke, T. Krauß, K. Wolter, Cluster-based fitting of phase-type distributions to empirical data. Comput. Math. Appl. **64**, 3840–3851 (2012)
36. J. Riascos-Ochoa, M. Sánchez-Silva, R. Akhavan-Tabatabaei, Reliability analysis of shock-based deterioration using phase-type distributions. Probab. Eng. Mech. **38**, 88–101 (2014)
37. Y. Mori, B. Ellingwood, Maintaining reliability of concrete structures. i: role of inspection/repair. J. Struct. ASCE **120**(3), 824–835 (1994)
38. K. Sobczyk, Stochastic models for fatigue damage of materials. Adv. Appl. Probab. **19**, 652–673 (1987)
39. S. Li, L. Sun, J. Weiping, Z. Wang, The paris law in metals and ceramics. J. Mater. Sci. Lett. **14**, 1493–1495 (1995)
40. J.M. Van Noortwijk, A survey of the application of gamma processes in maintenance. Reliab. Eng. Syst. Saf. **94**, 2–21 (2009)
41. T. Utsu, Y. Ogata, R.S. Matsuura, The centenary of the omori formula for a decay law of after shock activity. J. Phys. Earth **43**, 1–33 (1995)
42. A. Helmstetter, D. Sornette, Subcritical and supercritical regimes in epidemic models of earthquake aftershocks. J. Geophys. Res. **107**, 22–37 (2002)

Chapter 7
A Generalized Approach to Degradation

7.1 Introduction

In Chaps. 5 and 6, we presented and discussed a set of degradation models commonly used in engineering practice. However, more often than not, degradation is the result of a combination of various damaging mechanisms and, therefore, the use of any of these models in isolation is not necessarily representative of the actual system behavior. Furthermore, as degradation mechanisms are more complex, there are generally no tractable analytical models available to describe these processes. In this chapter, we present a general framework that allows modeling complex degradation behaviors based on the theory of *Lévy processes*. The compound Poisson process presented in Chap. 3 and the widely used gamma process are special cases of *Lévy processes*. Although this approach implies some important assumptions about the process, in our opinion, it is as far as analytical models can currently go to describe degradation. This framework allows, for example, the combination of various mechanisms; furthermore, it can be used to find computable expressions for the reliability quantities, avoiding some difficult computational issues such as convolutions, infinite sums, and integrals [1]. In the first part of the chapter, we present the basics of Lévy processes; afterward, we describe how they can be used for modeling degradation and we finalize with some illustrative examples. Proofs of the general properties of Lévy processes are not presented here, but are available in [2, 3].

7.2 Definition of a Lévy Process

Lévy processes are continuous-time stochastic processes with independent and stationary increments and with right continuous sample paths having left limits. Formally, a Lévy process is defined as follows [4]:

M. Sánchez-Silva and G.-A. Klutke, *Reliability and Life-Cycle Analysis of Deteriorating Systems*, Springer Series in Reliability Engineering,
DOI 10.1007/978-3-319-20946-3_7

Definition 47 Given a filtered probability space $(\Omega, \mathscr{F}, \mathbb{F}, P)$, an adapted process $\{X_t, t \geq 0\}$, with $X_0 = 0$ almost surely (a.s.), is a *Lévy process* if

1. $\{X_t, t \geq 0\}$ has increments independent of the past; that is, $X_t - X_s$ is independent of $\mathscr{F}_s$, $0 \leq s < t < \infty$;
2. $\{X_t, t \geq 0\}$ has stationary increments; that is, $X_t - X_s$ has the same distribution as X_{t-s}, $0 \leq s < t < \infty$; and
3. $\{X_t, t \geq 0\}$ is continuous in probability; that is, $\lim_{t \to s} P(X_t \in \cdot) = P(X_s \in \cdot)$.

In succinct terms, Lévy processes are stochastically continuous processes with stationary, independent increments. Note that based on this definition, and within the context of degradation, any compound Poisson process (CPP) shock model, and progressive models in the form of stationary gamma processes or linear deterministic models (see Chap. 5) are examples of Lévy processes. In modeling degradation, we consider only Lévy processes on $\mathbb{R}$ (one-dimensional Lévy processes) that have nondecreasing sample paths a.s.; such processes are known as *subordinators* (see Sect. 7.3.1). As we will see, the fact that these processes have independent increments leads to a very specific characterization, as well as to a relatively tractable model for degradation.

An important property of Lévy processes is that the sum of independent Lévy processes is also Lévy; therefore, it is possible to combine independent Lévy damage models with no additional difficulty. In this sense, the proposed framework can be used to describe many cases reported in the literature, such as cumulative CPP with linear drift [5] or the stationary gamma process combined with a CPP with shock sizes distributed gamma [6].

Existing papers that have used also Lévy processes for modeling degradation include [1, 7–9]. For a very readable introduction to Lévy processes, see [10, 11], and for a complete mathematical exposition, see [2].

7.2.1 Characteristic Function and Characteristic Exponent

The formalism presented in this chapter to describe degradation requires the definition of the *characteristic function* of the Lévy process $\{X_t, t \geq 0\}$ on $\mathbb{R}^d$. The characteristic function $\phi_Y(z)$ of a random variable Y is given by the following transformation (defined in terms of the Lebesgue integral) [12]:

$$\phi_Y(z) := \mathbb{E}\left[e^{i\langle z, Y\rangle}\right] = \int_{\mathbb{R}^d} e^{i\langle z, x\rangle} P(Y \in dx), \qquad z \in \mathbb{R}^d \tag{7.1}$$

where $i = \sqrt{-1}$ is the imaginary unit and $\langle \cdot, \cdot \rangle$ is the inner product in $\mathbb{R}^d$. Note that the characteristic function contains all the probabilistic information of Y. Some useful properties related to the characteristic function are

1. The characteristic function ϕ_Y uniquely determines the probability distribution $P(Y \in \cdot)$, and vice versa; they are related through the *Fourier inversion formula*, which is discussed in Sect. 7.6.

2. The characteristic function $\phi_Y(z)$ is uniformly continuous in $\mathbb{R}^d$.
3. For the characteristic function $\phi_Y(z)$:

$$|\phi_Y(z)| < 1, \quad \text{for} \quad z \in \mathbb{R}^d \quad \text{and} \quad \phi_Y(0) = 1. \tag{7.2}$$

Now for a Lévy process $\{X_t, t \geq 0\}$, consider the expression

$$X_n = (X_1 - X_0) + (X_2 - X_1) + \cdots + (X_n - X_{n-1})$$

for integer n. Since increments $X_j - X_{j-1}, j = 1, \ldots, n$ (with $X_0 \equiv 0$) are independent and identically distributed, the characteristic function of X_n can be expressed as

$$\phi_{X_n}(z) = [\phi_{X_1}(z)]^n. \tag{7.3}$$

In general, equation (7.3) holds for any $t \geq 0$, i.e.,

$$\phi_{X_t}(z) = [\phi_{X_1}(z)]^t, \tag{7.4}$$

and since X_t can be divided into an infinite number of independent, identically distributed increments, we say that X_t has an *infinite divisible* distribution. For infinitely divisible distributions, the characteristic function ϕ_{X_1} can be expressed as [2]

$$\phi_{X_1}(z) = e^{-\Psi(z)}, \tag{7.5}$$

where Ψ is a unique continuous function from $\mathbb{R}^d \to \mathbb{C}$, called the *characteristic exponent* of the distribution of X_1. Using Eqs. 7.4 and 7.5, the characteristic function of the distribution of X_t can then be written as

$$\phi_{X_t}(z) = [e^{-\Psi(z)}]^t = e^{-t\Psi(z)}, \tag{7.6}$$

and Ψ is known as the *characteristic exponent* of the Lévy process $\{X_t, t \geq 0\}$. Many of the results that are presented here are based on the form of the characteristic exponent for specific cases of the Lévy process, and on the evaluation of the probability law $P(X_t \in \cdot)$.

7.2.2 The Lévy–Khintchine Formula

Because the distribution of X_t is infinitely divisible, the characteristic exponent of the Lévy process can be expressed in terms of the triplet (γ, Q, Π) through the famous *Lévy–Khintchine formula* [2]:

$$\Psi(z) = -i\langle z, \gamma\rangle + \frac{1}{2}Q(z) + \int_{\mathbb{R}^d} \left(1 - e^{i\langle z,x\rangle} + i\langle z, x\rangle \mathbf{1}_{\{|x|<1\}}\right) \Pi(dx), \qquad (7.7)$$

where $\gamma \in \mathbb{R}^d$, Q is a positive semi-definite form on $\mathbb{R}^d$, and Π is a measure on $\mathbb{R}^d \setminus \{0\}$ with

$$\int_{\mathbb{R}^d} (1 \wedge |x|^2)\Pi(dx) < \infty. \qquad (7.8)$$

Expression 7.7 provides the basis for understanding the probabilistic structure of the Lévy process. The parameter γ is known as the *drift* parameter, the quadratic form Q is known as the *Gaussian coefficient*, and the measure Π is known as the *Lévy measure*. Their roles in the probabilistic evolution of the Lévy process will be clarified shortly.

7.2.3 Decomposition of a Lévy Process

As suggested by the Lévy–Khintchine formula (Eq. 7.7), a Lévy process can be characterized as the superposition of three independent Lévy processes [2], i.e.,

$$X_t = X_t^{\{1\}} + X_t^{\{2\}} + X_t^{\{3\}} \qquad t \geq 0 \qquad (7.9)$$

This formulation is referred to as the *Lévy-Ito decomposition*. The first process, $\{X_t^{\{1\}}, t \geq 0\}$, corresponds to the first term of the Lévy–Khintchine formula, and represents a deterministic drift process with parameter γ. The second process, $\{X_t^{\{2\}}, t \geq 0\}$, represents a Brownian motion with parameter Q. The third process, $\{X_t^{\{3\}}, t \geq 0\}$ (i.e., third term of the formula), represents a pure jump process with parameter Π. The Lévy measure Π governs the timing and sizes of the jumps of the process, and as we shall see, allows for further characterization of the jump part of the Lévy process.

7.2.4 The Lévy Measure Π and the Pure Jump Component of the Lévy Process

Some explanation of this measure, and in particular about the restrictions on the jump sizes given by Eq. 7.8, is in order.

Let ΔX represent the size of a jump and denote $N_t(B)$ as the number of jumps with sizes ΔX in a set B (i.e., $\Delta X \in B$) that occur by time t; note that $N_t(B)$ is a random variable. Then the Lévy measure, evaluated at the set B, is equal to the expected number of jumps in a unit time interval with sizes in B:

$$\Pi(B) = \mathbb{E}\left[N_1(B)\right]. \tag{7.10}$$

Due to the stationary property of Lévy processes, the expected number of jumps with sizes in B in an arbitrary time interval $[0, t]$ is given by

$$\mathbb{E}\left[N_t(B)\right] = \Pi(B)t. \tag{7.11}$$

For the process to be well defined (right continuous with left hand limits), it is necessary that the accumulated jump process does not explode (i.e., become arbitrarily large on finite-time intervals). The condition in Eq. 7.8 ensures that this does not happen. To see this, note that the condition is always satisfied if the Lévy measure is finite, in which case the jump process is simply a compound Poisson process with measure Π. On the other hand, if the Lévy measure is infinite, let us separate the jumps into those of size one or greater (the "large jumps") and those of size less than one (the "small jumps"). That is, the third term in Eq. 7.7 can be written as

$$\int_{\mathbb{R}^d} \left(1 - e^{i\langle z,x\rangle}\right) \mathbf{1}_{\{|x|\geq 1\}}\Pi(dx) + \int_{\mathbb{R}^d} \left(1 - e^{i\langle z,x\rangle} + i\langle z, x\rangle\right) \mathbf{1}_{\{|x|<1\}}\Pi(dx) \tag{7.12}$$

Condition 7.8 ensures that $\int_{[1,\infty)} \Pi(dx) < \infty$; that is, only finitely many jumps may exceed the cutoff value (taken to be one, but actually arbitrary). This implies that if Π is infinite, and there will be infinitely many jumps, but they will be of arbitrarily small size. In this case, we can consider the jump process as the independent superposition of a compound Poisson process having jumps of size 1 or greater, and a pure jump process (in fact, a martingale) having jumps of size less than 1. The decomposition of the jump process in expression 7.12 is unique.

7.2.5 Mean and Central Moments of a Lévy Process

The nth-moment of a random variable X can be computed from its characteristic function $\phi_X(z)$ as follows [13]:

$$\mathbb{E}\left[X^n\right] = (-i)^n \phi_X^{(n)}(0), \tag{7.13}$$

where $\phi_X^{(n)}(0)$ denotes the n-th derivative of $\phi_X(z)$ evaluated at $z = 0$. Therefore, it is possible to obtain expressions for the moments of the Lévy process X_t, for each t, by replacing Eq. (7.6) into (7.13):

$$\mathbb{E}\left[X_t^n\right] = (-i)^n \frac{d^n}{dz^n} e^{-t\Psi(z)}\bigg|_{z=0}. \tag{7.14}$$

Setting $n = 1$, differentiating (7.14) and noting that $\Psi(0) = 0$ (Sect. 7.2.1), the mean of X_t becomes

$$\mathbb{E}[X_t] = ie^{-t\Psi(0)}\Psi'(0)t = i\Psi'(0)t, \tag{7.15}$$

where $\Psi'(0)$ denotes the derivative of $\Psi(z)$ evaluated at $z = 0$.

In the same way, the n-central moments $\mu_n(t) = \mathbb{E}\left[(X_t - \mathbb{E}(X_t))^n\right]$ of X_t can be obtained from Eq. (7.14). The expressions for $n = 2, 3, 4$ are

$$\mu_2(t) = \Psi^{(2)}(0)t \tag{7.16}$$

$$\mu_3(t) = -i\Psi^{(3)}(0)t \tag{7.17}$$

$$\mu_4(t) = (3t\Psi^{(2)}(0)^2 - \Psi^{(4)}(0))t \tag{7.18}$$

Note that in Eqs. 7.15–7.17, the mean of X_t, its variance $\mu_2(t)$, and third central moment $\mu_3(t)$ vary linearly with time. These results are important for modeling degradation and will be used in Sects. 7.4 and 7.5 to compare different Lévy deterioration models.

7.3 Modeling Degradation via Subordinators

In this section, we present a degradation framework, which will be referred in the following as the *Lévy degradation formalism* [1]. We will first define the concept of *subordinators* and then state the basic assumptions of the model.

7.3.1 Subordinators

Formally, *subordinators* are Lévy processes that take values in $\mathbb{R}^+ := [0, \infty)$ with increasing sample paths [2]. Therefore, the Gaussian (Brownian) component $X_t^{\{2\}}$ of the Lévy process (Eq. 7.9) must be zero—i.e., $Q \equiv 0$. In addition, the Lévy measure Π has support on $[0, \infty)$ (i.e., the process has no negative jumps) and satisfies

$$\int_{(0,\infty)} (1 \wedge x)\, \Pi(dx) < \infty, \tag{7.19}$$

which is necessary for the sum of jumps $\sum_{s \le t} \Delta X_s$ to be finite. In addition, the term $i\langle z, x\rangle \mathbf{1}_{|x|<1} = izx\mathbf{1}_{|x|<1}$ in the integral in Eq. (7.7) can be integrated and included as part of the deterministic term $-i\langle z, \gamma\rangle = -iz\gamma$, thanks to the condition (7.19). This defines the *drift coefficient* of the subordinator as

$$q = \gamma - \int_{(0,1)} x \; \Pi(dx); \tag{7.20}$$

which must satisfy $q \geq 0$. Under these conditions, the characteristic exponent $\Psi(z)$ in (7.7) takes the special form:

$$\Psi(z) = -iqz + \int_{(0,\infty)} (1 - e^{izx}) \, \Pi(dx), \quad z \in R. \tag{7.21}$$

In summary, a subordinator X_t has the general form $X_t = qt + X_t^{\{3\}}$; it is uniquely determined by (q, Π) (with $q \geq 0$ and Π with support on $[0, \infty)$); and has a characteristic exponent given by the Lévy–Khintchine formula for subordinators presented in Eq. (7.21).

7.3.2 Assumptions of the Model

The Lévy degradation formalism is based on the following assumptions [1]:

1. Deterioration is described by a one-dimensional stochastic process $\{X_t, t \geq 0\}$.
2. The deterioration process $\{X_t, t \geq 0\}$ has independent and stationary increments.
3. Without maintenance, deterioration is an increasing (i.e., not decreasing) function.
4. Multiple sources of degradation (i.e., different shock and/or progressive processes) act independently.

An important assumption of the model is that the process requires independent stationary increments. This can be justified in many practical applications. For example, van Noortwijk et al. [14] argue that when there is only partial information about the initial and final state of the system, there is no way to know how the accumulated degradation was reached. It follows that the deterioration increments are *exchangeable*, which means that the order in which they appear does not matter, and independence of increments therefore follows. In the specific case of earthquake engineering, Iervolino et al. [6] showed that structural deterioration satisfy the independence condition because earthquakes arrivals follow a homogeneous Poisson process. Furthermore, if aging (progressive) deterioration, described by a stationary gamma process, is also included, this condition is also satisfied since both degradation mechanisms are increasing and act independently on the structure.

In some cases, the stationarity property may not be fulfilled, i.e., processes for which the probability law of deterioration in a time interval depends not only on the length of the interval but also on the initial time. A classical example is the case of aftershock sequences after a main-shock, whose arrival rate is nonlinear (i.e., nonstationary) [15]. Relaxing the stationarity condition in the proposed Lévy formalism defines a more general process known as *nonhomogeneous (NH) Lévy process*, which will not be discussed here, but more information can be found in [16].

7.4 Specific Models

In order to show the versatility of the Lévy formalism, in this section we show how it can be used to describe two important degradation models, the compound Poisson process and the gamma process. We also present a general framework to construct models that describe the combined effect of both progressive and shock-based degradation. These examples all assume that the Brownian coefficient Q is zero.

7.4.1 Compound Poisson Process (CPP)

Consider the case of shock-based degradation, which is caused by discrete events in time that remove finite amounts of capacity from the system. In the Lévy degradation formalism, the subordinator W_t can be used to model shock-based deterioration if the following conditions are satisfied:

1. The process $X_t^{\{1\}}$ is zero (Eq. 7.9), i.e., the drift term $q = 0$. Therefore, W_t is only a jump process $X_t^{\{3\}}$.
2. The Lévy measure Π_W of the process W_t has support on $\mathbb{R}^+$ and it is finite.

Under these assumptions, the process W_t constitutes a compound Poisson process and can be written as

$$W_t = X_t^{\{3\}} = \sum\nolimits_{s \leq t} \Delta X_s = \sum\nolimits_{i=0}^{N_t} Y_i, \tag{7.22}$$

where N_t is the number of shocks until time t, which is a Poisson process with rate λ. The sequence $\{Y_i\}_{i \geq 1}$ corresponds to *iid* shock sizes with distribution $G(\cdot)$ supported on $[0, \infty)$. Therefore, the Lévy measure is given by

$$\Pi_W(dx) = \lambda G(dx). \tag{7.23}$$

Note that Π_W is finite because $G(\cdot)$ is a distribution (i.e., $G(\mathbb{R}^+) = 1$); therefore, $\Pi_W(\mathbb{R}^+) = \lambda G(\mathbb{R}^+) = \lambda$. Under these conditions, the characteristic exponent is given by

$$\begin{aligned} \Psi_W(z) &= \lambda \int_{(0,\infty)} \left(1 - e^{izx}\right) G(dx) \\ &= \lambda \left(\int_{(0,\infty)} G(dx) - \int_{(0,\infty)} e^{izx}\, G(dx) \right) \end{aligned} \tag{7.24}$$

Note that the first integral in Eq. 7.24 is equal to 1 since $G(\mathbb{R}^+) = 1$; and the second integral corresponds to the characteristic function $\phi_Y(z)$ of the shock sizes; then,

$$\Psi_W(z) = \lambda\left(1 - \phi_Y(z)\right). \tag{7.25}$$

Therefore, the characteristic function of W_t becomes

$$\phi_{W_t}(z) = e^{-t\Psi_W(z)} = e^{-\lambda t(1-\phi_Y(z))}. \tag{7.26}$$

The mean, second, and third central moments of W_t are given by Eqs. 7.15–7.17:

$$\mathbb{E}[W_t] = \lambda t(-i)\phi_Y'(0) = \lambda t\,\mathbb{E}[Y] \tag{7.27}$$

$$\mu_n(t) = \lambda t(-i)^n\phi_Y^{(n)}(0) = \lambda t\,\mathbb{E}[Y^n] \quad n = 2, 3. \tag{7.28}$$

These results come from Eq. 7.13, which corresponds to Wald's equation [17].

7.4.2 Progressive Lévy Deterioration Models

The subordinator Z_t for modeling progressive deterioration can be constructed as the sum of two independent processes: a linear deterministic drift (LD) $X_t^{\{1\}} = qt$ (with $q > 0$) and a jump process $X_t^{\{3\}}$ with infinite and positive Lévy measure $\Pi_Z(\mathbb{R}^+) = \infty$. Thus,

$$Z_t = qt + X_t^{\{3\}}, \tag{7.29}$$

with the additional condition described by Eq. 7.19. Note that the second term in Eq. 7.29, i.e., $X_t^{\{3\}}$, describes a jump process with infinite number of small jumps in any finite-time interval (Sect. 7.2.4); which is used to model the randomness of the process. Then, the characteristic exponent of the progressive degradation process is given by Eq. (7.21):

$$\begin{aligned}\Psi_Z(z) &= -iqz + \int_{(0,\infty)} (1 - e^{izx})\,\Pi_Z(dx)\\ &= -iqz + \Psi_p(z),\end{aligned} \tag{7.30}$$

where

$$\Psi_p(z) = \int_{(0,\infty)} (1 - e^{izx})\,\Pi_Z(dx) \tag{7.31}$$

is the characteristic exponent of the component corresponding to the jump process $X_t^{\{3\}}$. Note that the integral in Eq. 7.31 cannot be split into two terms, like we did with Eq. 7.24 for the compound Poisson process because both terms are infinite.

The mean, second, and third central moments of the process have the following general form:

$$\mathbb{E}[Z_t] = qt + i\Psi_p'(0)t \tag{7.32}$$

$$\mu_n(t) = -t\Psi_p^{(n)}(0)(-i)^n \quad n = 2, 3. \tag{7.33}$$

An example of a Lévy process with infinite measure that has been used extensively for modeling progressive degradation is the stationary gamma process [18] (see Chap. 5).

A nonstationary gamma process X_t with shape function $v(t) > 0$ and scale parameter $u > 0$ has the following probability density (see also Chap. 5):

$$P(X_t \in dx) = \frac{u^{v(t)}}{\Gamma(v(t))} x^{v(t)-1} e^{-ux} dx, \quad x \geq 0. \tag{7.34}$$

Thus, if the shape parameter is linear with $v(t) = vt$ for $v > 0$, the gamma process is a Lévy process. Under the Lévy formalism, this stationary gamma process with rate c and scale parameter u is defined as a jump process with Lévy measure density:

$$\Pi_Z(dx) = vx^{-1}e^{-ux}dx. \tag{7.35}$$

Note that Π_Z is an infinite positive measure that satisfies the requirement of Eq. 7.19 for a subordinator. The characteristic exponent and function are given, respectively, by evaluating Eqs. 7.31 and 7.6:

$$\begin{aligned} \Psi_Z(z) &= \Psi_p(z) \\ &= v \ln\left(1 - \frac{iz}{u}\right), \end{aligned} \tag{7.36}$$

because the exponent of the characteristic function depends only on $\Psi_p(z)$ since the drift is zero, and

$$\phi_{Z_t}(z) = e^{-t\Psi_p(z)} = \left(1 - \frac{iz}{u}\right)^{-vt}. \tag{7.37}$$

The mean, second, and third central moments are given by Eqs. 7.15–7.17:

$$\mathbb{E}[Z_t] = \frac{vt}{u} \tag{7.38}$$

$$\mu_n(t) = \frac{(n-1)!vt}{u^n} \quad n = 2, 3. \tag{7.39}$$

Note that these expressions are also proportional to t as in the CPP case.

7.4.3 Combined Degradation Mechanisms

It is not uncommon to find a system whose degradation depends on both extreme events (shocks) and environmental (progressive) conditions. For example, this could be the case of structures such as bridges located in environments that combine both aggressive climatic conditions and high seismicity. Under the assumption that both damage accumulation processes are independent, which is easy to justify in many practical cases (see [5, 6, 19]), the Lévy degradation formalism described above can be used with advantage to model degradation.

Let us define W_t and Z_t as the two independent processes that describe shock-based and progressive degradation, respectively. Then, the combined degradation process K_t can be obtained by superposition, i.e.,

$$K_t = W_t + Z_t = \sum_{i=1}^{N_t} Y_i + \left(qt + X_t^{\{3\}}\right), \tag{7.40}$$

where $X_t^{\{3\}}$ is a jump process representing the progressive random deterioration with infinite Lévy measure Π_Z and characteristic exponent Ψ_Z; Y_i is the ith shock size with distribution $G(\cdot)$, for all $i = 1, 2, \ldots$. Since K_t is a Lévy process, its Lévy measure is given by the sum of the measures of the component processes Π_Z and Π_W:

$$\Pi(dx) = \Pi_W(dx) + \Pi_Z(dx) = \lambda G(dx) + \Pi_Z(dx), \tag{7.41}$$

where the first term comes from Eq. 7.23 with λ the shocks arrival rate of the Poisson process. Furthermore, the characteristic exponent is given by the sum of the corresponding characteristic exponents (Eqs. 7.25 and 7.30), i.e.,

$$\begin{aligned}\Psi_K(z) &= \Psi_W(z) + \Psi_Z(z)\\ &= \Psi_W(z) + (\Psi_p(z) - iqz)\\ &= \lambda(1 - \phi_Y(z)) + (\Psi_p(z) - iqz)\end{aligned} \tag{7.42}$$

and the characteristic function by the product of the corresponding characteristic functions:

$$\begin{aligned}\phi_{K_t}(z) &= e^{-t\Psi_K(z)} \\ &= e^{-\lambda t(1-\phi_Y(z))} e^{-t\Psi_p(z)} e^{iqtz}.\end{aligned} \quad (7.43)$$

Finally, the mean, second, and third central moments of K_t are computed as the sum of their values for each mechanism:

$$\mathbb{E}[K_t] = \mathbb{E}[W_t] + \mathbb{E}[Z_t] \quad (7.44)$$

$$\mu_n(t) = \mu_{n,W}(t) + \mu_{n,Z}(t) \quad n = 2, 3. \quad (7.45)$$

7.5 Examples of Degradation Models Based on the Lévy Formalism

In order to illustrate the applicability of the Lévy degradation formalism, in this section we provide explicit expressions for the characteristic exponent $\Psi(z)$; the mean, second, and third central moments ($\mathbb{E}[\cdot]$, $\mu_2(t)$ and $\mu_3(t)$) of the following three degradation models;

1. shock-based (compound Poisson process) (Tables 7.1 and 7.2);
2. progressive (Table 7.3); and
3. combined (Table 7.4).

For the case of shock degradation based on the compound Poisson process, different distributions for the shock sizes were evaluated. In Table 7.3, information about

Table 7.1 Examples of shock-based Lévy degradation process W_t(CPP with rate of shock occurrences λ)

Shock-based (CPP) models		
Quantities for W_t	Delta $Y_i \sim \delta(y)$	Uniform $Y_i \sim U(y-a, y+a)$
$\phi_Y(z)$	e^{izy}	$\frac{e^{iz(y+a)}-e^{iz(y-a)}}{iz2a}$
$\mathbb{E}[Y]$	y	y
$\text{cov}(Y)$	0	$a/\sqrt{3}y$
$\mathbb{E}[Y^2]$	y^2	$y^2 + a^2/3$
$\mathbb{E}[Y^3]$	y^3	$ya^2 + y^3$
$\Psi_W(z)$	$\lambda(1-\phi_Y(z))$	$\lambda(1-\phi_Y(z))$
$\mathbb{E}[W_t]$	λty	λty
$\mu_2(t)$	λty^2	$\lambda t(y^2 + a^2/3)$
$\mu_3(t)$	λty^3	$\lambda t(ya^2 + y^3)$

Table 7.2 Examples of shock-based (CPP with rate of shock occurrences λ) Lévy degradation process W_t

Shock-based (CPP) models			
Quantities for W_t	Exponential $Y_i \sim Exp(\nu)$	Lognormal $Y_i \sim LN(\mu, \sigma)$	PH-type $Y_i \sim PH(\boldsymbol{\tau}, \mathbf{T})$
$\phi_Y(z)$	$\frac{1}{1-(iz/\nu)}$	$\sum_{n=0}^{\infty} \frac{(iz)^n}{n!} e^{n\mu+n^2\frac{\sigma^2}{2}}$	$-\boldsymbol{\tau}(\mathbf{T}+iz\mathbf{I})^{-1}\mathbf{t}$
$\mathbb{E}[Y]$	$y = 1/\nu$	$y = e^{\mu+\sigma^2/2}$	$y = -\boldsymbol{\tau}\mathbf{T}^{-1}\mathbf{1}$
$cov(Y)$	1	$\sqrt{e^{\sigma^2}-1}$	$\frac{\sqrt{2\boldsymbol{\tau}\mathbf{T}^{-2}\mathbf{1}-(\boldsymbol{\tau}\mathbf{T}^{-1}\mathbf{1})^2}}{-\boldsymbol{\tau}\mathbf{T}^{-1}\mathbf{1}}$
$\mathbb{E}[Y^2]$	$2y^2$	$y^2\left(cov(Y)^2+1\right)$	$2\boldsymbol{\tau}\mathbf{T}^{-2}\mathbf{1}$
$\mathbb{E}[Y^3]$	$6y^3$	$y^2\left(cov(Y)^2+1\right)^3$	$-6\boldsymbol{\tau}\mathbf{T}^{-3}\mathbf{1}$
$\Psi_W(z)$	$\lambda(1-\phi_Y(z))$	$\lambda(1-\phi_Y(z))$	$\lambda(1-\phi_Y(z))$
$\mathbb{E}[W_t]$	λty	λty	λty
$\mu_2(t)$	$2\lambda ty^2$	$\lambda ty^2\left(cov(Y)^2+1\right)$	$\lambda t(2\boldsymbol{\tau}\mathbf{T}^{-2}\mathbf{1})$
$\mu_3(t)$	$6\lambda ty^3$	$\lambda ty^2\left(cov(Y)^2+1\right)^3$	$\lambda t(-6\boldsymbol{\tau}\mathbf{T}^{-3}\mathbf{1})$

Table 7.3 Examples of the progressive Lévy degradation process Z_t

Progressive models		
Quantities for Z_t	LD (drift q)	Gamma process $GP(v, u)$
$\Psi_Z(z)$	$-iqz$	$v\ln(1-iz/u)$
$\mathbb{E}[Z_t]$	qt	vt/u
$\mu_2(t)$	0	vt/u^2
$\mu_3(t)$	0	$2vt/u^3$

Table 7.4 Examples of the combined Lévy degradation process K_t

Combined models		
Quantities for K_t	Shocks + LD	Shock + gamma process $GP(v, u)$
$\Psi_K(z)$	$\lambda(1-\phi_Y(z))-iqz$	$\lambda(1-\phi_Y(z))+v\ln(1-iz/u)$
$\mathbb{E}[K_t]$	$\lambda ty+qt$	$\lambda ty+vt/u$
$\mu_2(t)$	$\lambda t\mathbb{E}[Y^2]$	$\lambda t\mathbb{E}[Y^2]+vt/u^2$
$\mu_3(t)$	$\lambda t\mathbb{E}[Y^3]$	$\lambda t\mathbb{E}[Y^3]+2vt/u^3$

two cases of progressive degradation are presented, including the gamma process, which is the most common model used for this type of problems. Finally, in Table 7.4 there is a description of two models for the combined effect of shock-based and progressive degradation.

7.6 Expressions for Reliability Quantities

7.6.1 Computational Aspects: Inversion Formula

In order to derive the probability law $P(X_t \in \cdot)$ of the process and other key reliability quantities, it is necessary to invert Eq. 7.1 to obtain the probability law $P(X_t \in \cdot)$ from the characteristic function $\phi_{X_t}(z)$ of X_t. Then, given $a < x$, (for more details see [13]):

$$P(X_t \in (a, x]) = \frac{1}{2\pi i} \int_{-\infty}^{\infty} \frac{e^{-iza} - e^{-izx}}{z} \phi_{X_t}(z) dz. \tag{7.46}$$

Based on this expression, it can be proved [20] that the cumulative distribution function $P(X_t \in (-\infty, x])$ is

$$P(X_t \in (-\infty, x]) = \frac{1}{2} - \frac{1}{2\pi i} \int_{-\infty}^{\infty} \frac{e^{-izx}}{z} \phi_{X_t}(z) dz. \tag{7.47}$$

7.6.2 Reliability and Density of the Time to Failure

Equation 7.47 corresponds to the reliability function $R(t)$ in which x is the threshold that differentiate between failure and survival states. Based on the notation in Chaps. 4 to 5, $x = V_0 - k^*$. For practicality, we will write $R_x(t)$ to indicate that x is the deterioration to be surpassed for the system to fail; thus, the reliability is given by

$$\begin{aligned} R_x(t) &= \frac{1}{2} - \frac{1}{2\pi i} \int_{-\infty}^{\infty} \frac{e^{-izx}}{z} \phi_{X_t}(z) dz \\ &= \frac{1}{2} - \frac{1}{2\pi i} \int_{-\infty}^{\infty} \frac{e^{-izx}}{z} e^{-t\Psi(z)} dz. \end{aligned} \tag{7.48}$$

Differentiating equation (7.48) with respect to t, we obtain the lifetime density

$$\begin{aligned} f_x(t) &= -\frac{dR_x(t)}{dt} \\ &= -\frac{1}{2\pi i} \int_{-\infty}^{\infty} \frac{e^{-izx}}{z} \Psi(z) e^{-t\Psi(z)} dz \end{aligned} \tag{7.49}$$

7.6.3 Numerical Solution

The expressions for the reliability quantities must be evaluated numerically by approximating the improper integrals (7.48) and (7.49) as an infinite sum (i.e., *discretization*), which is *truncated* when convergence has been achieved [21]. In the literature, there are several rules of discretization out of which the most common is the *trapezoidal rule* [21–24]. Then, the expression for the reliability function (7.48) can be approximated by [22]

$$R_x(t) \approx R_x(t;h) := \frac{1}{2} - \frac{1}{2\pi i}\sum_{m=-\infty}^{\infty} \frac{e^{-ix(m-1/2)h}}{(m-1/2)} e^{-t\Psi((m-1/2)h)} \tag{7.50}$$

where z has been replaced by $((m-1)/2)h$ and $h > 0$ is the discretization step size. For computing the sum in Eq. 7.50, it is necessary to truncate it at a maximum/minimum index $\pm\ell$; then,

$$R_x(t) \approx R_x(t;h,\ell) := \frac{1}{2} - \frac{1}{2\pi i}\sum_{m=-\ell}^{\ell} \frac{e^{-ix(m-1/2)h}}{(m-1/2)} e^{-t\Psi((m-1/2)h)} \tag{7.51}$$

Similar expressions are obtained for the pdf of the lifetime (Eq. 7.49):

$$f_x(t) \approx f_x(t;h,\ell) := -\frac{1}{2\pi i}\sum_{m=-\ell}^{\ell} \frac{e^{-ix(m-1/2)h}}{(m-1/2)} \Psi((m-1/2)h) e^{-t\Psi((m-1/2)h)} \tag{7.52}$$

Clearly, the discretization step size h is critical for the model; Riascos-Ochoa et al. [1] proposed the following step size:

$$h = r\frac{2\pi}{x + \mathbb{E}[X_t] + \mathbb{E}[X_1]} = r\frac{2\pi}{x + (t+1)\Psi'(0)i} \tag{7.53}$$

The numerical examples that will be presented in the following sections will use a value of $r = 1/20$. Experimental and analytical results have shown that a good approximation to ℓ is $\ell \approx 10^5$ [1].

Finally, the moments of the system's lifetime, i.e.,

$$\mathbb{E}[L^n] = \int_0^\infty t^n f_x(t)dt \tag{7.54}$$

can be approximated numerically using, for example, the trapezoidal rule. The procedure consists of two steps:

1. Define a time increment $\Delta t > 0$ and the set of times $t_1, t_2, ..., t_N$ with $t_i = t_{i-1} + \Delta t$ and $t_0 = 0$ at which the density $f_x(t)$ of the lifetime L is evaluated by using the approximation $f_x(t; h, \ell)$ from Eq. (7.52). The final time t_N and the increment Δt are set in order to have the following trapezoidal approximation

$$\int_0^{\infty} f_x(t)dt \approx F_x(\Delta t, t_N) := \frac{t_n - t_0}{2N}\left[f_x(t_0) + 2f_x(t_1) + 2f_x(t_2) + \cdots + 2f_x(t_{N-1}) + f_x(t_N)\right] \tag{7.55}$$

which approaches 1 with an absolute error $|1 - F_x(\Delta t, t_N)| \leq \epsilon$; where ϵ is a predefined value.
2. Approximate the moments $\mathbb{E}[L^n]$ by applying the trapezoidal rule, i.e.,

$$\mathbb{E}[L^n] \approx \frac{t_N - t_0}{2N}\left[t_0^n f_x(t_0) + 2t_1^n f_x(t_1) + 2t_2^n f_x(t_2) + \cdots + 2t_{N-1}^n f_x(t_{N-1}) + t_N^n f_x(t_N)\right]. \tag{7.56}$$

7.6.4 Construction of Sample Paths Using Simulation

Sample paths of different Lévy deterioration processes can be simulated from its probability law $P(X_t \in \cdot)$ using, for example, the increment-sampling method described in [18]. Thus, considering that Lévy processes have independent and identically distributed increments, the procedure consists of two steps:

1. Define a time increment $\Delta t > 0$ and the set of times $t_0, t_1, t_2, ..., t_n$ with $t_i = t_{i-1} + \Delta t$ and $t_0 = 0$, at which damage increments will be evaluated. This means that $X_{\Delta t} = (X_{t_i} - X_{t_{i-1}})$, with $X_{t_0} = 0$, is *iid* for all t_i.
2. Randomly draw independent damage increments $\hat{X}_i$ (associated to every t_i) from the cumulative distribution function (CDF) of $X_{\Delta t}$.

Note that the cumulative distribution function of $X_{\Delta t}$ is numerically computed as in Sect. 7.6.3, i.e., Eq. 7.51 for fixed Δt and several values of x. In summary, the sample path is constructed as a series of successive increments $\hat{X}_i$ occurring at times t_i.

Example 7.36 Construct several sample paths of a system subjected to two progressive degradation processes $Z_t^{\{1\}}$ and $Z_t^{\{2\}}$ using the Lévy formalism. Both degradation mechanisms are modeled using a gamma process with the following parameters:

1. $\text{GP}_1(v_1 = 1, u_1 = 1/2)$;
2. $\text{GP}_2(v_2(t) = 0.02t^2, u_2 = 1/2)$.

Note that the mean of the degradation processes are $\mathbb{E}_1[Z_t^{\{1\}}] = 2t$ and $\mathbb{E}_2[Z_t^{\{2\}}] = 0.04t^2$. The sample paths of this degradation processes are shown in Figs. 7.1 and 7.2, where the mean of the degradation process is indicated with a dashed line.

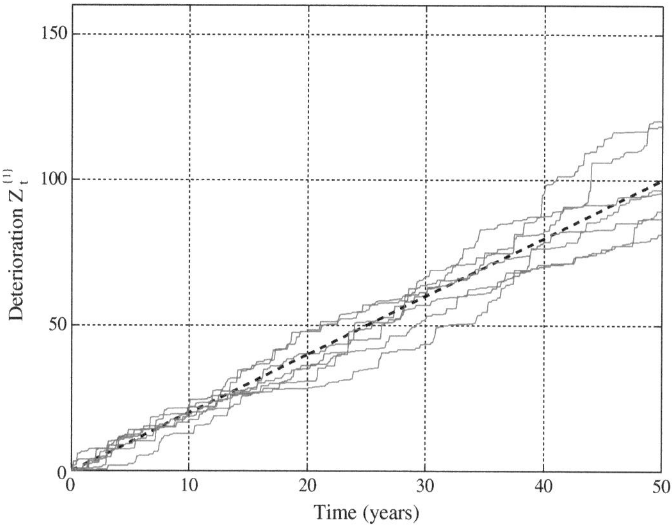

Fig. 7.1 Sample paths of the progressive degradation model described by a gamma process with $\mathrm{GP}_1(v_1 = 1, u_1 = 1/2)$

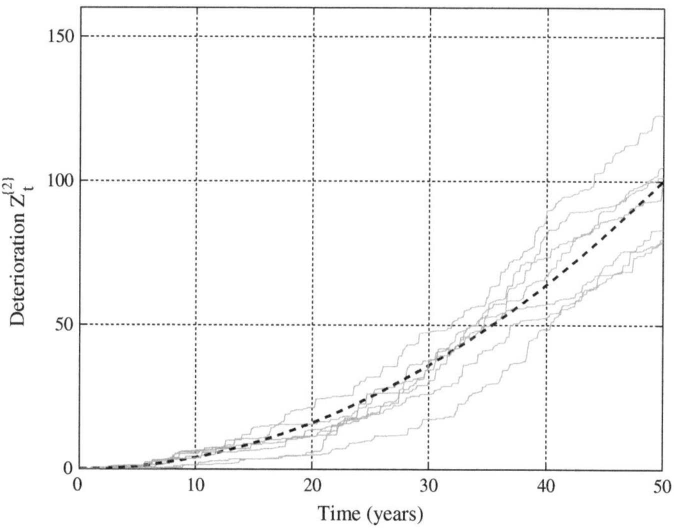

Fig. 7.2 Sample paths of the progressive degradation model described by a gamma process with $\mathrm{GP}_2(v_2(t) = 0.02t^2, u_2 = 1/2)$

Simulation was implemented using the increment-sampling method described above, in which the time intervals selected for the simulations were $\Delta t = 0.1$. An important observation is that for the second case, the process is not homogeneous and some additional considerations are required for the evaluation; these can be found in [16].

Example 7.37 Using the Lévy formalism, draw several realizations of two shock-based degradation models described by a compound Poisson process with the following shock size distributions:

1. $Y_i \sim \delta(y = 10)$; and
2. $Y_i \sim \exp(1/\nu)$ with rate $\nu = 10$.

In both cases, the rate of shock occurrence is $\lambda = 0.2$. Also both models have the same mean deterioration $\mathbb{E}[X_t] = 2t$.

The sample paths of the two processes are shown in Figs. 7.3 and 7.4. The mean of the degradation process is indicated with a dashed line. It can be observed that while the CPP-delta model has always shocks with identical sizes, i.e., $y = 10$, in the realization of the CPP-exp, shocks have different sizes. As expected, in both cases the sample paths are distributed around the dashed line that represents the mean. It is interesting to note that the dispersion around the mean is greater for the CPP-exp model, which is explained from the fact that its second central moment ($\mu_2(t) = 2\lambda t y^2 = 40t$) is larger than the one for the CPP-Delta model ($\mu_2(t) = \lambda t y^2 = 20t$) (see Tables 7.1 and 7.2).

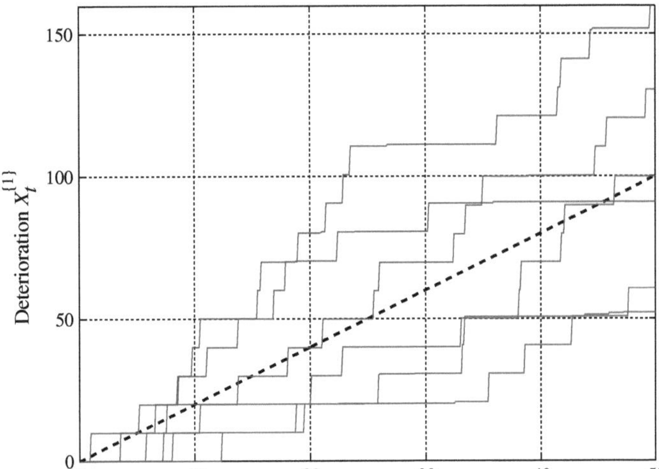

Fig. 7.3 Sample paths for a CPP model with Poisson rate $\lambda = 0.2$; and shock sizes distributed $Y_i \sim \delta(y = 10)$

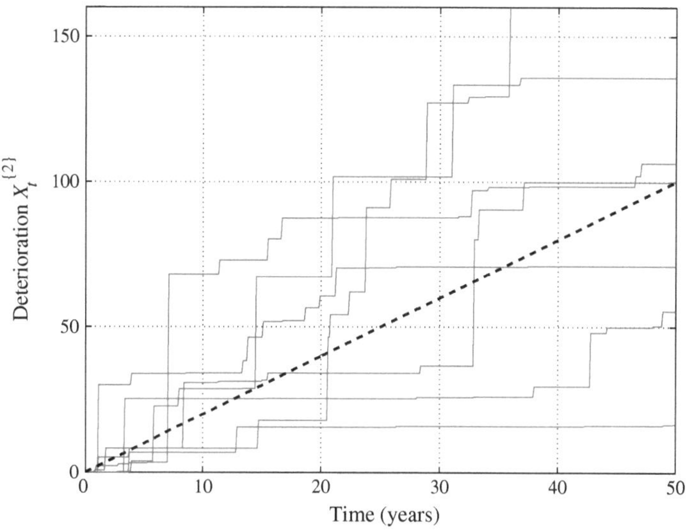

Fig. 7.4 Sample paths for a CPP model with Poisson rate $\lambda = 0.2$; and shock sizes distributed $Y_i \sim \exp(1/10)$

Example 7.38 In this example, we are interested in the sample path of a combined degradation process K_t. The shock-based model component corresponds to the CPP-exp presented in the previous example. The progressive deterioration Z_t is given by the gamma process $GP_1(v_1 = 1, u_1 = 1/2)$.

Several realizations of the progressive deterioration process were already shown in Fig. 7.1, while Fig. 7.5 presents various sample paths for the combined case—i.e., K_t. Note that both models have the same mean, i.e., $\mathbb{E}[W_t] = \mathbb{E}[Z_t] = 2t$, while the mean of the combined process is $\mathbb{E}[K_t] = 4t$. As expected, the variance of the combined model is greatly controlled by the CPP-exp model.

Example 7.39 Consider a system that degrades with failure threshold $x = v_0 - k^* = 100$. We are now interested in obtaining the lifetime density for different degradation models.

The system is subjected to progressive degradation, modeled as a GP with parameters $GP(v = 0.1, u = 1/20)$. For the case of shocks, we considered a CPP with rate $\lambda = 0.1$ and the following shock size distributions:

1. $Y_i \sim \delta(y = 20)$;
2. $Y_i \sim \exp(\nu = 1/20)$;
3. $Y_i \sim U(0, 40)$; and
4. $Y_i \sim LN(\mu, \sigma)$.

For the particular case of the CPP-LN, the parameters (μ, σ) are determined according to Table 7.1 such that the mean of shock sizes is $\mathbb{E}[Y] = 20$ with a coefficient

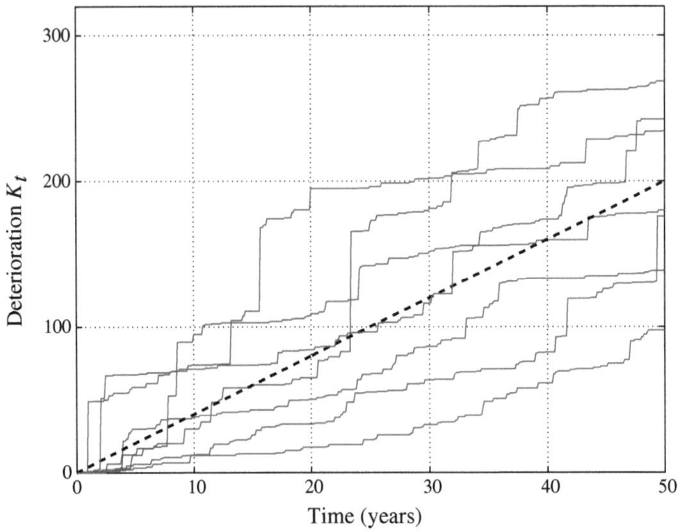

Fig. 7.5 Sample paths for combined model of $\text{GP}_1(v_1 = 1, u_1 = 1/2)$ and CPP-exp, with $\lambda = 0.2$ and $Y \sim \exp(1/10)$

of variation $COV(Y) = 2$. The mean deterioration in all of the models considered is $\mathbb{E}[X_t] = 2t$. The results of the analysis are shown in Figs. 7.6 and 7.7. Furthermore, it can be observed that, as expected, the processes with greater variance produce greater dispersions in their lifetime. The second central moments are $\mu_2(t) = 40t$, $(160/3)t$, $80t$, $120t$ for the CPP-delta (and GP model), CPP-U, CPP-exp, and CPP-LN, respectively. Finally, Fig. 7.7 shows the density for the combined cases. In this case, each CPP model was combined with a progressive gamma degradation $GP(v = 0.1, u = 1/20)$. Note that the combined models lead to smaller failure times, which is expected since we have added an additional source of degradation.

These results can be compared with available analytical expressions for the GP model (given in [18]) and the CPP-Delta and CPP-Exp models; these are

$$f_x^{GP}(t) = \frac{v}{\Gamma(vt)} \int_{xu}^{\infty} \left[\log(z) - \Gamma'(vt)/\Gamma(vt)\right] z^{vt-1} e^{-z} dz \tag{7.57}$$

$$f_x^{\delta}(t) = \lambda e^{-\lambda t} \frac{(\lambda t)^{\lfloor x/y \rfloor}}{\lfloor x/y \rfloor !} \tag{7.58}$$

$$f_x^{Exp}(t) = -\lambda e^{-\lambda t} \left[\sum_{k=1}^{\infty} \frac{\gamma(k, \nu x)}{(k-1)!} \frac{(\lambda t)^k}{k!} \left(\frac{k}{\lambda t} - 1 \right) - 1 \right], \tag{7.59}$$

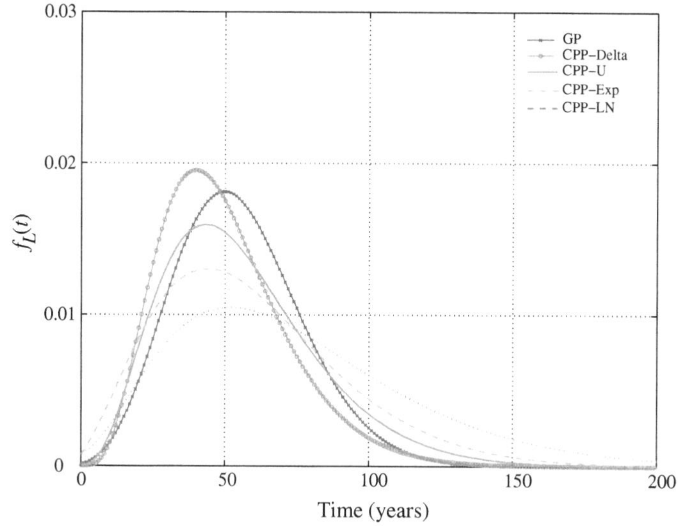

Fig. 7.6 PDF $f_x(t)$ of the lifetime L of a system with threshold level $x = 100$ for not combined GP and CPP's models; $\lambda = 0.1$

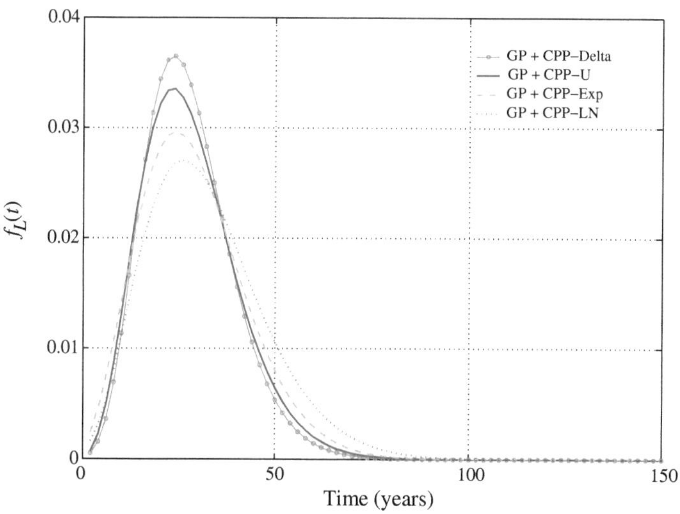

Fig. 7.7 PDF of the lifetime L of a system with threshold level $x = 100$ for combined degradation $GP(v = 0.1, u = 1/20)$ with several CPP's models; $\lambda = 0.1$

with $\lfloor \cdot \rfloor$ the integer part function, $\Gamma(x)$ the Gamma function, and $\gamma(k, \nu x)$ the lower incomplete gamma function. The densities obtained for these cases match exactly to the numerically computed curves with the formalism presented in this chapter; they are superimposed on the densities shown in Figs. 7.6 and 7.7.

7.7 Summary and Conclusions

This chapter presents a general framework within which it is possible to accommodate most degradation models used in practical applications (Chap. 5). Degradation is modeled as an increasing Lévy process known as subordinator, i.e., a process with independent, stationary, and nonnegative increments. A subordinator is specified by its Lévy measure, characteristic function, and characteristic exponent. We show how these quantities are used to obtain analytical expressions for the mean and the moments of the degradation process. In addition, expressions for the important reliability quantities, namely the reliability function, the probability density of lifetime, and its mean and moments, can also be easily obtained. The assumption of independence among different degradation processes allows the superposition, and therefore, modeling combined degradation mechanisms. An important advantage of the proposed formalism is that it overcomes analytical difficulties that appear frequently when modeling degradation such as infinite sums and convolutions. In fact, at this moment, this approach is as far as any analytical solution can go to model the complexity of degradation.

References

1. J. Riascos-Ochoa, M. Sánchez-Silva, G-A. Klutke, Modeling and reliability analysis of systems subject to multiple sources of degradation based on Lévy processes (2015) (Under review)
2. J. Bertoin, *Lévy Processes* (Cambridge University Press, Cambridge, U.K., 1996)
3. K.-I. Sato. *Lévy processes and infinitely divisible distributions* (Cambridge University Press, Cambridge, 1999)
4. P.E. Protter, *Stochastic Integration and Differential Equations* (Springer, Germany, 2004)
5. G.-A. Klutke, Y. Yang, The availability of inspected systems subject to shocks and graceful deterioration. IEEE Trans. Reliab. **51**(3), 371–374 (2002)
6. I. Iervolino, M. Giorgio, E. Chioccarelli, Gamma degradation models for earthquake-resistant structures. Struct. Saf. **45**, 48–58 (2013)
7. M. Abdel-Hameed, Life distribution properties of devices subject to a pure jump damage process. J. Appl. Probab. **21**, 816–825 (1984)
8. M. Abdel-Hameed, *Lévy Processes and their Applications in Reliability and Storage* (Springer, New York, 2014)
9. Y. Yang, G.-A. Klutke, Lifetime-characteristics and inspection-schemes for lévy degradation processes. IEEE Trans. Reliab. **49**(4), 377–382 (2000)
10. D. Applebaum, Lévy process—from probability theory to finance and quantum groups. Not. AMS **51**(11), 1336–1347 (2004)

11. D. Applebaum, *Lévy Processes and Stochastic Calculus* (Cambridge University Press, Cambridge, U.K., 2004)
12. S. Resnick, *A Probability Path* (Birkhauser, Boston, 1999)
13. R. Durret, *Probability: Theory and Examples* (Cambridge University Press, USA, 2010)
14. J.M. van Noortwijk, R.M. Cooke, M. Kok, A bayesian failure model based on isotropic deterioration. Eur. J. Oper. Res. **82**, 270–282 (1995)
15. I. Iervolino, M. Giorgio, E. Chioccarelli, Closed-form aftershock reliability of damage-cumulating elastic-perfectly-plastic systems. Earthq. Eng. Struct. Dyn. **43**, 613–625 (2014)
16. J. Riascos-Ochoa, M. Sánchez-Silva, G-A. Klutke, Degradation modeling and reliability estimation via non-homogeneous Lévy processes (2016) (Under review)
17. S. Ross, *Introduction of Probability Models* (Academic Press, San Diego, CA, 2007)
18. J.M. Van Noortwijk, A survey of the application of gamma processes in maintenance. Reliab. Eng. Syst. Saf. **94**, 2–21 (2009)
19. M. Sánchez-Silva, G.-A. Klutke, D. Rosowsky, Life-cycle performance of structures subject to multiple deterioration mechanisms. Struct. Saf. **33**(3), 206–217 (2011)
20. J. Gil-Pelaez, Note on the inversion theorem. Biometrika Trust **38**(3/4), 481–482 (1951)
21. H. Bohman, Numerical inversions of characteristic functions. Scand. Actuarial J. **2**, 121–124 (1975)
22. L. Feng, X. Lin, Inverting analytic characteristic functions and financial applications. SIAM J. Financ. Math. **4**, 372–398 (2013)
23. L.A. Waller, B.W. Turnbull, J.M. Hardin, Obtaining distribution functions by numerical inversion of characteristic functions with applications. Am. Stat. **49**(4), 346–350 (1995)
24. R.B. Davies, Numerical inversion of a characteristic function. Biometrika Trust **60**(2), 415–417 (1973)

Chapter 8
Systematically Reconstructed Systems

8.1 Introduction

In Chaps. 4–7, we addressed the problem of modeling systems that degrade over time and that are abandoned after failure. However, frequently, once systems reach a serviceability threshold, or experience failure, they are updated or reconstructed so as to be put back in service. In these cases, some additional considerations are needed to describe the system's performance over time. Since models for systematically reconstructed systems are based on renewal theory (under specific assumptions; see Chap. 3), one of the modeling challenges in this chapter is the study and evaluation of the distribution function for the times between renewals. We also integrate the degradation models presented in Chaps. 4 and 7 with renewal theory to build models able to describe the long-term performance of large engineering systems. The chapter is divided into two parts. The first part presents models that do not explicitly take deterioration into account, while the second part considers explicit characterizations of deterioration over time. The models presented in this chapter will be used later to carry out life-cycle analysis (Chap. 9) and to define maintenance policies (Chap. 10).

8.2 Systems Renewed Without Consideration of Damage Accumulation

The problem of systematically reconstructed systems has been studied for many years, but has received increasing attention as life-cycle analysis has become more important. In particular, it has impacted the way in which long-term decisions related to the management and operation of most large infrastructure projects are made. The first papers addressing this subject in civil engineering were presented by Rosemblueth and Mendoza [1] and Rosemblueth [2] and by Hasofer [3].

M. Sánchez-Silva and G.-A. Klutke, *Reliability and Life-Cycle Analysis of Deteriorating Systems*, Springer Series in Reliability Engineering,
DOI 10.1007/978-3-319-20946-3_8

Rackwitz [4] presents a critical review of these papers and extends the concepts to failures under normal and extreme conditions, serviceability failures, obsolescence, and other failure mechanisms. In the pioneering work of Rackwitz and his colleagues [5–10], the main concepts associated with this problem are discussed in depth. These works have opened a large spectrum of research opportunities in many areas with important applications in practice. Much of this section is based on this body of work, which will lead into out discussion of life-cycle analysis in Chap. 9.

8.2.1 Description of the Process

In the simplest case of a systematically reconstructed system, the system condition is not observed or monitored over time (it is assumed to be operating satisfactorily), until it suddenly fails and is taken out of service due to an anomalous internal characteristic or an extreme event. After failure, the system is repaired and put back into service immediately (instantaneous interventions) and the process of operation, failure, and repair continues indefinitely or until the system is abandoned (Fig. 8.1). It is important to stress that the assumption that interventions take the system to a satisfactory operating condition is justified under the presumption that the first design was already optimal and, therefore, there is no reason to change the design rules [5].

8.2.2 Successive Reconstructions at Shock Times

In this section, we consider the case in which failures, and the corresponding instantaneous interventions, occur randomly with inter-arrival times X_i; $i = 1, 2, \ldots$. The

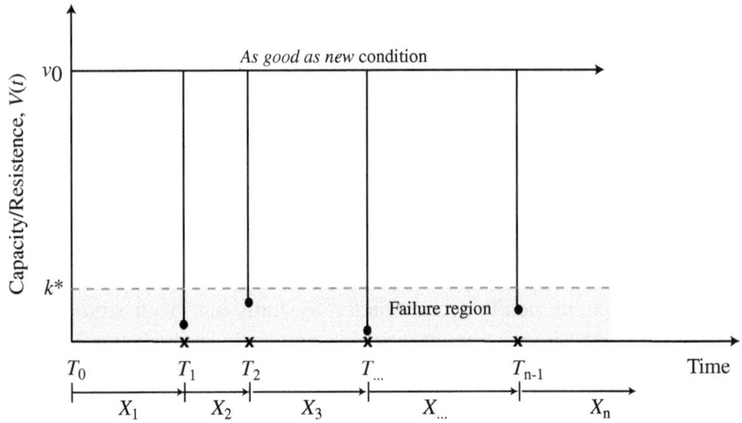

Fig. 8.1 Description of a system subject to systematic reconstruction with instant failures and repairs

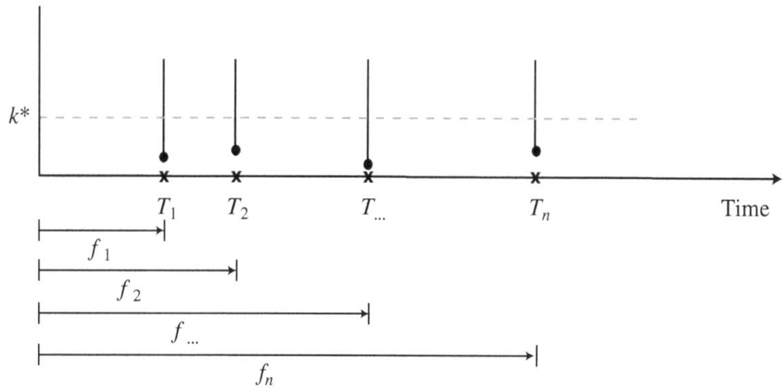

Fig. 8.2 Description of the probability density to the nth intervention

times X_i are assumed to be independent and identically distributed random variables with distribution function $F(t)$ and density $f(t)$. In this case, the time to the nth event, T_n, has distribution

$$T_n = \sum_{i=1}^{n} X_i \sim F_n(t) \tag{8.1}$$

where $F_n(t)$ is the distribution of the time to the nth intervention (renewal) and is computed as the nth convolution of F with itself. The corresponding density of F_n is f_n, which can be expressed as (Fig. 8.2)

$$f_n(t) = \int_0^t f_{n-1}(t-\tau)f(\tau)d\tau; \quad n = 2, 3, \ldots \tag{8.2}$$

For convolution integrals, the Laplace transform can be used with advantage [4]. The Laplace transform of $f(t)$ is

$$\mathscr{L}[f(t)] = f^*(\theta) = \int_0^\infty f(t)e^{-\theta t}dt \tag{8.3}$$

For the case in which $f(t)$ is a probability density, $f^*(0) = 1$ and $0 < f^*(\theta) \leq 1$ for all $\theta > 0$. The analytical solution for the Laplace transform is not always available; however, a list of common probability models for which it exists is shown in Table 8.1.

The Laplace transform of $f_n(t)$ is

$$\mathscr{L}[f_n(t)] = f_n^*(\theta) = \int_0^\infty f_n(t)e^{-\theta t}dt. \tag{8.4}$$

In addition, convolutions have the following property:

$$f_n^*(\theta) = f_1^*(\theta)f_{n-1}^*(\theta) = f_1^*(\theta)[f^*(\theta)]^{n-1} \tag{8.5}$$

where it may be the case that $f_1^*(\theta) = f^*(\theta)$.

Table 8.1 Analytic Laplace transform expressions for selected distributions

Name	Density function $f(t)$	Laplace transform $f^*(\theta)$
δ-Spike	$\delta(a)$	$\exp(-\theta a)$
Exponential	$\lambda\exp(-\lambda t)$	$\frac{\lambda}{\lambda+\theta}$
Uniform	$\frac{1}{b-a}$	$\frac{\exp(-a\theta)-\exp(-b\theta)}{\theta(b-a)}$
Beta	$\frac{y^{r-1}(1-y)^{s-1}}{B(r,s)}$	$F_1(r, r+s; -\theta)$
Rayleigh	$\frac{1}{w^2}\exp\left[-\left(\frac{t}{w}\right)^2\right]$	$1-\theta w\sqrt{\frac{\pi}{2}}\exp\left(\frac{1}{2}\theta^2 w^2\right)\text{erfc}\left(\frac{\theta w}{\sqrt{2}}\right)$
Gamma	$\frac{\lambda^k}{\Gamma(k)}t^{k-1}\exp(-\lambda t)$	$\left(\frac{\lambda}{\theta+\lambda}\right)^k$

Example 8.40 Consider a system where shocks occur according to a stationary Poisson process with rate λ (i.e., rate at which failures and immediate repairs occur). Compute the Laplace transform of the process.

By definition, the inter-arrival times of events that follow a Poisson process are independent and exponentially distributed (i.e., $f(t) = \lambda\exp(-\lambda t)$). Then, according to Eq. 8.4, the Laplace transform of the time between events (e.g., shocks) can be computed as

$$f^*(\theta) = \int_0^\infty \lambda\exp(-\lambda t)e^{-\theta t}dt = \frac{\lambda}{\theta+\lambda} \tag{8.6}$$

which is an important result when modeling the occurrence of extreme events such as earthquakes or storms [7].

If the probability function of the time to the nth failure is known (Eq. 8.1), it is now possible to compute the expected number of failures in time t. This is carried out by evaluating the renewal function (see Chap. 3)

$$M(t) = \mathbb{E}[N(t)] = \sum_{n=1}^{\infty} F_n(t) \tag{8.7}$$

where $N(t)$ is the number of renewals in $[0, t]$. The derivative of the renewal function $M(t)$ is called the renewal density $m(t)$ and is defined as

$$m(t) = \sum_{n=1}^{\infty} f_n(t) \tag{8.8}$$

where, as mentioned before, f_n is the density of the time to the nth renewal (Eq. 8.2). For ordinary renewal processes,[1] the property of the Laplace transform shown in

[1]In an ordinary renewal process, all times between renewals are *iid*.

Eq. 8.5 can be used conveniently to obtain [5]

$$m^*(\theta) = \sum_{n=1}^{\infty} f_n^*(\theta) = \sum_{n=1}^{\infty} [f^*(\theta)]^n = \frac{f^*(\theta)}{1 - f^*(\theta)} \tag{8.9}$$

since $\sum_{n=1}^{\infty} x^n = \sum_{n=0}^{\infty} x^n - 1 = 1/(1-x) - 1 = x/(1-x)$. Similarly, for modified renewal processes (i.e., when the time to first failure is different, $f_1 \neq f_i$; for $i > 1$), the density to the nth failure is computed as [5]

$$m_1^*(\theta) = \sum_{n=1}^{\infty} f_n^*(\theta) = \sum_{n=1}^{\infty} f_1^*(\theta)[f^*(\theta)]^{n-1} = \frac{f_1^*(\theta)}{1 - f^*(\theta)} \tag{8.10}$$

Note that the solutions presented in Eqs. 8.9 and 8.10 constitute an expression for the density of the expected number of failures and immediate repairs for a system that is *successively reconstructed.*

Example 8.41 Consider a system that is successively reconstructed after failures, which occur according to a Poisson process with rate $\lambda = 0.5$. If the cost of future repairs is discounted to time $t = 0$ with a continuous discounting function $\delta(t) = \exp(-\theta t)$; $\theta = 0.05$, compute the expected net present value (NPV) of the all investments for a system with infinite lifetime.
The expected discounted[2] total cost of investments is

$$\mathbb{E}[C_T] = \sum_{n=1}^{\infty} \left[\int_0^{\infty} C_n \delta(t) f_n(t) dt \right] = \sum_{n=1}^{\infty} \left[\int_0^{\infty} C_n f_n(t) e^{-\theta t} dt \right]$$

where C_n indicates the cost of the nth failure and repair with $n = 1, 2, \ldots$. If the cost of interventions is assumed to be equal, i.e., $C_n = C$, and taking advantage of the form of the discount function, this equation can be written as (see Eq. 8.9)

$$\mathbb{E}[C_T] = \sum_{n=1}^{\infty} C_n f_n^*(t) = C \sum_{n=1}^{\infty} [f^*(\theta)]^n = C \frac{f^*(\theta)}{1 - f^*(\theta)}$$

Because failures are exponentially distributed, there is an analytical expression for Laplace transform; then

$$\mathbb{E}[C_T] = C \frac{f^*(\theta)}{1 - f^*(\theta)} = \frac{\frac{\lambda}{\lambda+\theta}}{1 - \frac{\lambda}{\lambda+\theta}} = \frac{\lambda}{\theta} C = \frac{0.5}{0.05} C = 10C.$$

[2] A detailed discussion about the problem of discounting will be provided in Chap. 9.

8.2.3 Systems Subject to Random Failures—Extreme Overloads

Consider now a system subjected to random external demands such that there may exist events (demands) that make the system to fail (with probability P_f), and other events that do not cause failure (with probability $1 - P_f$). As in the previous case, if the system does not fail, it continues operating in a satisfactory condition, and once it fails, it is immediately repaired and taken to its original condition (Fig. 8.3).

In order to model this case, we need to make a distinction between two processes that occur simultaneously. Let us first assume that the events that may (or may not) cause the failure follow a renewal process with the time to the first event having distribution F_1, and the times between any two successive events having distribution F. Furthermore, let us define G_1 as the distribution function to the first failure and G as the distribution of the time between failures. The densities of F and G will be denoted as f and g, respectively (Fig. 8.4).

The density of the time to the *first* failure can be written as [4]

$$g_1(t) = \sum_{n=1}^{\infty} f_n(t) P_f (1 - P_f)^{n-1} \tag{8.11}$$

where $f_n(t)$ is the nth convolution of f with itself and describes the density function of the time to the nth event (not necessarily a failure) (Fig. 8.4).

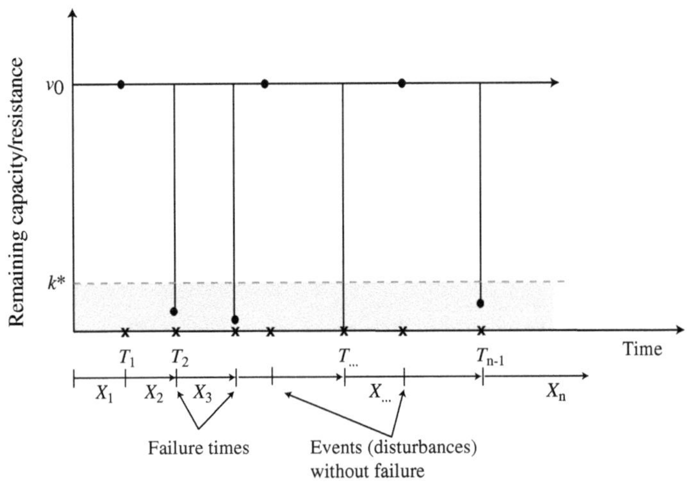

Fig. 8.3 Systematic reconstruction after failure—failures due to extreme overloads

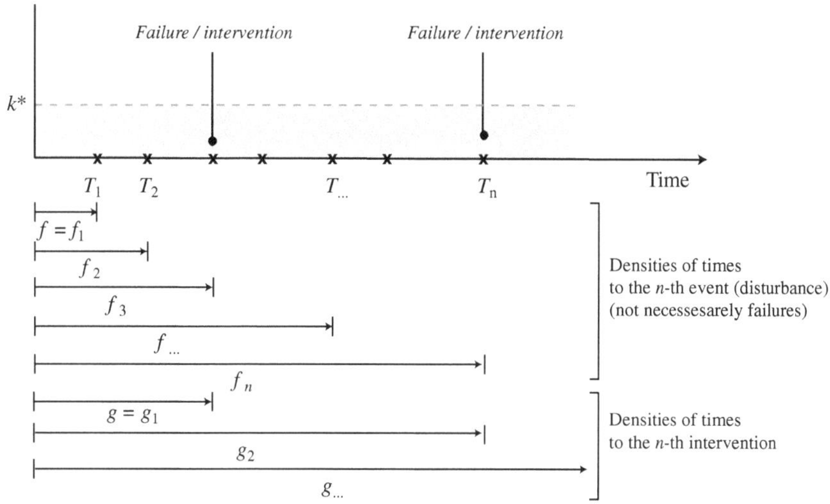

Fig. 8.4 Description of the probability density to the nth intervention

By taking advantage of the Laplace transform and Eq. 8.5, it is possible to rewrite the function of the time to first failure (Eq. 8.11) as follows [4]:

$$
\begin{aligned}
g_1^*(\theta) &= \sum_{n=1}^{\infty} f_1^*(\theta) f_{n-1}^*(\theta) P_f (1-P_f)^{n-1} \\
&= \sum_{n=1}^{\infty} f_1^*(\theta) [f^*(\theta)]^{n-1} P_f (1-P_f)^{n-1} \\
&= \frac{P_f f_1^*(\theta)}{1-(1-P_f) f^*(\theta)}
\end{aligned}
\tag{8.12}
$$

where $g_1^*(\theta) = \mathscr{L}[g_1(t)]$ is the Laplace transform of the probability density of the time to first failure. Note that this expression is defined in terms of the Laplace transform of the inter-arrival event densities f.

Let us now evaluate the density of the time between any two failures as function of the density of the time between disturbances. It should be clear that if the system is at a time just right after a reconstruction, the density to the next failure is the same as between any other two failures; then,

$$g(t) = \sum_{n=1}^{\infty} f_n(t) P_f (1-P_f)^{n-1} \tag{8.13}$$

Then, by taking the Laplace transform, i.e., $\mathscr{L}[f^{(n)}(t)] = f_n^*(\theta)$, and considering Eq. 8.5 [4],

$$g^*(\theta) = \sum_{n=1}^{\infty} f^*(\theta)[f^*(\theta)]^{n-1} P_f (1 - P_f)^{n-1}$$
$$= \frac{P_f f^*(\theta)}{1 - (1 - P_f) f^*(\theta)} \tag{8.14}$$

Note that in Eqs. 8.12 and 8.14, it is assumed that the system is abandoned after the first failure. Consider now that the system is subject to shocks that may or may not cause the failure with certain probability P_f, and that it is systematically reconstructed immediately after every failure; furthermore, we assume that the system operates over an infinite time horizon. Then, we can apply the same rationally as in previous derivations to obtain the discounted expected value of losses. Again, the density between failures would be g (Eq. 8.14) for the case in which the times between failures are *iid*, and g_1 (Eq. 8.12) for the case in which the time to first failure is different from the rest (which are all identically distributed). Then, $\mathbb{E}[C_T] = Ch^*(\theta)$ such that

$$h^*(\theta) = \frac{g^*(\theta)}{1 - g^*(\theta)} \tag{8.15}$$

or

$$h_1^*(\theta) = \frac{g_1^*(\theta)}{1 - g^*(\theta)} \tag{8.16}$$

where $h^*(\theta)$ and $h_1^*(\theta)$ are the Laplace transform of the probability density of the times between failures. Hasofer [3] called $h^*(\theta)$ and $h_1^*(\theta)$ the discount factor.

Example 8.42 Consider a system is subjected to events that occur randomly in time with exponential distribution F and density f. Every time there is an event, the system may fail with probability P_f (or survive with probability $1 - P_f$). If the cost of failure of the system is C, and the discounting function $\delta(t) = \exp(-\theta t)$ with θ the discount rate, compare the expected discounted value of losses, for the following cases:

1. A system that starts operating right after an event has occur and therefore the rate of occurrence of all disturbances is λ_1. The system is abandoned after failure.
2. A system that starts operating sometime after an event has occur and therefore the rate of occurrence of the first disturbance is $\lambda_2 = \alpha\lambda_1$, with $\alpha \geq 1$, the rest of occurrences have rate λ_1. The system is abandoned after failure.
3. A system that starts operating right after an event has occur and therefore the rate of occurrence of all disturbances is λ_1. The system is systematically reconstructed for and infinite time horizon.

In the first case, and keeping in mind Eq. 8.14, we get

$$\begin{aligned}\mathbb{E}[C_T] &= C\int_0^{\infty} g(t)\delta(t)dt = Cg^*(t)\\ &= C\frac{P_f f^*(\theta)}{1-(1-P_f)f^*(\theta)}\\ &= CP_f\frac{\lambda_1}{\theta+\lambda_1 P_f}.\end{aligned}$$

For the second case, the discounted expected total cost $\mathbb{E}[C_T]$ can be computed as

$$\mathbb{E}[C_T] = C\int_0^{\infty} g_1(t)\delta(t)dt = Cg_1^*(t) = C\frac{P_f f_1^*(\theta)}{1-(1-P_f)f^*(\theta)} \tag{8.17}$$

therefore,

$$\begin{aligned}\mathbb{E}[C_T] &= CP_f\frac{\frac{\lambda_2}{\theta+\lambda_2}}{1-(1-P_f)\frac{\lambda_1}{\theta+\lambda_1}}\\ &= CP_f\frac{\frac{\alpha\lambda_1}{\theta+\alpha\lambda_1}}{1-(1-P_f)\frac{\lambda_1}{\theta+\lambda_1}}\\ &= CP_f\frac{\alpha\lambda_1(\theta+\lambda_1)}{(\theta+P_f\lambda_1)(\theta+\alpha\lambda_1)}\end{aligned}$$

Note that for $\alpha = 1$ the solution becomes $\mathbb{E}[C_T] = CP_f\cdot\lambda_1/(\theta+\lambda_1 P_f)$, which is the same result obtained in the first case.

Finally, for the third case, we have that

$$\begin{aligned}\mathbb{E}[C_T] &= Ch^*(\theta)\\ &= C\frac{g^*(\theta)}{1-g^*(\theta)}\\ &= CP_f\frac{f^*(\theta)}{1-f^*(\theta)}\\ &= CP_f\frac{\lambda_1}{\theta}\end{aligned}$$

8.3 Renewal Models Including Repair Times

The performance of many engineered systems can be modeled as a two-state system; for example, operating/nonoperating, safe/unsafe, etc. Furthermore, in some cases, immediate reconstruction (instantaneous) cannot be assumed and repair times become important in the analysis. In this section, we present models that include repair times and, in particular, we focus on the problem of system availability.

8.3.1 System Availability

Consider a system that starts operating and remains is a satisfactory condition until failure. Once it fails, some time is required for the system to be repair and put back into service. After repaired, the system continues operating satisfactorily until next failure. These cycles of failures and repairs continue over an infinite time horizon (see Fig. 8.5).

Let us define X_i as the time between the ith and the $i-1$th failures, and Y_i as the associated repair time (Fig. 8.5). Both X and Y are *iid* random variables with probability distribution $F(t)$ and $H(t)$, respectively. Let us further define a cycle as $Z = X + Y$, which corresponds to the length of time between two consecutive failures. Then, the probability distribution of the length of the cycle is

$$G(t) = P(Z \leq t) = F(t) + H(t) = \int_0^{\infty} F(t-\tau)H(\tau)d\tau \tag{8.18}$$

and the time to the nth renewal has a probability distribution:

$$T_n = \sum_{i=1}^{n} Z_i \sim G_n(t) \tag{8.19}$$

where $G_n(t)$ is the nth Stieljes convolution of G with itself.

A quantity of particular interest in operational decision making for this type of problems is the *system availability*. Availability is defined as the long-run proportion of time that the system is operating. Then, the asymptotic availability of the system can be computed as [11]

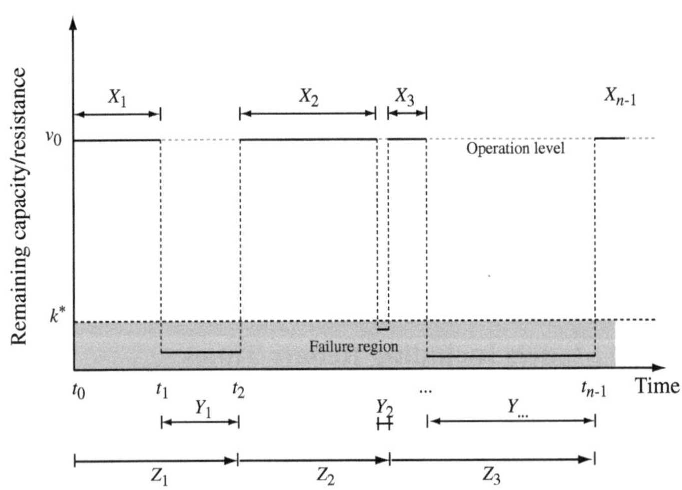

Fig. 8.5 Definition of a cycle for systems with repair times

$$A(\infty) = P(\text{System is operating as } t \to \infty) = \frac{\mathbb{E}[X]}{\mathbb{E}[X] + \mathbb{E}[Y]} \tag{8.20}$$

where the operator $\mathbb{E}[\cdot]$ indicates the expected value.

Example 8.43 Consider a bridge that may be only in two-state service or out of service. Both the times it spends in service and out of service are exponentially distributed. If the bridge is operating, it becomes out of service with a rate $\lambda_1 = 0.01$, and the time for it to be repaired has a rate $\lambda_2 = 0.2$. Then, we are interested in computing the long-term availability of the bridge.

Because the times in service and out of service are exponentially distributed, the long-run availability can be computed as

$$A(\infty) = \frac{1/\lambda_1}{1/\lambda_1 + 1/\lambda_2} = \frac{100}{100 + 5} = 0.95$$

which means that, on average, the bridge will be in operation 95 % of the time.

Although it is not shown in Fig. 8.5, the condition of the system when in operation does not necessary mean that it is functioning in *as good as new* state permanently. In actual problems, the system condition decreases as a result of different degradation mechanisms (see Chap. 5). Thus, when damage accumulates, the terms in Eq. 8.20 describe the expected time the system operates above or below a certain threshold (e.g., failure threshold). This problem is illustrated with the following example.

Example 8.44 Consider a bridge in a seismic region such that every time an extreme event occurs (e.g., earthquake) it suffers some damage (e.g., loss of stiffness). The inter-arrival times of the extreme events are assumed to be random with distribution F, and the amount of damage caused by the event i will be D_i, which is also a random variable. Furthermore, we will assume that the damages accumulated at every shock and the occurrence of shocks are independent.

Let us assume that the condition of the structure at time $t = 0$ is v_0. Furthermore, in order to characterize the operation, two capacity thresholds are defined. The threshold level y^* defines the serviceability limit state; this means that as long as its condition is above y^*, the system is considered to be in a level of service which is *acceptable*. In addition, the ultimate limit state k^*, defines the actual failure of the system, which necessarily leads to reconstruction (Fig. 8.6). It is assumed that the authorities will not make an intervention unless the system's condition falls below k^*. Then, although the operation within the range between y^* and k^* is considered *not acceptable*, the authorities are willing to allow the system to operate under these circumstances.

The objective is to compute the long-run proportion of time (availability) that the system is operated above a threshold value y^* (*acceptable* condition).

In order to compute the availability, we need first to compute the length of cycle. A cycle is defined by the amount of time the system is operating above k^*, i.e.,

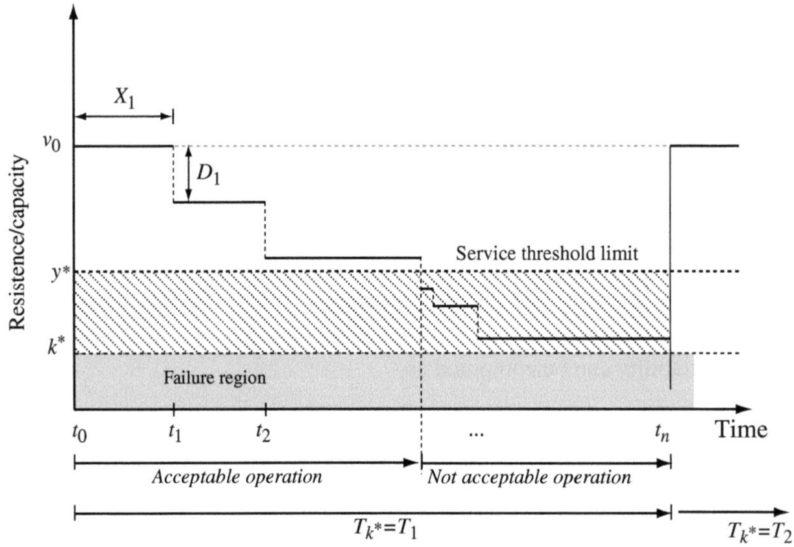

Fig. 8.6 Systematic reconstruction after failure or maintenance

$$T_{k^*} = \sum_{i=1}^{N_{k^*}} X_i$$

where $N_{k^*} = \min\{n : \sum_{i=1}^{n} D_i > v_0 - k^*\}$ and the time the bridge is in service above the limit y^* is

$$T_{y^*} = \sum_{i=1}^{N_{y^*}} X_i$$

where $N_{y^*} = \min\{n : \sum_{i=1}^{n} D_i > v_0 - y^*\}$. The expected values of T_{k^*} and T_{y^*} are computed as

$$\mathbb{E}\left[\sum_{i=1}^{N_{k^*}} X_i\right] = \mathbb{E}[X]\mathbb{E}[N_{k^*}] \quad \text{and} \quad \mathbb{E}\left[\sum_{i=1}^{N_{y^*}} X_i\right] = \mathbb{E}[X]\mathbb{E}[N_{y^*}].$$

Therefore, the long-run proportion of time that the system will perform over a limit y^* is computed as

$$A(\infty) = \frac{\mathbb{E}[N_{y^*}]}{\mathbb{E}[N_{k^*}]}$$

If the damage caused by the events is independent and identically distributed random variables with probability distribution G, it can be proven that [12]

$$\mathbb{E}[N_{y^*}] = m_G(v_0 - y^*) + 1 \quad \text{and} \quad \mathbb{E}[N_{k^*}] = m_G(v_0 - k^*) + 1$$

where m_G is the renewal function of G, i.e., $m_G(t) = \sum_{n=1}^{\infty} G_n(t)$. Therefore,

$$A(\infty) = \frac{m_G(v_0 - y^*) + 1}{m_G(v_0 - k^*) + 1}; \quad k^* \leq y^* \leq v_0.$$

8.3.2 Markov Processes

A way of modeling problems in which the system may take only two states (e.g., operation and failure) is by using Markov processes (Fig. 8.7). In this case, the Markov chain model is defined by a 2×2 transition probability matrix $\mathbf{P}$, which, for the case shown in Fig. 8.7, has the following form:

$$\mathbf{P} = \begin{bmatrix} P_{11} & P_{12} \\ P_{21} & P_{22} \end{bmatrix} \tag{8.21}$$

If state 1 indicates operation and state 2 failure, the probability P_{21} indicates the probability that the system will go back from a failure state to an operation state (i.e., reconstruction). Note also that P_{22} is the probability that the system remains in state 2 (failure state in Fig. 8.7). For Markov chains, the probability that the system is in a given state $\mathbf{S} = \{S_1, S_2\}$ (i.e., operation or failure) after n transitions can be computed as (see Chap. 6)

$$\mathbf{p}^n = \mathbf{p}^0\mathbf{P}^n = \mathbf{p}^0 \begin{bmatrix} P_{11} & P_{12} \\ P_{21} & P_{22} \end{bmatrix}^n \tag{8.22}$$

where $\mathbf{p}^0$ is the initial state probability vector and $\mathbf{p}^n$ is the probability vector after n transitions.

Example 8.45 Consider a system as the one shown in Fig. 8.7 with transition probability matrix:

$$\mathbf{P} = \begin{bmatrix} 0.9 & 0.1 \\ 0.75 & 0.25 \end{bmatrix}$$

Compute the long-term probability of being in every system state.

Note that the transition probability matrix implies that $P_f = 0.1$, which is the probability that the system moves from an operation state to a failure state. If the system starts operating at $n = 0$, with initial state probability vector $\mathbf{p}^0 = [1, 0]$, the probability of being in a given state after n transitions is computed using Eq. 8.22. The evolution of state probabilities is shown in Table 8.2. Note that in the long run, the probability of being in an operating state stabilizes to $P_{11} = 0.8824$, while the probability of being in a failure state to $P_{22} = 0.1176$. Note also that $P_{11} = 0.8824$ corresponds to the system availability.

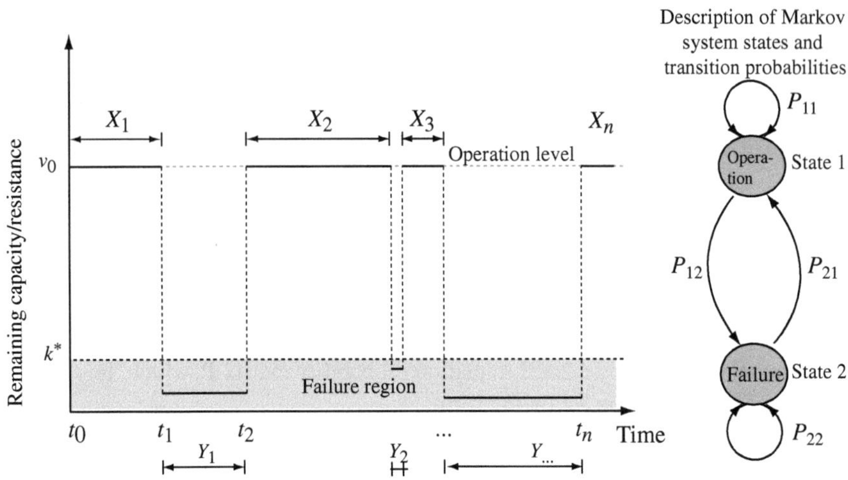

Fig. 8.7 Description of the alternating operation and repair system states

Table 8.2 Evolution of system state probabilities

	Transition - n							
Prob.	1	2	3	4	5	6	7	....
P_{11}	0.9	0.885	0.8828	0.8824	0.8824	0.8824	0.8824	0.8824
P_{22}	0.1	0.115	0.1173	0.1176	0.1176	0.1176	0.1176	0.1176

8.4 Models Including Damage Accumulation

So far, we have described various models for successive reconstruction in which the system condition alternates between operating and failure states. However, in practice the transition from a satisfactory operating condition to a failure state is not instantaneous but defined by the degradation process (see Chap. 4). In Sects. 8.2 and 8.3, the main interest was on obtaining the functions $f_1^*(\theta)$ and $f^*(\theta)$ (Eqs. 8.9 and 8.10); and $g_1^*(\theta)$ and $g^*(\theta)$ (Eqs. 8.12 and 8.14). For the case of systems that degrade, the methods to compute these functions were presented in Chaps. 5–7. In this section, we will discuss the renewal properties of systems for which damage accumulates with time (see Figs. 8.8 and 8.9) and on presenting a general formulation for the problem.

Consider a system that is systematically reconstructed and let us define a random variable, Z_i, as the time of the ith structural replacement (end of cycle i) with $Z_0 := 0$ (Figs. 8.8 and 8.9). Then, the system's failure probability at time t is then computed as

$$P_f(t) = P(V(t) < k^*) \cdot \mathbf{1}_{\{Z_i \leq t < Z_{i+1}\}}, \quad i = 0, 1, 2, \ldots \tag{8.23}$$

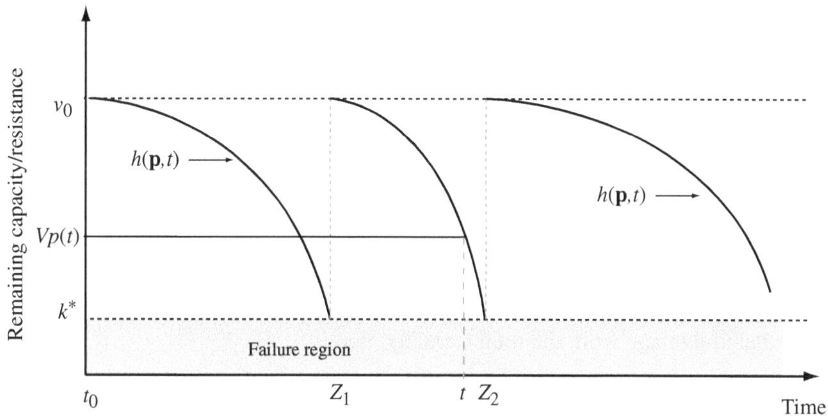

Fig. 8.8 Successively reconstructed process of a system subjected to progressive deterioration

where $V(t)$ is the state of the system at time t, Z_i with $i = 0, 1, 2, \ldots$ indicates the cycle the system is in at time of evaluation, and $\mathbf{1}_{\{Z_i \leq t < Z_{i+1}\}}$ is an indicator function. For progressive degradation, this evaluation is straightforward; however, for the case of systems that degrade as a result of shocks (see Fig. 8.9), some special considerations are needed. In what follows, we will focus on the later.

Then, let us assume that the shock inter-arrival times constitute a sequence of nonnegative independent random variables X_i with $i = 1, 2, \ldots$ and common distribution $F(t)$. Furthermore, assume that damage accumulates as a result of successive *iid* random shocks Y_i, with $i = 1, 2, \ldots$ and distribution $G(y)$. If no intervention takes place in the time interval $[0, t]$, the accumulated damage at time t is given by $D(t) = \sum_{i=1}^{N(t)} Y_i$, where $N(t)$ accounts for the number of shocks by time t. Then,

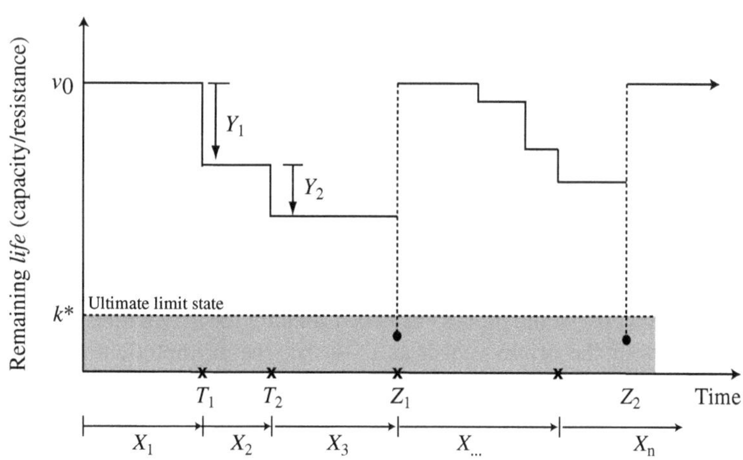

Fig. 8.9 Successively reconstructed process of system subjected to shock-based deterioration

the deterioration at time t, expressed in terms of the cycle the system is in, can be computed as

$$Q(t) = \sum_{j=1}^{N(t)} Y_j - \left\{ \sum_{j=1}^{N(Z_i)} Y_j \right\} \cdot \mathbf{1}_{\{Z_i \leq t < Z_{i+1}\}}, \quad i = 0, 1, 2, \ldots, \tag{8.24}$$

where the term $N(Z_i)$ is the number of shocks that have occurred to the end of cycle Z_i.

Consider now that at the beginning of cycle i with $i \geq 2$, the capacity is reset to a random value v_{i-1}, which may or may not be different from the initial state at $t = 0$ (i.e., v_0). Therefore, the capacity at time t is computed by subtracting the accumulated damage from the total capacity, that is,

$$V(t) = \sum_{j=0}^{\infty} v_j \cdot \mathbf{1}_{\{Z_j \leq t\}} - Q(t) \tag{8.25}$$

Let us now define $\{L(t), t \geq 0\}$ as the counting process of interventions, i.e., $L(t)$ is the number of interventions by time t with $L(0) = 0$. Then, the *instantaneous* intervention (*intensity*) can be written in infinitesimal terms as

$$\nu(t) := \mathbb{E}[dL(t)|\mathscr{L}_{t^-}] = P(dL(t) = 1|\mathscr{L}_{t^-}) = \lambda(t)\left(\int_{V(t,k^*)}^{\infty} dG(y)\right) \tag{8.26}$$

where $\mathscr{L}_{t^-}$ denotes the history of interventions and shock processes; $V(t, k^*) = V(t) - k^*$; f is the density of shock occurrences; and $\lambda(t)$ is the *intensity* or *hazard rate* of the shock process (see Chap. 5) within a cycle Z_i defined as

$$\lambda(t) = \sum_{n \geq 0} \frac{f(t - T_n)}{1 - \int_0^{t-T_n} f(x)dx} \cdot \mathbf{1}_{\{T_n < t \leq T_{n+1}\}} \tag{8.27}$$

where $\mathbf{1}_{\{T_n < t \leq T_{n+1}\}}$ is an indicator random variable. This indicator function is equal to 1 if the time t is between shocks n and $n+1$; and 0 otherwise [13, 14].

Because this section deals with systems that regenerate, the main interest is on estimating the expected number of failures in a infinite time horizon (successive reconstruction) or in a finite time T. The only difference with the cases presented in Sect. 8.2 is the way in which the failure probability is computed and the form of the density of the time to failure.

If a structure is systematically reconstructed (after failure or intervention), its performance with time can be modeled as a renewal process. In this case, the cycle within which the structure is at the time of evaluation becomes important in the assessment. However, if the process has been running for a long time and assuming that the effects of the origin vanish as $t \to \infty$, the asymptotic solution for the instantaneous failure probability of systems subject to shocks (see Chap. 5) can be expressed as [14]

$$\lim_{t\to 0}\frac{1}{\mathbb{E}[L]}\left(\sum_{n=0}^{\infty}\left(\int_{V(t,k^*,n)}^{\infty}\lambda(t)dG(y)\right)P(N(t)=n)\right) \tag{8.28}$$

where $\mathbb{E}[L]$ is the expected value of the length of a cycle. The length of one cycle is the expected time between interventions given that repair or reconstruction times are not significant with respect to the total life cycle. Note that in this case the delayed and ordinary processes converge asymptotically although the transient behavior is different.

8.5 Simulation of Systems Performance Over Time

For most type of problems presented in this chapter, an analytical solution cannot be found. Under these circumstances, numerical methods may be of great help; in particular, Monte Carlo simulations can be used to find quantities of particular interest such as the average number of renewals in a finite time T. Then, for the case of systems that deteriorate as a result of shocks only, a numerical solution, using Monte Carlo simulations, is presented in Algorithm 4. The basic assumption of the model is that shock sizes and shock occurrences are independent.

Algorithm 4 Monte Carlo simulation to compute the performance of systematically reconstructed systems subject to shocks only.

Require: T {Time window for the analysis}
  F {Probability distribution of shock times}
  G {Probability distribution of shock sizes}
  v_0 {Performance condition at time $t = 0$}
  k^* {Minimum performance condition}
  N {number of simulations}
for $i = 1 : N$ **do**
  $cont = 0$ {keeps track of the number of failures before T}
  $t = 0$;
  $s = 0$;
  Generate a random value of the shock time, $\widehat{t_r}$, from F;
  $t = t + \widehat{t_r}$;
  while $t \leq T$ **do**
    Generate a random value of the shock size, $\widehat{s_r}$, from G;
    $s = s + \widehat{s_r}$;
    if $s \geq (v_0 - k^*)$ **then**
      $cont = cont + 1$;
      $s = 0$;
    end if
    Generate a random value of the shock time, $\widehat{t_r}$, from F;
    $t = t + \widehat{t_r}$;
  end while
  $EF(i) = cont$;
end for;
Compute Expected number of failures in time T as $\frac{1}{N}\sum_{i=1}^{N} EF(i)$.

In addition, the Algorithm 5 uses simulation to manage the case of degradation that result from the combined effect of both shocks and deterministic progressive deterioration. Again, the main assumption is that shock sizes and shock times are independent, and also that shock sizes are independent of the progressive deterioration function $h(t)$.

In the Algorithm 5, a few aspects should be clarified. First, note that failures do not necessarily occur at shock times but may occur at any time. Thus, if the system is in a state $V(t_i)$ at time t_i, it will fail at time t_a if $V(t_i) - h(t_a - t_i) = k^*$, and solving for t_a

$$t_a = h^{-1}(V(t_i) - k^*) + t_i. \tag{8.29}$$

Algorithm 5 Monte Carlo simulation to compute the performance of systematically reconstructed systems subject to shocks and deterministic progressive deterioration.

Require: T {Time window for the analysis}
 F {Probability distribution of shock times}
 G {Probability distribution of shock sizes}
 h {Deterministic progressive deterioration function. Note that $h(t = 0) = 0$.}
 v_0 {Performance condition at time $t = 0$}
 k^* {Minimum performance condition}
 N {number of simulations}
for $i = 1 : N$ **do**
 $cont = 0$ {keeps track of the number of failures/repairs before T}
 $t = 0$;
 $t_q = 0$;{keeps track of the time that the system has spend in the cycle.}
 $s = 0$; {keeps track of the system state $V(t)$}
 while $t \leq T$ **do**
  Generate a random value of the shock time, $\widehat{t_r}$, from F;
  $t_a = h^{-1}(v_0 - s - k^*) + t$ {time at which the system would fail due to progressive deterioration}
  if $min(\widehat{t_r}, t_a) = t_a$ **then**
   $cont = cont + 1$;
   $s = 0$;
   $t = t + t_a$;
   $t_q = 0$;
  else
   Generate a random value of the shock size, $\widehat{s_r}$, from G;
   $s = s + (h(t_q + \hat{t_r}) - h(t_q)) + \widehat{s_r}$;
   $t = t + \widehat{t_r}$;
   $t_q = t_q + \widehat{t_r}$;
   if $s \geq (v_0 - k^*)$ **then**
    $cont = cont + 1$;
    $s = 0$;
    $t_q = 0$;
   end if
  end if
 end while
 $EF(i) = cont$;
end for;
Compute Expected number of failures in time T as $\frac{1}{N}\sum_{i=1}^{N} EF(i)$.

In the algorithm, this condition is checked after every shock. It is important to stress that the function h has to be positive and monotonically increasing. The algorithm keeps track of total time through variable t, and of the time in a cycle via the variable t_q. The algorithm assumes that the deterioration in a cycle keeps the same trend throughout until there is failure, i.e., the trend does not change after a shock. Finally, note that the Algorithm 5 can be used to evaluate the case of progressive degradation only by making the shock sizes equal to 0.

Example 8.46 Using Algorithms 4 and 5, compute the expected number of interventions that might be required for

1. a system subject to deterioration as a result of shocks only, where both inter-arrival times and shock sizes are distributed exponentially with rates $\lambda = 0.35$ and $\eta = 0.1$, respectively; and
2. a system subject to both shocks (with the same parameters as in case 1) and progressive deterioration function $h(t) = (t/\alpha T)^b(v_0 - k^*)$ with $b = 1.75$ and T the time window. Consider the following cases: $\alpha = 1$, $\alpha = 1/2$, and $\alpha = 1/4$.

The analysis should be carried out for various time windows that vary from $T = 0$ to $T = 100$ years. Finally, take the minimum acceptable performance level of the system as $k^* = 25$.

The results for both cases are shown in Fig. 8.10. Every expected value in the figure was computed using 10,000 simulations. It can first be observed that as the time window increases the expected number of interventions also increases. With the exception of the first 30 years approximately, this growth is almost linear. The results also show that when progressive deterioration is included in the model,

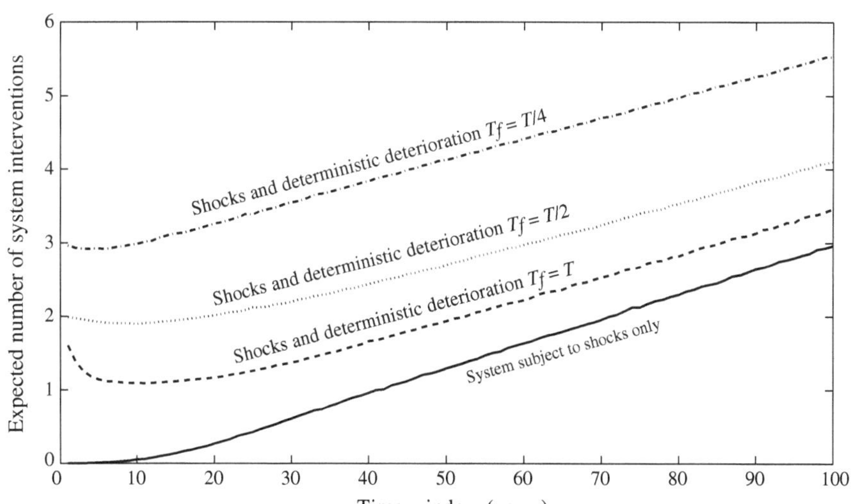

Fig. 8.10 Expected number of renewals (interventions) for different time windows and various progressive deterministic deterioration functions

the expected number of interventions becomes larger than when it is not. Then as the deterministic time to failure becomes smaller, the expected number of interventions becomes larger.

8.6 Summary and Conclusions

The chapter presents several models to deal with cases in which the system is systematically reconstructed after failure or as a result of any other intervention (e.g., maintenance). The results obtained are important for life-cycle analysis since they allow understanding the system behavior over time (see Chap. 9). In essence, the objective of regenerative models is to compute the expected number of failures for a finite or infinite time horizon. Due to the complexity of most models and the difficulty in finding a close form for which there is an explicit solution, at the end of the chapter various algorithms that use Monte Carlo simulations were presented as an alternative to the analytical complexity.

References

1. E. Rosemblueth, E. Mendoza, Optimization in isostatic structures. J. Eng. Mech., ASCE (EM6), 1625–1642 (1971)
2. E. Rosemblueth, Optimum design for infrequent disturbances. Struct. Div., ASCE **102**(ST9), 1807–1825 (1976)
3. A.M. Hasofer, Design for infrequent overloads. Earthq. Eng. Struct. Dyn. **2**(4), 387–388 (1974)
4. R. Rackwitz, Optimization—the basis of code making and reliability verification. Struct. Saf. **22**(1), 27–60 (2000)
5. R. Rackwitz, Optimization and risk acceptability based on the life quality index. Struct. Saf. **24**, 297–331 (2002)
6. R. Rackwitz, A. Lenz, M. Faber, Sustainable civil engineering infrastructure by optimization. Struct. Saf. **27**(3), 187–285 (2004)
7. M. Sánchez-Silva, R. Rackwitz, Implications of the high quality index in the design of optimum structures to withstand earthquakes. J. Struct., ASCE **130**(6), 969–977 (2004)
8. R. Rackwitz, A. Lentz, M.H. Faber, Socio-economically sustainable civil engineering infrastructures by optimization. Struct. Saf. **27**, 187–229 (2005)
9. R. Rackwitz, The effect of discounting, different mortality reduction schemes and predictive cohort life tables on risk acceptability criteria. Reliab. Eng. Syst. Saf. **91**, 469–484 (2006)
10. R. Rackwitz, A. Joanni, Risk acceptance and maintenance optimization of aging civil engineering infrastructures. Struct. Saf. **31**, 251–259 (2009)
11. S. Ross, *Introduction of Probability Models* (Academic Press, San Diego, 2007)
12. S.M. Ross, *Stochastic Processes*, 2nd edn. (Wiley, New York, 1996)
13. M. Sánchez-Silva, G.-A. Klutke, D. Rosowsky, Life-cycle performance of structures subject to multiple deterioration mechanisms. Struct. Saf. **33**(3), 206–217 (2011)
14. M. Sánchez-Silva, G.-A. Klutke, D. Rosowsky, Optimization of the design of infrastructure components subject to progressive deterioration and extreme loads. Struct. Infrastruct. Syst. **8**(7), 655–667 (2012)

Chapter 9
Life-Cycle Cost Modeling and Optimization

9.1 Introduction

The purpose of the previous chapters was to provide tools that can be used to predict the future performance of engineering systems. This is important since the economic and functional feasibility of large engineering projects depends mostly on their operation and management through time. In this chapter, we discuss the concept of life-cycle analysis, a modern project evaluation paradigm for assessing the impacts (e.g., environmental, economic) of a product (e.g., engineering project) or service from "cradle to grave." Up to Chap. 8 we focused on existing mathematical models to describe system degradation and the alternatives to derive lifetime distributions. In this and the following chapters, we will use these models within the context of life-cycle analysis. In the first part of the chapter, we discuss in some detail the problem of life-cycle analysis and describe all aspects involved in the evaluation. In the second part, we focus on the problem of defining optimum design parameters for systems with long lifetimes. Some of the concepts developed in this chapter will be used also in Chap. 10 to define maintenance strategies.

9.2 Definition and General Aspects

9.2.1 Importance of Life-Cycle Analysis

As mentioned before, life-cycle analysis (LCA) is a project evaluation strategy directed to assess the environmental and/or economic impacts of a product or service throughout its lifetime. For the particular case of large infrastructure projects the system life cycle includes extraction of raw materials; processing, manufacturing, and fabrication (construction); use and operation; and disposal or recovery after its useful life. By taking a broader view of what an engineering project is, LCA goes beyond

M. Sánchez-Silva and G.-A. Klutke, *Reliability and Life-Cycle Analysis of Deteriorating Systems*, Springer Series in Reliability Engineering,
DOI 10.1007/978-3-319-20946-3_9

the traditional idea that the central element in design is the physical (mechanical) behavior of the system (e.g., structure). This means that financial factors (e.g., cost of future investments, discount rates, etc.), inter-generational responsibility, environmental aspects and sustainability, among others, become relevant elements in the analysis and the definition of the project characteristics.

There are three forces driving the evolution and use of LCA during the last decade: first, government regulations all over the world are moving in the direction of life-cycle "accountability;" second, businesses of all sorts have recognized that LCA is key to fostering efficiency and continuous improvement; and third, continuous and long-term environmental protection has emerged as a criterion in both consumer markets and government procurement guidelines [1]. Thus, LCA has emerged as a valuable decision-support tool for both policy makers and industry in assessing the lifetime impacts of a product or process. It has also played an important role in defining environmental policies and strategies that contribute to sustainable development. In practice, LCA has been extensively used to assess the environmental impact of large projects, which includes estimating the effects on global climate change, natural resource depletion, ozone depletion, acidification, eutrophication, human health, and ecotoxicity [2, 3]. From the traditional infrastructure engineering perspective, LCA has been used mainly to obtain design parameters and to define maintenance strategies. Therefore, there is still a need for large engineering projects, especially civil infrastructure, to better integrate with their context and to participate more actively in sustainability development.

9.2.2 Definition of Basic Terms

The idea of *life-cycle analysis* has been used in many different contexts, which include, among others, social sciences, health, environmental impact and protection, biology and engineering. Although the basic idea of LCA is similar in all fields, the discussion and definitions presented in this section will focus on problems related to infrastructure systems.

The *life* (or *lifetime*) of a project is the time horizon during which it operates as planned (see Chap. 4); note that it can be finite or infinite. In many practical applications, the term *life* describes also the time span for which the system is planned or designed; this is also called *mission time*. The *life-cycle* is a term commonly used to describe the time span between the conception and the decommissioning of the project; however, it is a term used loosely, for example, to specify some time window that somewhat characterize the project performance.

The *life-cycle analysis* can be broadly defined as:

> a tool to evaluate the performance of a project throughout its *lifetime* in terms of some *utility* measure.

An *utility* measure that is commonly used in infrastructure projects is the economic worth (or equivalently, cost). However, recently, other measures are gaining attention. For example, carbon dioxide (CO_2) emissions and other measures to evaluate the environmental footprint and sustainability are becoming important and have started to be included in government regulations for the development of large infrastructure projects [4, 5].

If the analysis is restricted to a monetary evaluation, the total costs which the owner (or user) will incur, during its lifetime, to keep the system operating is referred to as the *life-cycle cost*. The US National Institute of Standards and Technology (NIST) Handbook 135 [6], defines *life-cycle cost* as

> "the total discounted dollar cost of owning, operating, maintaining, and disposing of a building or a building system over a period of time."

Therefore, a Life-cycle cost analysis (LCCA) can be defined as [7]:

> ".. an economic assessment of competing design alternatives, considering all costs of ownership over the economic life of each alternative, expressed in equivalent dollars."

Then, in essence LCCA can be seen as an economic alternative for project evaluation [6] and to support long-term cost-based decisions [8].

Additional definitions of life-cycle cost analysis in various contexts include: "the total cost to the owner of acquisition and ownership of a system over its useful life" (ACQuipedia.com); the "sum of all recurring and one-time (non-recurring) costs over the full life span or a specified period of a good, service, structure, or system. It includes purchase price, installation cost, operating costs, maintenance and upgrade costs, and remaining (residual or salvage) value at the end of ownership or its useful life." (Business Dictionary.com); and the "total cost throughout its life including planning, design, acquisition and support costs and any other costs directly attributable to owning or using the asset" [9]. For more references on LCCA see also the RMS Guidebook [10] for a life-cycle cost summary; the Reliability and Maintainability Guideline for Manufacturing Machinery and Equipment [11] optimum maintenance strategies; the Total Asset Management: Life Cycle Costing Guideline report prepared by the New South Wales Treasury [9]; the Infrastructure Planning Handbook [12]; and the life-cycle costing for design professionals [13].

9.2.3 Complexity of LCCA

The complexity of LCCA goes beyond the mathematical models used to describe the system performance over time (see Chaps. 4–9). It requires understanding the relationship of those models with the context. In Fig. 9.1 we show, in particular, the relationship between the stages (processes) of a project development, the actors that participate and the "mechanical"[1] performance of the system.

Note first that the execution and operation of a project consists of a set of processes (activities or tasks), that extend from the conceptual design to the decommissioning. Processes are related and executed by different actors, whose relationships and

---

[1] By a "mechanical" we mean a problem that can be fully described by physical laws.

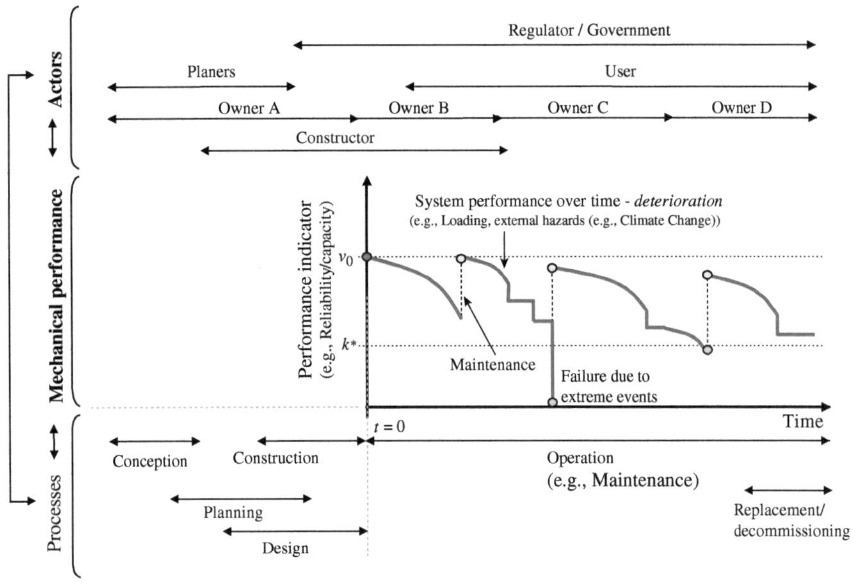

Fig. 9.1 Integration of deterioration and operation aspects within the different stages of an infrastructure project

interests govern operational decisions. In addition to the complexity of the interaction among different actors, all decisions are inevitably conditioned by the system's physical performance. Thus, they are strongly related with the design assumptions, the material properties, the operational constraints, and the relationship with the environment. It is important to stress that decisions about the operation cannot be related to the system's physical state alone since they involve the complexities of the interactions among different actors at different points in time (Fig. 9.1). This interaction of processes, actors, and the system's performance through time is at the heart of life-cycle models; important research developments on this subject can be found in [14–16].

9.2.4 LCCA and Sustainability

Large engineered systems with long life cycles (e.g., dams, large bridges, roadways) usually have an impact on the long-term socioeconomic development of a country. In these cases, the concept of *sustainability* becomes relevant and should be included as part of LCCA. Sustainability is a term that has been discussed in many different contexts and across many disciplines (e.g., economics, biology, engineering, social sciences). *Sustainable development* refers to the continued socioeconomic growth by the rational use of natural resources and the appropriate management of the environment. A widely accepted definition is given by the Brundtland Commission (1987) [17]:

> "Meeting the needs of the present without compromising the ability of future generations to meet their needs."

Note that based on this definition, sustainability is not in itself a fixed goal, but rather a continuous and long-term commitment. For the particular case of large physical infrastructure, LCCA is consistent with the Agenda 21 for Sustainable Construction in Developing Countries (CIB and UNEP-IETC, 2002), where sustainable construction is defined as:

> "... a holistic process aiming to restore and maintain harmony between the natural and built environments, and create settlements that affirm human dignity and encourage economic equity."

Against this backdrop, sustainable development should seek to provide people with opportunities for an acceptable quality of life by protecting the physical environment and its resources. It has been argued in various forums that current concerns about sustainability have become a new ethical standard related to intergenerational equity with implications for the development of civil infrastructure [18]. These implications have to do with aspects, such as pollution control, rational use of resources, and financial feasibility of engineering projects.

The importance of the relationship between LCCA and sustainability is an aspect that should be addressed in practice for major and long lasting projects. However, despite the importance of sustainability, we will discuss it marginally when addressing the problem of discounting; other aspects are beyond the scope of this book.

9.2.5 LCCA and Decision Making

In engineering, LCCA can be used for different purposes, among which the following are of special interest:

- as a criteria for comparing various project (system) investment alternatives;
- as a tool for establishing optimum management policies; and
- as a criteria for defining consistent and cost-effective design and operation parameters.

In summary, a LCCA is a cost-based evaluating strategy with the objective of selecting the design and management requirements that lead to the lowest cost of ownership (i.e., construction and operation); which is in turn consistent with the system quality and function specifications. Clearly, LCCA is a tool that is intended to lead to better investment decisions in the long term. This means that decisions should balance the economic investment, the benefits derived from the existence of the project and the consequences of poor performance or failure [19].

As a decision-making tool (see Chap. 1), LCCA should take into consideration the following aspects:

1. decisions about the system's performance and the associated costs (e.g., cost of interventions) are based on predictions with some degree of uncertainty;
2. decisions are influenced by the time-dependent variability in financial and economic parameters;
3. decisions should be made based on a cost and asset management policy and not simply on a mechanical performance model of the system;
4. decisions should be made taken into account the social, economic and political context.

9.3 Life-Cycle Cost Formulation

The life-cycle cost analysis integrates the benefits derived from the existence of the system with the costs associated to the process of construction, operation (i.e., inspection and maintenance), and decommissioning (i.e., removing the system from service). This relationship is illustrated in Fig. 9.2 and can be expressed as follows:

$$Z(\mathbf{p}, t_s) = B(\mathbf{p}, t_s) - C_O(\mathbf{p}) - C_L(\mathbf{p}, t_s) - C_D(t_s) \tag{9.1}$$

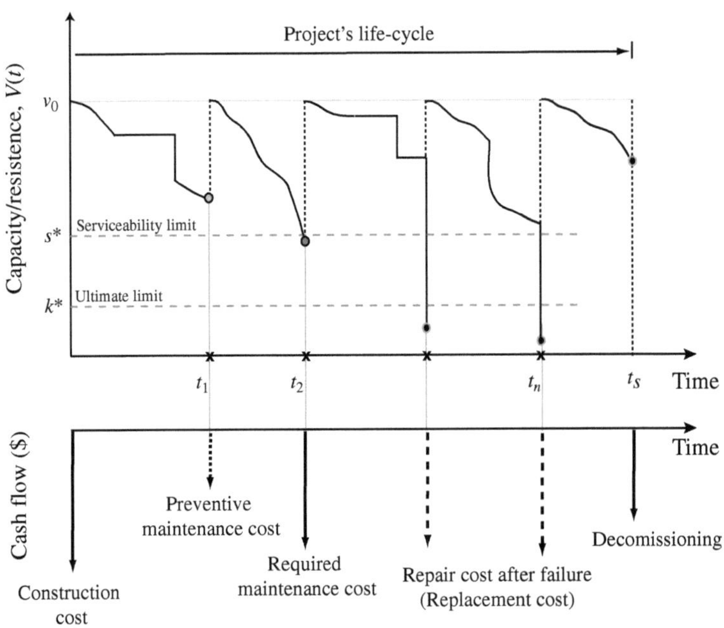

Fig. 9.2 Description of the life-cycle cost of a system

where t_s is the system lifetime (which might be finite or infinite); $\mathbf{p}$ is a vector of parameters that describes the decision variables, which include aspects, such as the design criteria; e.g., v_0 in Fig. 9.2 (e.g., geometry, material properties, external demands) and inspection and maintenance schedules; $B(\mathbf{p}, t_s)$ is the benefit expected from the investment and operation, $C_0(\mathbf{p})$ is the cost of planning, designing, and constructing the project, $C_L(\mathbf{p}, t_s)$ are all additional costs required for the system to operate as required; these may include:

- inspection and maintenance;
- insurance coverage;
- quality assurance measurements;
- financial costs (e.g., finance charges such as loan interest payments);
- loss of business opportunity;
- direct and indirect losses in case of failure;
- loss of life.

Finally, $C_D(t_s)$ describes the cost of decommissioning (when it exists) at the end of the life cycle t_s.

Equation 9.1 can be rewritten in many ways; for instance, by discretizing costs or by extending the problem to multiple hazards (e.g., environmental, earthquakes, hurricanes, climate change) [20, 21]. Closed-form solutions for the optimization (i.e., maximization of the benefit-cost relationship) of Eq. 9.1 can be obtained in a few specific cases; e.g., see [18–20] where solutions are based on strong assumptions about costs and the performance of the system. The main modeling difficulties are due to the fact that the life-cycle performance of the system and the corresponding decisions depend upon the unpredictable combination of the occurrence and magnitude of external events, the system degradation mechanisms, and the decisions about system operation.

9.4 Financial Evaluation and Discounting

9.4.1 LCCA Assessment Criteria

The challenge in LCCA, as in any other economic evaluation method, is to quantify the economic value of a set of possible designs to support the decision-making process. According to statistical decision theory [22], optimum decisions should be based on the expected value of the objective cost function (e.g., Eq. 9.1); i.e.,

$$\mathbb{E}[Z(\mathbf{p}, t_s)] = \mathbb{E}[B(\mathbf{p}, t_s) - C_0(\mathbf{p}) - C_L(\mathbf{p}, t_s) - C_D(t_s)] \tag{9.2}$$

Given that benefits and costs are distributed over a time horizon defined by the life cycle, they should be discounted to a given point in time, usually taken as $t = 0$ (see Chap. 1). This is to have a standard value representation for comparison purposes.

Table 9.1 Net present value for different cash-flow strategies [12]

Discounting equation	Description
Single amount	$P_v = F \cdot \frac{1}{(1+\gamma)^n}$
Uniform flow chart	$P_v = A \cdot \frac{(1+\gamma)^n - 1}{\gamma(1+\gamma)^n}$
Geometric gradient $g \neq \gamma$	$P_v = A_1 \cdot \frac{1-\left[\frac{1+g}{1+\gamma}\right]^n}{\gamma - g}$
Geometric gradient $g = \gamma$	$P_v = A_1 \cdot \frac{n}{1+\gamma}$

γ—discount factor (time-independent);
P_v—Net Present Value;
F—Future value in the n-time unit;
A—Cash-flow equally distributed $\{A, A, \ldots, A\}$;
A_1—Cash-flow distributed as: $\{A_1, A_1(1+g), \ldots, A_1(1+g)^{n-1}\}$;

This approach is called net present value (NPV) evaluation, and it is widely used as tool to choose among various alternatives; as an example, in Table 9.1 we present a set of NPV expressions for various cash-flow structures.

For a project to be feasible, the expected discounted objective function at $t = 0$ must be positive: i.e., $\mathbb{E}[Z(\mathbf{p}, t_s)] \geq 0$; otherwise the owner (or stakeholders) will incur a loss. Thus, the optimal technical solution is the one for which the system's parameters, i.e., $\mathbf{p} = \mathbf{p}_{opt}$ satisfy:

$$\max_{\mathbf{p}} \{\mathbb{E}[Z(\mathbf{p}, t_s)] \geq 0\}. \tag{9.3}$$

The components of the objective function $\mathbb{E}[Z(\mathbf{p}, t_s)]$ (Eq. 9.2), as function of the vector parameter $\mathbf{p}$ are illustrated in Fig. 9.3. Note that since decommissioning costs usually do not depend of $\mathbf{p}$, they are not included in the figure.

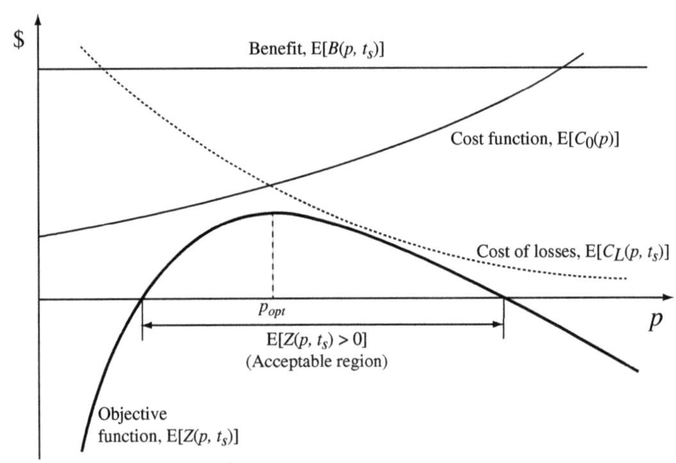

Fig. 9.3 Description of the life-cycle cost objective function

9.4.2 Discounting

Evaluation of the Discount Rate

In order to compute the NPV of future investments, costs, and benefits should be discounted to time $t = 0$. The general form of the discounting function $\delta(t)$ for the first cash-flow model presented in Table 9.1, which is the most widely used model, can be approximated as follows:

$$\delta(t) = \frac{1}{(1+\gamma_1)^t} \approx \exp(-\gamma t) \quad \text{for } \gamma \ll 1 \tag{9.4}$$

where $\gamma \approx \gamma_1$ is called the *discount rate*. Other expressions of the discount function with the corresponding implications, can be found in [23].

For projects in the public interest, the discount rate is frequently associated with the so-called *social discount rate* (SDR). This rate reflects the value that society assigns to its current condition (well-being) compared with possible future states. Some of the main approaches for discounting future benefits and costs will be briefly presented here; a more extensive discussion can be found elsewhere; e.g., [24].

The first and most common approach is the *social rate of time preference* (SRTP), which establishes that there are two main effects that have to be considered when selecting the discount rate:

- pure time consumption; and
- economic growth.

The pure time consumption (also called "utility discount rate") is purely psychological and accounts for the weight that an individual assigns to future utility compared with present utility. In other words, it captures possibly nonrational behavior through which individuals compare present with future experiences. Then, future investments are discounted at rate ρ indicating that there is a preference for current consumption over any future expenditure. On the other hand, the criteria of economic growth accounts for the fact that as access to resources increases with time, the marginal utility of future investments (costs) becomes smaller. This reduction in marginal utility is discounted at rate η.

Then, the discount rate (Eq. 9.4) should combine the effect of both economical growth and pure time preferences; i.e., [25],

$$\gamma = \rho + \theta\eta \tag{9.5}$$

where ρ is the discount rate associated to the pure time preference; η is the annual rate of growth per capita real consumption; and θ is a constant that takes into consideration the elasticity of marginal utility of consumption. Note that the elasticity of a variable is a measure of how much a variable changes (in percentage) in response to a change in a second variable:

$$\theta = \frac{\%\text{ change in variable 1}}{\%\text{ change in variable 2}} \tag{9.6}$$

For instance when evaluating the elasticity of demand in the price of a product, the variable 1 is the quantity demanded and the variable 2 the price of the product. In most engineering projects, $\theta > 1$, which implies that the demand responds more than proportionally to changes in variable 2. Empirical evidence suggests that values of θ vary from 1.5 to 2 % [24]. As an example, for Japan $\gamma = \rho + \theta\eta = 1.5 + 1.3 \cdot 2.3 = 4.5$ [24].

Based on this description, the social discount function can then be computed as:

$$\delta(t) = \exp(-(\rho + \theta\eta)t) = \exp(-\gamma t) \tag{9.7}$$

The second approach to obtain the SDR is to use the *Social Opportunity Cost* of capital (SOC), which is based on the idea that resources are always scarce and both the government and the private sector should compete for the same funds. Under these circumstances, both public and private sectors should have the same return on investment. Then, the SOC is a measure of the marginal earning rate for private business investments.

An intermediate alternative is the *weighted average approach*, which recognizes that rates and funds may come from different sources. Therefore, the rate should be computed as the weighted average of the rates coming from SOC and SRTP; i.e., [26],

$$\gamma = \alpha \cdot \text{SOC} + (1 - \alpha) \cdot \text{SRTP} \tag{9.8}$$

where the weighting factor α defines the proportion of funds from each source. This approach can be extended to include resources that need to be obtained from private or public sectors as well as international markets. In this approach, also known as the Harberger approach, the discount rate can be expressed as [27]:

$$\gamma = \alpha \cdot \text{SOC} + \beta \cdot \text{SRTP} + (1 - \alpha - \beta)r_j \tag{9.9}$$

where r_j is the government long-term foreign borrowing rate. In Eq. 9.9, α is the share of funds for public investment obtained at the expense of private investment; and β is the proportion of funds obtained from current consumption [24]. Clearly the factor $(1 - \alpha - \beta)$ is the percentage of funds that should be obtained from foreign markets. Note that the terms SOC and SRTP are rates.

A detailed and deeper discussion on the methods for selecting discount rates is beyond the scope of this book but an extensive and critical review can be found in [24].

Selection of Discount Rates

The selection of the discount rate is a matter of great debate and there is not a unique way to select a value. In fact, the selection of discount rates in many projects is not justified as part of the design process despite the importance it has on the outcome [28].

For the particular case of infrastructure systems, and given its importance on society, discount rates have been calculated, for instance, as the long-term average of the economic growth per capita [29]. In this case, values vary between 0.9 % (Africa) and 2.5 % (USA and Canada). On the other hand, if discount rates are associated to the financial market, in industrialized countries interest rates may vary between 2 and 8 %, while in moderate and low developed countries interest rates may vary from 8 to 18 % and from 15 to 30 % respectively. The differences observed in financial and social discount rates, in terms of the level of development, reflect also the "differences in the perceived social opportunity cost of public funds across countries and in the extent to which the issue of intergenerational equity is taken into consideration" [24].

When selecting discount rates, it is also important to distinguish between private and public investments. For example, Wen [20] states that for the public sector these rates vary between 4 and 6 % while for the private sector, they vary between 6 and 10 %. Typical values of the discount rate for selected countries are shown in Table 9.2. This table presents also the approach that was used to define them. An additional discussion on this topic can be found in [30].

Other related issues that are beyond the scope of this book but that ought to be considered are the time-dependent variability of the rates, the variation of the discount rate depending upon the nature of the discounted value (e.g., monetary costs versus cost of saving lives), and the differences between financial and economic discounting [29, 31–33]. An interesting discussion on public and private discounting for life-cycle cost analysis can be found in [29, 34]. Finally, some actual data can be found in [24].

9.4.3 Inter- and Intra-generational Discounting

When the consequences of decision affect future generations, a distinction between discounting within the same generation (i.e., intra-generational) and discounting for future generations (i.e., intergenerational) is required. In the first case, the effects of pure time preference rate should be included because today's generation suffers both effects; i.e., they prefer present utility to future utility and they are subject of the diminishing marginal utility effect. However, when discounting utility for different generations only the effect of economical growth has to be considered. If the pure time preference is considered in inter-generational discounting, the preferences of future generation would be less valued in relation to present preferences [35].

Although there is still debate about whether today's societal decisions should take into account the effects on future generations, from the sustainability point of view, a reasonable argument can be made to support this premise. If a societal

Table 9.2 Typical social discount rates for selected countries (taken from [24])

Country	Disc. rate, γ (%)	Observations
Australia	8	1991 (SOC-approach)
Canada	10	(SOC-approach)
China	>8	Short term projects
	<8	Long-term projects
France	8	Before 1985
	4	After 1985
Germany	4	Before 1999
	3	2004
Norway	7	1978
	3.5	1998
Italy	5	(SRTP-approach)
Spain	6	Transportation project (SRTP-approach)
	4	Water-related projects (SRTP-approach)
United Kingdom	8	1967 (SOC-approach)
	10	1969
	5	1978
	6	1989
	<3.5	2003 (Long term) (SOC-approach)
USA	8	Before 1992 (Off. Management & Budget) (SOC-approach)
	7	After 1992 (SRTP-approach)
	0.5–3	EPA-Intergenerational discounting
		(SRTP-approach)
India	12	(SOC-approach)
Pakistan	12	(SOC-approach)
Philippines	15	(SOC-approach)

SOC—Social Opportunity Cost of Capital
SRTP—Social Rate of Time Preference

decision aims to be sustainable, "the influences of our decision have to consider future consequences, looking in a scheme where the altruism of present society is not consider as a constraint of governmental duties" [35].

For an extended discussion of this topic see [28, 35, 36].

9.5 Assessment of Benefits and Costs

As mentioned in Sect. 9.1, for a project to be feasible, it is necessary to compute the expected value of all (current and future) discounted investments. In this section, we will present and discuss the general form of the quantities in Eq. 9.1.

9.5.1 Evaluation of Benefits

Often the benefit function $B(\mathbf{p}, t_s)$ as described in Eq. 9.1 is assumed to be independent of the vector parameter $\mathbf{p}$ and constant over time, thus $B(\mathbf{p}, t_s) = b$. According to Rackwitz [19], it is reasonable to assume $b = \beta C_0$ with $0 < \beta \leq 0.3$, where C_0 is the part of the construction costs (i.e., initial investment) that is independent of $\mathbf{p}$. Under this assumption, the discounted benefits derived from the existence and operation of the project can be computed as:

$$B(t_s) = \int_0^{t_s} b\delta(\tau)d\tau = \int_0^{t_s} b\exp(-\gamma\tau)d\tau = \frac{b}{\gamma}[1 - \exp(-\gamma t_s)] \tag{9.10}$$

for a reference time t_s, which is the length of the life-cycle—i.e., the service lifetime. The asymptotic solution of Eq. 9.10; i.e., $t_s \to \infty$ is

$$B(\infty) = \frac{b}{\gamma} \tag{9.11}$$

Note that the benefit is independent of all other costs and of the "mechanical" performance of the system (i.e., degradation process).

Example 9.47 Consider a system for which the construction cost is $C_0 = \$1000$. Build a table of the benefit for various discount rates and lifetimes.

In large engineering projects, the benefit factor β derived from the construction and operation the project is in the order of $\beta \approx 0.1$. Then, the constant benefit over time is $b = \beta \times C_0 = \$100$. For finite lifetimes, the benefit is computed using Eq. 9.10. The results for various discount rates and lifetimes are presented in Table 9.3.

It can be observed that, as expected, for larger discount rates the benefit becomes smaller. Also, the benefit increases with time but converges to a maximum value

Table 9.3 Benefit value for various discount rates and lifetimes

Discount rate γ	Time window t						b/γ (Eq. 9.11)
	5	10	25	50	100	200	
0.01	487.7	951.6	2212.0	3934.7	6321.2	8646.6	10000.0
0.03	464.3	863.9	1758.8	2589.6	3167.4	3325.1	3333.3
0.05	442.4	786.9	1427.0	1835.8	1986.5	1999.9	2000.0
0.07	421.9	719.2	1180.3	1385.4	1427.3	1428.6	1428.6
0.1	393.5	632.1	917.9	993.3	1000.0	1000.0	1000.0
0.125	371.8	570.8	764.9	798.5	800.0	800.0	800.0
0.15	351.8	517.9	651.0	666.3	666.7	666.7	666.7
0.25	285.4	367.2	399.2	400.0	400.0	400.0	400.0

at large lifetimes. This convergence depends on the time window but also on the discount rate. For example, for a discount rate of 0.05, convergence is reached at 200 years; while for a discount rate of 0.15, the limiting solution is achieved in 50 years.

9.5.2 Intervention Costs

The cost of interventions during the structure's life-cycle, C_L, can be divided into direct and indirect costs. Direct costs are those imputed to the owner; for instance, costs associated with inspection, maintenance and reconstruction after failure. On the other hand, indirect costs are all those imposed on the user; i.e., costs derived from the impossibility to use the system (e.g., a bridge closure). Further details and a discussion on cost-related issues in life-cycle analysis can be found in [37, 38].

Consider, the case of a system subjected to systematic interventions or reconstructions; and let's denote by X_i the time between interventions $i-1$ and i. Then, the time to the mth intervention is (Fig. 9.4),

$$T_m = \sum_{i=1}^{m} X_i \tag{9.12}$$

Furthermore, if the times between interventions, X_i, are *iid* random variables with pdf $F(t) = P(X \leq t)$, the probability distribution of the time to the nth intervention is the nth convolution of F with itself; i.e., $F_n(t)$.

On the other hand, if $C(T_i)$ describes the cost in which the owner incurs in the ith intervention, which occurs at time T_i (Fig. 9.4), the total discounted cost of interventions for an infinite time horizon can be computed as:

$$C_T = \sum_{i=1}^{\infty} C(T_i)e^{-\gamma T_i} \tag{9.13}$$

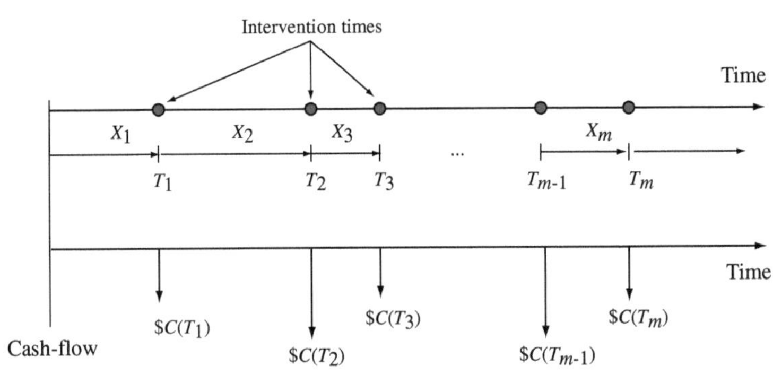

Fig. 9.4 Description of the life-cycle cost of a system

where γ is the discount rate, which is assumed to be constant. If the discount rate is not time-invariant,

$$C_T = \sum_{i=1}^{\infty} C(T_i)e^{-\int_0^{T_i} \gamma(\tau)d\tau} \tag{9.14}$$

Although infrequent, in the case of a continuous cash-flow structure, or any other structure for that matter, appropriate modifications to Eqs. 9.13 and 9.14 must be made. Furthermore, in cases where the cost of inspections is small compared to the cost of interventions, it can be left out of the analysis; however, if it is included, it should be discounted and an expression similar to Eq. 9.13 should be derived. It is important to notice that inspections do not always occur at the same time as interventions. A detailed description about the nature of inspections is presented in Chap. 10.

Finally, if the mth intervention occurs at time T_m with probability distribution $F_m(t)$, the total expected discounted costs needed to evaluate $\mathbb{E}[Z(\mathbf{p}, t_s)]$ can be computed as:

$$\mathbb{E}[C_T(t_s)] = \sum_{m=1}^{\infty} \int_0^{t_s} C(\tau_m)e^{-\gamma\tau_m} dF_m(\tau_m) \tag{9.15}$$

where $dF_m(t)$ is the density of the time where the cost $C(T_m)$ is executed. The details of this calculations were presented in Chap. 8. Note that the upper limit of the integral in Eq. 9.15 can be finite of infinite (i.e., $t_s \to \infty$) depending on the time window selected for the analysis.

Example 9.48 Consider a system that needs to be reconstructed over time at a fixed cost of \$100 for each intervention. Compare the long term (i.e., $t_s \to \infty$) total discounted cost for three deterministic and three random intervention policies. Interventions are carried out at fixed time intervals: $T_1 = 5$ (case 1), $T_2 = 10$ (case 2), and $T_3 = 25$ (case 3) years; while the random intervention policies assume times between events to be exponentially distributed with rates $\lambda_1 = 0.2$ (case 4), $\lambda_2 = 0.1$ (case 5), and $\lambda_3 = 0.04$ (case 6).

In order to compare various intervention policies for several discount rates, Monte Carlo simulation was used to compute the total cost for every case considered. The values reported in the table correspond to mean values. Note that every case of the deterministic policies corresponds, on average, to a random case; for example, in Case 1, there is one event every 5 years, while in case 4 there is one event every 5 years on average. The results show that the models with deterministic intervention times have slightly smaller total costs (Table 9.4).

Table 9.4 Total discounted costs for every policy

Case	Intervention policy	Discount rate γ					
		0.01	0.03	0.05	0.10	0.15	0.25
1	Every 5 y.	1950.4	617.9	352.1	154.1	89.5	40.2
2	Every 10 y.	950.8	285.8	154.1	58.2	28.7	8.9
3	Every 25 y.	352.1	89.5	40.2	8.9	2.4	0.2
4	$\lambda = 0.2$	1997.9	668.7	396.2	199.1	133.1	79.0
5	$\lambda = 0.1$	994.0	333.5	201.9	100.9	66.3	40.1
6	$\lambda = 0.04$	400.6	134.3	80.5	40.2	26.4	13.3

9.5.3 End of Service Life Considerations

At the end of the service life, the owner (or stakeholders) is presented with various options, which typically involve either major upgrading (i.e., extending the service life) or demolition. Service life extensions may include rehabilitation of the structure to extend the use for its initial purpose, or may enable an extended structural life with a change in purpose. An example of such service life extension is the common rehabilitation and conversion of industrial or commercial space for residences, typical of modern urban regeneration projects. Modeling and understanding lifetime extensions of large infrastructure is still a topic for which there is a need for further research. If, on the other hand, the LCCA does not consider extensions of the system's lifetime after it has accomplished its time mission, the expected discounted decommissioning costs can be computed as:

$$\mathbb{E}[C_D(t_s)] = \int_0^{t_s} C_D(\tau)e^{-\gamma\tau} dF_D(\tau) \tag{9.16}$$

where $dF_D(t)$ is the density of the time to decommissioning. Note that the existence of decommissioning implies that the time horizon for the analysis is finite. If the system is upgraded, instead of demolished, the system can be treated as systematically reconstructed (see Chap. 8).

End-of-life decisions are an important part of infrastructure management; however, their contribution relative to other life-cycle phases (see Fig. 9.1) vary greatly on a case-by-case basis depending upon the system of interest and scope of analysis [39, 40]. However, their consideration in a life-cycle analysis is essential for completeness and informed decision making.

On a final note, it is important to mention that recent research (e.g., see [2, 4, 41]) has also shown that, for large infrastructure systems, the environmental impact of decommissioning may significantly influence the initial design decisions and the selection of materials. Thus, if a structure is deconstructed and demolished, the end-of-life stage entails decisions regarding waste generation and management, as well as recovery and recycle or reuse of the structure's contents, components, and material constituents [42–44].

9.6 Cost of Loss of Human Lives

In traditional engineering design problems, and in LCCA in particular, including the potential loss of human lives in the analysis is a difficult and controversial subject; however, the problem of accounting for these losses is an issue that cannot be avoided. In this section, we discuss some existing approaches to model the loss of lives and their implementation within LCCA.

9.6.1 Approaches to the Problem of Life Loss Evaluation

The failure of large engineering systems, may frequently involve risk to human life and limb. Over the last decades, the question of risk to human live has moved from making monetary estimations of the value of the human losses to finding ways of assessing the cost of saving lives; i.e., the cost to reduce the risk to life. Although there is still a great deal of debate over this topic, recently, the work in many different disciplines, such as economics, social sciences, health-related sciences, engineering, etc., has moved in similar directions.

Before approaching the problem it is necessary to establish the socioeconomic context within which the evaluation of human losses is carried out. According to Rackwitz [18, 29], this discussion can only take place within the context of

> "our moral and ethical principles as laid down in our constitutions and elsewhere including everyone's right to live, the right of a free development of her/his personality and the democratic equality principle."

This means that the approach to the cost of saving lives can only be formulated for involuntary risks [29], which are those to which an anonymous member of society is exposed. In other words, it cannot be used to economically assess the life of a particular individual; it can only be used as a criteria for decisions in the public interest (e.g., public policies for risk reduction). Within this context, the standard approach to placing a monetary value on the life-saving benefits of regulations is frequently referred to as the Societal Willingness to Pay (SWTP) for mortality risk reductions [45–49].

Within this context, there are two basic approaches for estimating the future costs associated to possible life-losses that have been used in practice:

1. Cost of saving lives and
2. Cost of saving life-years.

In problems that involve the possibility of instantaneous death (e.g., building collapse, traffic accidents) the analysis is often carried out using the concept of *lives-saved*. On the other hand, in problems where preventive measures may have a long-term impact on the life of an individual, the concept of *life years* saved has been the metric preferred; this application is of common use in areas of public health including medicine, vaccination, and disease screening [50].

The cost associated to saving lives is commonly evaluated by using the Value of Statistical Life (VSL), while the cost of saving life-years uses the value per statistical life-year (VSLY). Clearly, neither of them is constant over an individual's life and vary with age, health, socioeconomic standards, wealth, gender, and other factors; overall, an accurate evaluation requires using values that depend on characteristics of the affected individuals.

The discussion in the following will focus mainly on the cost of saving lives given the nature and type of consequences of most large engineering systems (i.e., future casualties as a result of failures). However, it is important to keep in mind that the approach of cost of saving lives is still a matter of great debate. This discussion is beyond the scope of this book but some interesting reflexions can be found in [28, 30, 35, 50–52].

9.6.2 The Cost of Saving Lives Within LCCA

The costs associated to the loss of lives can be included in LCCA in two ways. The direct alternative consists on estimating the potential number of casualties and assigning them a value, usually based on the VSL. It can be interpreted as the value assigned for compensation to the relatives of the victims in case of an event [18]. This value can be entered in Eq. 9.1 as part of the cost of losses C_L. This approach, however, has many criticisms, specially because it has the connotation that it is a way to assign value to life. The second approach is to use the life quality index (see subsections below) as a criterion to define a threshold that separates efficient from inefficient life saving investments [53]. In this case, the cost of saving lives is included as a restriction in the analysis and not as a direct cost [18, 54, 55].

In the following subsections we will present a discussion on the Life Quality Index [18, 56] and its use in LCCA.

LQI Formulation

The *Life Quality Index* (LQI) is a socioeconomic composite indicator developed by Nathwani et al. [56] as a general principle for supporting decision making concerning activities with an impact on health and life safety in the public domain. The LQI addresses the question of how much society is willing to pay and can afford to reduce the probability of premature death by some intervention changing the behavior of individuals or organizations and/or technology [57]. It is important to stress that the LQI makes sense only for social and administrative units (e.g., country) with common beliefs represented in documents such as a constitution [18, 48]. Thus, imbedded in the nature of the LQI is the idea that it is derived for an anonymous person. The LQI principles have been also discussed and expanded by Rackwitz [18, 29, 30, 54] and others [55, 58, 59].

The original derivation of the LQI can be found in [56] while the derivation from a utility function perspective is presented in [29, 48]. The LQI can be interpreted as a utility function consisting of three main components [29, 48]:

1. life expectancy;
2. consumption (income); and
3. the time necessary to rise the total income.

It has the following general form:

$$L(a) = g^{w} e(a)^{1-w} (1-w)^{1-w} \approx g^{q} e(a) \tag{9.17}$$

where g is the GDP per capita; $e(a)$ is the life expectancy at age a; and w is the fraction of time devoted to rise g. Statistical data for selected countries is presented in Table 9.5. The term $(1-w)^{1-w}$ is constant and can be dropped to get the approximation shown in Eq. 9.17, where the constant $q = w/(1-w)$ is a measure of the trade-off between the resources available for consumption and the value of the time of healthy life [29]. In later developments, Rackwitz [30, 54] suggests the following modification: $q = w/(\beta(1-w))$, where the term β is added to represent the fraction of GDP that is produced through labor and not as return on investments; typical values of β are between 0.6 for developed countries and 0.8 in underdeveloped countries.

Table 9.5 Basic statistics used to evaluate the life quality index (LQI)

Region	g($) [60]*	w [61]	q [62]
Australia	36,570	0.182	0.318
Brazil	10,214	0.193	0.342
Canada	35,241	0.179	0.311
China	6,714	0.232	0.432
Colombia	9,592	0.204	0.366
Dem. Republic of Congo	398	0.195	0.346
France	29,661	0.162	0.276
Germany	33,423	0.150	0.253
Japan	30,579	0.187	0.329
Mali	1,099	0.195	0.346
Mexico	12,991	0.202	0.361
Mozambique	1,083	0.195	0.346
Sierra Leone	844	0.195	0.346
South Africa	9,469	0.195	0.346
United Kingdom	32,449	0.173	0.299
United States	41,976	0.183	0.320
World (World Life Table)[61]	9,042	0.160	0.318

*2010-g-GDP per capita (2005 PPP USD);
OECD and IMF statistics available online

Societal Willingness to Pay

The so-called *Willingness to Pay* (WTP) is a concept widely used in economics as measure of what an individual is willing to exchange for a particular object or goal. This term can be extended to a societal level expressing what a society is willing to invest, for instance, in risk reduction. If this is the case, a change in life expectancy and the corresponding investment should balance each other to keep the LQI constant. Thus, by differentiating $L(a)$ (Eq. 9.17) [30],

$$dL(a) = \frac{dL(a)}{de(a)}de(a) + \frac{dL(a)}{dg}dg \geq 0 \tag{9.18}$$

then,

$$dL(a) = g^q de(a) + qg^{q-1}e(a)dg = 0 \tag{9.19}$$

Taking the expectation and rearranging the terms in Eq. 9.19, the societal willingness to pay (SWTP) can be expressed as:

$$SWTP = dg = -\mathbb{E}\left[\frac{g}{q}\frac{de(a)}{e(a)}\right] \tag{9.20}$$

Then, by expressing $de(a)/e(a)$ in terms of a small change in mortality, dm, [29, 63],

$$SWTP = dg = -\mathbb{E}\left[\frac{g}{q}\frac{de_d(a)}{e_d(a)}\right] \approx -\frac{g}{q}C_x dm = G_x dm \tag{9.21}$$

where the term $de(a)/e(a)$ in Eq. 9.20 has been replaced by $de_d(a)/e_d(a)$ in Eq. 9.21. The term e_d is the age averaged discounted life expectancy. This discounting follows the same form described in Sect. 9.4.2 and is defined in terms of an intergenerational discounting rate (typical values 4 to 7 %) [29, 30, 54].

Furthermore, in Eq. 9.21 the small change in discounted life expectancy is replaced by a small change in mortality; i.e., $C_x dm = de_d(a)/e_d(a)$. In this case, C_x is a demographical constant for a specific mortality reduction scheme x, which is associated to a safety-related intervention (e.g., maintenance, retrofitting). Then, the constant $G_x = (g/q)C_x$ depends also on the mortality reduction scheme x of a particular intervention.

A mortality reduction regime defines the way in which the intervention affects the survival curve (i.e., survival probability by age) of a society. Typical mortality reduction schemes include:

- proportional to age;
- only at certain age ranges;
- constant at all ages.

A detailed discussion and formulation of various mortality regimes can be found in [54]. The key concept behind the formulation of Eqs. 9.18–9.21 is that they provide

Table 9.6 SWTP for a *unitary* change in mortality proportional over the age a distribution for year 2010

Region	G_{Π} in US$(millions)			
	1 %	2 %	3 %	4 %
Australia	0.942	1.121	1.321	1.625
Brazil	0.259	0.308	0.363	0.463
Canada	1.230	1.464	1.725	2.164
China	0.136	0.161	0.190	0.231
Colombia	0.242	0.288	0.340	0.388
Dem. Republic of Congo	0.009	0.011	0.013	0.016
France	1.112	1.324	1.560	1.910
Germany	1.358	1.616	1.904	2.225
Japan	0.887	1.056	1.244	1.523
Mali	0.028	0.033	0.039	0.050
Mexico	0.331	0.394	0.465	0.577
Mozambique	0.030	0.035	0.042	0.052
Sierra Leone	0.022	0.027	0.031	0.037
South Africa	0.263	0.313	0.369	0.467
United Kingdom	1.175	1.399	1.649	1.993
United States	1.430	1.702	2.006	2.467
World (World Life Table) [61]	0.501	0.422	0.359	0.308

SWTP values are expressed in 2005 PPP US Dollars (millions) for different discount rates

a way to estimate the impact that a marginal investment on a safety measure (i.e., dg) may have on risk reduction (i.e., reduction of mortality, dm) [29]. According to Nathwani et al. [56] the acceptable criteria presented in Eq. 9.18 is "necessary, affordable and efficient from a societal point of view; also, it is inter-generationally equitable."

The SWTP for countries with diverse socioeconomical conditions, and for the world [61], are presented in Tables 9.6 and 9.7. In Table 9.6, the SWTP is computed for a mortality reduction scheme that is proportional over the age distribution; while in Table 9.7 the SWTP is evaluated with a mortality reduction scheme uniformly distributed over all ages. The details of these calculations are not presented inhere but can be found in [30].

A complete discussion on clear guidelines for a consistent application of the LQI net benefit criterion in a variety of practical applications can be found in [30, 53].

Societal Value of Statistical Life

Because the impact of a safety measure does not discriminate with respect to the characteristics of the individuals, i.e., mortality reduction scheme, the SWTP can be

Table 9.7 SWTP for *unitary* change of mortality uniformly distributed over all ages for year 2010

Region	G_Δ			
	1 %	2 %	3 %	4 %
Australia	1.765	2.101	2.476	3.099
Brazil	0.402	0.478	0.564	0.656
Canada	1.472	1.753	2.066	2.413
China	0.199	0.237	0.279	0.355
Colombia	0.329	0.392	0.462	0.570
Dem. Republic of Congo	0.015	0.017	0.020	0.024
France	1.220	1.452	1.711	2.054
Germany	1.320	1.571	1.851	2.196
Japan	0.933	1.111	1.309	1.560
Mali	0.039	0.046	0.054	0.062
Mexico	0.452	0.538	0.634	0.780
Mozambique	0.038	0.045	0.053	0.068
Sierra Leone	0.018	0.021	0.025	0.031
South Africa	0.337	0.401	0.473	0.602
United Kingdom	1.330	1.583	1.866	2.142
United States	1.449	1.724	2.032	2.312
World (World Life Table)	0.654	0.582	0.517	0.462

SWTP values are expressed in 2005 PPP US Dollars (millions) for different discount rates

replaced by what is known as the statistical value of societal life (SVSL). The SVSL can be derived from Eq. 9.20 as follows:

$$SVSL = -\mathbb{E}\left[\frac{g}{q}\frac{de_d(a)}{e_d(a)}\right] \approx -\frac{g}{q}\bar{e}_d \tag{9.22}$$

where $\bar{e}_d$ is the discounted expected life of the society, which usually is in the order of $e_d \approx 0.65e$. Note that, this is the value that society is willing to pay to save the life of an anonymous individual. The SVSL has been used extensively, in particular, in environmental risk-related problems [32]. The SVSL for selected countries and for various discount rates is presented in Table 9.8.

It is important to stress the difference between the meaning of the SVSL and the SWTP. The SVSL correspond to the amount which must be compensated for each fatality, regardless of the age. On the other hand, the SWTP is the amount that society is willing to pay for a reduction in mortality dm; i.e., it depends on the marginal change that the investment in the safety measure has on the discounted life expectancy.

In summary, both the SVSL and the SWTP are the maximum value that society as a whole is willing to invest for saving lives. Therefore, these values are constraints in LCCA and particularly in cost-based optimization problems.

Table 9.8 SVSL for the year 2010 expressed in US million in 2005 (PPP) for different discount rates

Region	SVSL			
	1 %	2 %	3 %	4 %
Australia	1.98	2.36	2.78	3.42
Brazil	0.48	0.57	0.68	0.86
Canada	2.53	3.01	3.55	4.45
China	0.28	0.34	0.40	0.48
Colombia	0.56	0.67	0.78	0.90
Dem. Republic of Congo	0.02	0.02	0.02	0.03
France	2.05	2.43	2.87	3.51
Germany	2.72	3.23	3.81	4.45
Japan	1.73	2.06	2.42	2.97
Mali	0.05	0.06	0.07	0.09
Mexico	0.63	0.75	0.89	1.10
Mozambique	0.06	0.07	0.08	0.10
Sierra Leone	0.05	0.06	0.07	0.08
South Africa	0.40	0.48	0.56	0.71
United Kingdom	2.08	2.47	2.92	3.53
United States	2.66	3.16	3.73	4.58
World (World Life Table)	0.98	0.78	0.63	0.52

9.6.3 Use of the LQI as Part of LCCA

As discussed at the beginning of the chapter, the result of a life-cycle cost analysis is to determine the discounted expected value of all investments throughout the life of the project. This value is defined based on some project specifications (e.g., design resistance/capacity, maintenance program); described in previous sections as the vector parameter $\mathbf{p}$. If the analysis is carried out on an existing project, the value of $\mathbf{p}$ is already defined. Then the LCCA will determine if $\mathbb{E}[Z(\mathbf{p}, t)] > 0$ or not; and the LQI evaluation can be used to find if the actual $\mathbf{p}$ complies with the safety requirements from a societal point of view. On the other hand, for new projects, the objective of using a LCCA is to take into consideration, during the design phase, the performance of the project throughout its lifetime. The objective is to find the optimum value of $\mathbf{p}$ that maximizes $\mathbb{E}[Z(\mathbf{p}, t)]$. This optimization should be clearly restricted by the LQI evaluation. In summary, in the case of new projects the LQI, and the derived SWTP, enter as restriction in the optimization process. For recent publications on the application of the LQI in practical cases see [64].

9.7 Models for LCCA in Infrastructure Projects

9.7.1 Background

The life-cycle performance of civil infrastructure projects is a topic that has been discussed widely during the last decades. The first works on this topic were published by Rosemblueth and Mendoza [65, 66] in the context of earthquake-resistant design optimization. Their ideas were reconsidered by Hasofer [67] and later by Rackwitz [19] to propose a general framework for optimal design and reliability verification. Further developments on this topic can be found in [8, 23, 29, 54, 56, 68]. A particular application to the relevant problem of structures subjected to extreme loads (i.e., earthquakes and winds) can be found in [20, 69–71]. Some documents that include a review of LCCA current practice in civil engineering are [72–74]; and additional relevant reference documents in other areas include [2, 6, 9, 42]. Analytical developments have been complemented with the development of specialized software. Several commercial reliability analysis software packages have been developed that manage the combined problem of degradation and extreme events. In particular, it is important to mention the software COMREL [75].

Performing life-cycle analysis on infrastructure projects requires making certain assumptions about the manner in which the system will be operated. In the models that follow, we consider, the cases of systems that are abandoned after first failure and systems that are systematically reconstructed for a finite or infinite time horizon. Figure 9.5 illustrates common system's life-cycle performances. In the following sections, we will develop formulations for the LCCA, which can serve as a foundation in building more complex models.

9.7.2 Systems Abandoned After First Failure

Consider a system that starts operating at time $t = 0$. Sometime after it is put in service it fails, and once it fails, it is abandoned (i.e., it is not reconstructed) (see Fig. 9.5a). The time to failure of the system is modeled as a random variable with density $f(\mathbf{p}, t)$, which can be obtained based on a specific physical degradation mechanism and according to the methods presented in Chaps. 5–7.

In this case, for a specific service lifetime t_s, the expected discounted benefit over the life of the project is given by

$$\mathbb{E}[B(t_s)] = \int_0^{t_s} b(\tau)\delta(\tau)(1 - F_1(\mathbf{p}, \tau))d\tau \tag{9.23}$$

where $b(t)$ is the benefit at time t, $\delta(t)$ is the discount function, and $F_1(\mathbf{p}, t)$ is the distribution of the time to first failure. Furthermore, assuming that the cost of losses

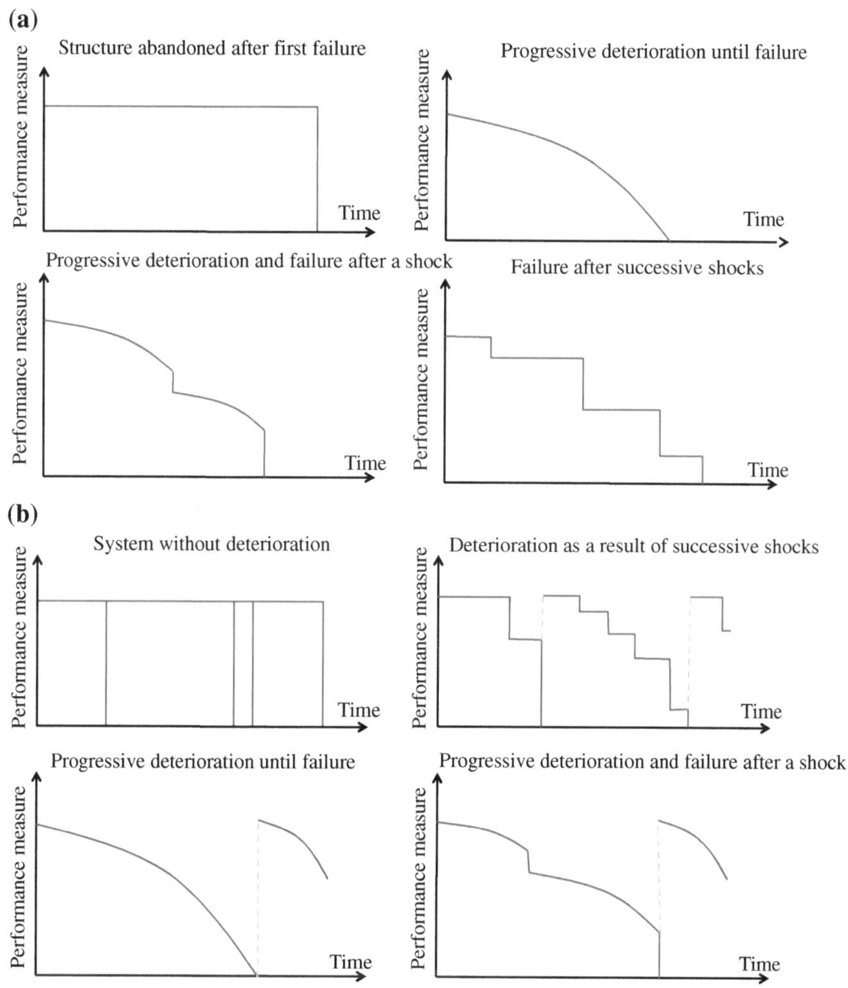

Fig. 9.5 Basic life-cycle performance cases. **a** Systems abandoned after first failure. **b** Systems systematically reconstructed

due to failure do not depend on t, for all t—i.e., $C_L(\mathbf{p})$, the total expected discounted cost of losses is computed as follows:

$$\mathbb{E}[C_T(\mathbf{p}, t_s)] = C_L(\mathbf{p}) \int_0^{t_s} f_1(\mathbf{p}, \tau)\delta(\tau)d\tau \tag{9.24}$$

Therefore, the expected discounted cost-benefit relationship is described by the following life-cycle cost function:

$$\begin{aligned} \mathbb{E}[Z(\mathbf{p}, t_s)] = \int_0^{t_s} b(\tau)\delta(\tau)(1 - F_1(\mathbf{p}, \tau))d\tau - C_0(\mathbf{p}) \\ -C_L(\mathbf{p}) \int_0^{t_s} f_1(\mathbf{p}, \tau)\delta(\tau)d\tau \end{aligned} \tag{9.25}$$

In order to solve Eq. 9.25 several considerations are important. First, Laplace transform has the form

$$\mathscr{L}(f(\mathbf{p},t)) = f^*(\mathbf{p},\gamma) = \int_0^\infty f(\mathbf{p},\tau)e^{-\gamma\tau}d\tau. \tag{9.26}$$

Then, we can conveniently assume a discount function of the form $\delta(t) = \exp(-\gamma t)$ (with γ the discount rate), and take $t_s \to \infty$ in Eq. 9.25. Under these conditions, we get the following expression for the expected life cycle cost:

$$\mathbb{E}[Z(\mathbf{p})] = \lim_{t_s\to\infty} \mathbb{E}[Z(\mathbf{p},t_s)] = \frac{b}{\gamma}(1 - f_1^*(\mathbf{p},\gamma)) - C_0(\mathbf{p}) - C_L(\mathbf{p})f_1^*(\mathbf{p},\gamma) \tag{9.27}$$

where $b(t) = b$.[2]

In time-dependent problems, the probability of failure is defined by the deterioration mechanism. In many industrial equipment and major infrastructure such as pipelines, failure probability is computed based on the number of failures observed and through the so-called failure rate. This approach was presented in Chap. 2 but it has been discussed widely in the literature; see for example [76–78].

Example 9.49 Consider, the case of a system for which the time to failure is exponentially distributed with constant parameter $\lambda(\mathbf{p}) = \lambda = 0.1$. The construction cost is $C_0 = 10^3$; the benefits are computed as: $b = 0.3C_0$; and the costs of losses in case of failure $C_L(\mathbf{p}) = C_L = 1.1 \times C_0$. Compute the discounted expected life-cycle cost for various discount rates and for an infinite time horizon.

For events whose time to failure is exponentially distributed with parameter λ, the Laplace transform is:

$$\mathscr{L}(f(t)) = f^*(\gamma) = \frac{\lambda}{\gamma+\lambda} \tag{9.28}$$

Therefore, the discounted expected life-cycle cost becomes,

$$\mathbb{E}[Z] = \frac{b}{\gamma+\lambda} - C_0 - C_L\frac{\lambda}{\gamma+\lambda} \tag{9.29}$$

[2]Note that based on the following Laplace transform property $F_1^*(\mathbf{p},\gamma) = f_1^*(\mathbf{p},\gamma)/\gamma$, the form of the benefit for an infinite lifetime can be derived as follows [18]:

$$B(\mathbf{p},\gamma) = \int_0^\infty b(\tau)\delta(\tau)(1 - F_1(\mathbf{p},\tau))d\tau = b\int_0^\infty \exp(-\gamma t) - F_1(\mathbf{p},\tau)\exp(-\gamma t)d\tau = \frac{b}{\gamma}(1 - f_1^*(\mathbf{p},\gamma)).$$

Then, with the cost data given, the values of $\mathbb{E}[Z]$ for various discount rates are:

γ (%)	$\frac{b}{\gamma+\lambda}$	C_0	$C_L\frac{\lambda}{\gamma+\lambda}$	$\mathbb{E}[Z]$
1	2727	1000	1000	727
3	2308	1000	846	461
5	2000	1000	733	267
10	1500	1000	550	−50
15	1200	1000	440	−240

Note that, interestingly, as the discount rate becomes larger (e.g., $\gamma > 10\,\%$) the objective function shows that the project is not feasible (i.e., $\mathbb{E}[Z] < 0$)

9.7.3 Systematically Reconstructed Systems

Systems that are successively repaired after failure are called *systematically reconstructed* (see Chap. 8). In this section, we will present several important cases.

Successive Reconstructions

Consider, a system that starts operating at an initial state, say $V(0) = v_0$ (in suitable units) and degrades until failure. Failure times are random and their distribution is defined using any of the methods presented in Chaps. 5–7. The system is systematically reconstructed after every failure. Assume further that the time between failures, and immediate interventions, are independent, so that failure times constitute a (possibly delayed) renewal process. Let the density of the time to first failure be given by $f_1(\mathbf{p}, t)$ and the density of the time between any other two successive failures $f(\mathbf{p}, t)$. These functions clearly depend on the system mechanical properties and other parameters comprised in the vector $\mathbf{p}$. Then, for constant benefits per time unit $b(t) = b$, the expected discounted life-cycle cost is:

$$\mathbb{E}[Z(\mathbf{p}, t_s)] = \int_0^{t_s} be^{-\gamma\tau}d\tau - C_0(\mathbf{p}) - C_L(\mathbf{p})\sum_{n=1}^{\infty}\int_0^{t_s} f_n(\mathbf{p}, \tau)e^{-\gamma\tau}d\tau \quad (9.30)$$

where $f_n(\mathbf{p}, t)$ is the probability density of the time to the nth failure/intervention. For the particular case where $t_s \to \infty$, Eq. 9.30 becomes (see Sect. 8.2.2) [18],

$$\mathbb{E}[Z(\mathbf{p}, t_s)] = \frac{b}{\gamma} - C_0(\mathbf{p}) - C_L(\mathbf{p})\sum_{n=1}^{\infty}\int_0^{\infty} f_n(\mathbf{p}, \tau)e^{-\gamma\tau}d\tau$$

$$= \frac{b}{\gamma} - C_0(\mathbf{p}) - C_L(\mathbf{p})\frac{f_1^*(\mathbf{p},\gamma)}{1 - f^*(\mathbf{p},\gamma)}$$
$$= \frac{b}{\gamma} - C_0(\mathbf{p}) - C_L(\mathbf{p})h_1^*(\gamma,\mathbf{p}) \qquad (9.31)$$

where $h_1^*(\gamma, \mathbf{p})$ is the Laplace transform of the renewal density. For ordinary renewal processes where the distribution between all failure occurrences are *iid* with density $f(\mathbf{p}, t)$, the last term of Eq. 9.31 is slightly modified and

$$\mathbb{E}[Z(\mathbf{p}, t_s)] = \frac{b}{\gamma} - C_0(\mathbf{p}) - C_L(\mathbf{p})\frac{f^*(\mathbf{p},\gamma)}{1 - f^*(\mathbf{p},\gamma)}$$
$$= \frac{b}{\gamma} - C_0(\mathbf{p}) - C_L(\mathbf{p})h^*(\gamma,\mathbf{p}) \qquad (9.32)$$

where $f_1^*(\mathbf{p}, t)$ in $h_1^*(\gamma, \mathbf{p})$ is replaced by $f^*(\mathbf{p}, t)$ and therefore $h_1^*(\gamma, \mathbf{p})$ replaced by $h^*(\gamma, \mathbf{p})$. For the renewal density and the Laplace transform there is an important asymptotic result [79]:

$$\lim_{t\to\infty} h(t,\mathbf{p}) = \lim_{\gamma\to 0} \gamma h^*(\gamma,\mathbf{p}) = \frac{1}{\bar{T_f}(\mathbf{p})} \qquad (9.33)$$

where $\bar{T_f}(\mathbf{p})$ is the mean time between renewals (failures) [18].

Systems Subjected to Extreme Events

Consider now a slight modification of the previous case in which we assume a system that does not deteriorate; i.e., the system remains in its initial condition, e.g., $V(0) = v_0$, through time. The system is exposed to extreme events (e.g., earthquakes, hurricanes, explosions and floods) that occurs randomly in time, and that may cause the failure of the system with probability $P_f(\mathbf{p})$. The system is immediately and systematically reconstructed after every failure. The density of the times between events (non necessarily failures) are $f_1(\mathbf{p}, t)$ (to the first event) and $f(\mathbf{p}, t)$ (between successive events).

In this case, the expected discounted cost of losses become (see Sect. 8.2.3) [66]:

$$\mathbb{E}[C_L(\mathbf{p})] = C_L(\mathbf{p})\sum_{n=1}^{\infty}\left[\int_0^{\infty} f_n(\mathbf{p},\tau)e^{-\gamma\tau}d\tau\right]P_f(\mathbf{p})(1-P_f(\mathbf{p}))^{n-1}$$
$$= C_L(\mathbf{p})\sum_{n=1}^{\infty} f_n^*(\mathbf{p},\gamma)P_f(\mathbf{p})(1-P_f(\mathbf{p}))^{n-1}$$
$$= C_L(\mathbf{p})\sum_{n=1}^{\infty} f_1^*(\mathbf{p},\gamma)P_f(\mathbf{p})(f^*(\mathbf{p},\gamma)(1-P_f(\mathbf{p})))^{n-1} \qquad (9.34)$$

where $f_n^*(\mathbf{p},\gamma) = f_1^*(\mathbf{p},\gamma) * f_{n-1}^*(\mathbf{p},\gamma)$ and $f_n^*(\mathbf{p},\gamma) = [f^*(\mathbf{p},\gamma)]^n$ (see Eq. 8.5). Then, the value of $h_1(\gamma,\mathbf{p})$ (Eq. 9.31) becomes

$$h_1^*(\mathbf{p},\gamma) = \frac{P_f(\mathbf{p}) f_1^*(\mathbf{p},\gamma)}{1-(1-P_f(\mathbf{p}))f^*(\mathbf{p},\gamma)} \tag{9.35}$$

and for an ordinary renewal process,

$$h^*(\mathbf{p},\gamma) = \frac{P_f(\mathbf{p}) f^*(\mathbf{p},\gamma)}{1-(1-P_f(\mathbf{p}))f^*(\mathbf{p},\gamma)} \tag{9.36}$$

The expressions in Eqs. 9.35 and 9.36 should then be replaced in Eq. 9.31 and 9.32 accordingly to model renewed systems subject to random external events.

Example 9.50 The occurrence of most natural extreme events (e.g., earthquakes) can be described as a stationary Poisson process. If every time there is one of such events the system may fail with probability $P_f(\mathbf{p})$, find an expression for the renewal density h^*.

The expression for h^* was derived in Eq. 9.36; i.e.,

$$h^*(\mathbf{p},\gamma) = \frac{P_f(\mathbf{p}) f^*(\mathbf{p},\gamma)}{1-(1-P_f(\mathbf{p}))f^*(\mathbf{p},\gamma)}$$

If the events occur with a Poisson intensity, λ, and remembering that $f^*(\mathbf{p},\gamma) = \lambda/(\gamma+\lambda)$, we get

$$\begin{aligned} h^*(\mathbf{p},\gamma) &= \frac{P_f(\mathbf{p})\frac{\lambda}{\gamma+\lambda}}{1-(1-P_f(\mathbf{p}))\frac{\lambda}{\gamma+\lambda}} \\ &= \frac{P_f(\mathbf{p})\lambda}{\gamma + P_f(\mathbf{p})\lambda} \end{aligned} \tag{9.37}$$

Example 9.51 Consider, the basic case of a system subject to extreme events (e.g., earthquakes) that occur according to a Poisson process with rate $\lambda = 2$/year. For the purpose of this example, a single parameter p will describe the system's remaining capacity/resistance of the system; note that p should be measured in appropriate system capacity units. The probability of failure in case of an event is function of the system parameter p and follows a lognormal distribution with mean $\mu = p$ and COV$= 0.35$.

The cost assumptions of the problem are the following: $C_0(p) = \$2 \cdot 10^7 + \$8 \cdot 10^3 p^2$ and $C_L(p) = \$2 \cdot 10^3(100-p)^{2.5}$ (includes direct and indirect losses) for $0 \leq p \leq 100$. The discount rate is $\gamma = 0.035$; and the constant benefit is calculated as $b = 0.15 \cdot \$2 \cdot 10^7$, which in the long run leads to: $b/\gamma = 8.571 \cdot 10^7$ (Eq. 9.11).

The objective function, benefit, construction cost, and cost of losses, as function of the system's vector parameter p, are presented in Fig. 9.6. It is observed that the construction cost increases with p, while the cost of losses decreases. The latter is

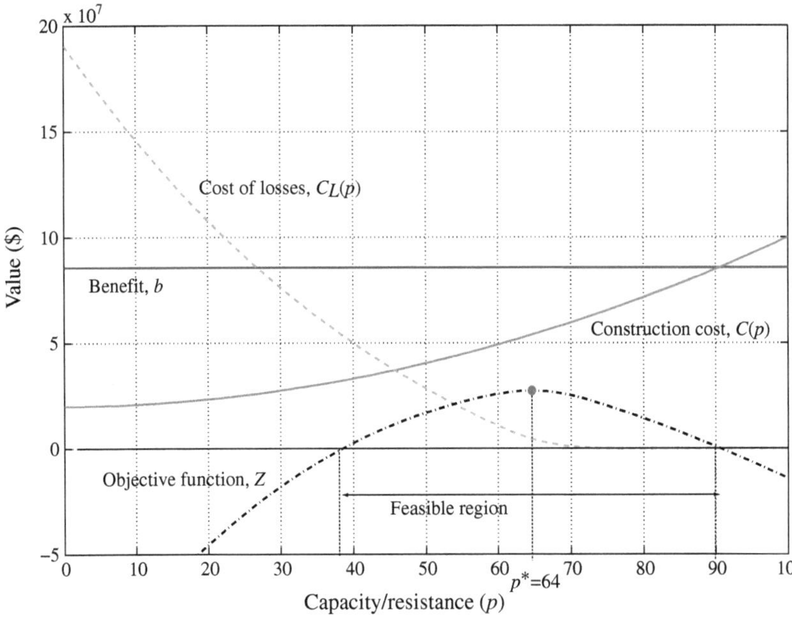

Fig. 9.6 Objective function, benefit, construction cost and cost of losses as function of the system's vector parameter p

clearly justified by the fact that as p increases, enhancing the system performance, the failure probability decreases, and therefore, the expected value of losses becomes smaller. The Laplace transform of the renewal density used to evaluate the expected value of losses is computed based on Eq. 9.37:

$$h^*(p) = \frac{P_f(p)\lambda}{\gamma + P_f(p)\lambda} = \frac{2P_f(p)}{0.035 + 2P_f(p)}$$

It can be observed in Fig. 9.6 that the objective function has a positive region within the interval [38.5, 90]. This means that, for the given financial conditions and cost structure, the project should be designed for a capacity resistance within this range; otherwise, the investment is not cost-effective. Finally, the optimum design parameter is $p^* = 64$, which will lead to a failure probability of $P_f(64) = 9.9 \cdot 10^{-3}$.

Systems Subject to Multiple Extreme Events

Many systems, especially large infrastructure projects, are designed to operate for long periods of time, and may be exposed to multiple hazardous events. Furthermore, in practice the system performance may be characterized by multiple limit states, which are used to define different intervention measures. In this section, we present an approximation to this case based on the work presented in [20, 70].

Consider a system (e.g., bridge) subject to extreme events and whose performance is defined by multiple limit states (Fig. 9.7). Under these conditions, the discounted expected value of the investments throughout the systems lifetime t_s can be written as [20]:

$$\mathbb{E}[Z(\mathbf{p}, t_s)] = \mathbb{E}\left[B(\mathbf{p}, t_s) - C_0(\mathbf{p}) - \sum_{i=1}^{N(t)} \sum_{j=1}^{k} [C_L(\mathbf{p})]_j P_{ij}(\mathbf{p}, t_i) e^{-\gamma t_i} \right] \tag{9.38}$$

where $[C_L(\mathbf{p})]_j$ is the cost of exceeding the j limit state, with $j = 1, 2, \ldots, k$, and $P_{ij}(\mathbf{p}, t_i)$ is the probability of exceeding the limit state j, given the ith occurrence of the extreme event. The term $e^{-\gamma t_j}$ describes the discount function with γ being the constant discount rate and t_j the time at which the j limit state is exceeded. External events are assumed to occur randomly in time and $N(t)$ describes the number of events that have occurred in time t (Fig. 9.7). Note that implicitly in Eq. 9.38 is the idea that the system is restored to its initial contain after each hazard occurrence (every intervention).

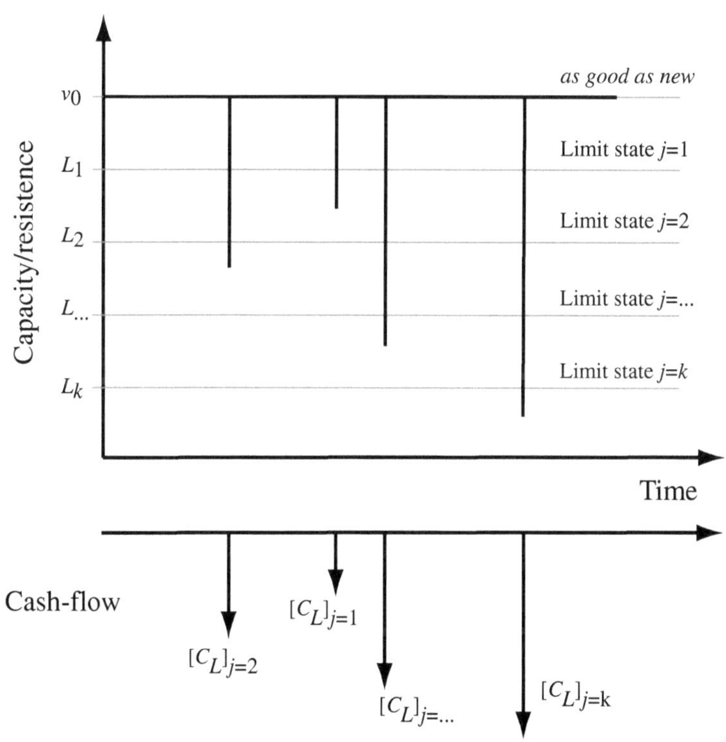

Fig. 9.7 Realization of the performance of a system with multiple limit states and subject to extreme events

Let us consider the case of a system subject to a single event whose occurrence is modeled by a Poisson process with rate ν. If the system does not deteriorate with time (i.e., the probability $P_{ij}(\mathbf{p}) = P_j(\mathbf{p})$ remains constant) the total discounted expected cost for the system's lifetime t_s becomes (see [20] for the derivation):

$$\mathbb{E}[Z(\mathbf{p}, t_s)] = \mathbb{E}[B(\mathbf{p}, t_s)] - C_0(\mathbf{p}) - \left(\sum_{j=1}^{k} [C_L(\mathbf{p})]_j P_j(\mathbf{p}) \right) \frac{\nu}{\gamma}(1 - e^{-\gamma t_s}) \tag{9.39}$$

where $P_j(\mathbf{p})$ is the probability of exceeding the limit state j.

Consider now the case of a system exposed to multiple extreme events, where all events follow also a Poisson process with rate ν_x. In this case, the join occurrence of two Poisson processes is also a Poisson process with join occurrence rate [80]:

$$\nu_{ij} = \nu_i \nu_j (\mu_{d_i} + \mu_{d_j}) \tag{9.40}$$

where ν_i and ν_j are the rates of the individual events and μ_{d_x} is the mean duration of the event x; similarly, for three extreme events,

$$\nu_{ijk} = \nu_i \nu_j \nu_k (\mu_{d_i}\mu_{d_j} + \mu_{d_i}\mu_{d_k} + \mu_{d_j}\mu_{d_k}) \tag{9.41}$$

In this case, the losses associated with exceeding a limit state w may result from the action of individual events, plus the case of two events occurring at the same time, etc. Then, the discounted expected cost of losses can be computed as [20]:

$$\begin{aligned} \mathbb{E}[C_L(\mathbf{p})] = \sum_{w=1}^{k} [C_L(\mathbf{p})]_w & \left[\sum_{i=1}^{n} \nu_i P_w^i + \sum_{i=1}^{n-1} \sum_{j=i+1}^{n} \nu_{ij} P_w^{ij} \right. \\ & \left. + \sum_{i=1}^{n-2} \sum_{j=i+1}^{n-1} \sum_{k=j+1}^{n} \nu_{ijk} P_w^{ijk} + \cdots \right] \frac{(1 - e^{-\gamma t_s})}{\gamma} \end{aligned} \tag{9.42}$$

where ν_{ij} and ν_{ijk} are obtained from Eqs. 9.40 and 9.41. The terms P_w^i, P_w^{ij} and P_w^{ijk} correspond to the probabilities of exceeding limit state w under the action of event i, or the combined action of events i and j; or i, j and k respectively.

Several interesting and complete examples with practical applications of this model can be found in [20, 70, 80].

9.8 Optimal Design Parameters

9.8.1 Problem Definition

The structure of the LCCA is frequently used to find the optimal set of design or operational parameters, i.e., the vector **p**, that maximizes the profit or minimizes cost. This approach constitute a new design paradigm in engineering in which special engineering systems should not be necessarily designed according to the requirements specified in codes of practice (or any type of regulation or that matter), but should be designed and operated using criteria based on optimum life-cycle cost evaluations. This means that safety and risk control strategies should be defined within a cost-effectiveness framework, and not only as arbitrary measures based on the system's physical performance.

Then, if $\mathbf{p} = \{p_1, p_2, \ldots, p_k\}$ is the vector that contains the system design and operation parameters, the optimal design is obtained by finding the value of **p** that solves the following objective function:

$$\max_{\mathbf{p}} \mathbb{E}[Z(\mathbf{p}, t_s)] \tag{9.43}$$

or which is the same,

$$\max_{\mathbf{p}} \mathbb{E}[B(\mathbf{p}, t_s) - C_0(\mathbf{p}) - C_L(\mathbf{p}, t_s) - C_D(t_s)] \tag{9.44}$$

As it was mentioned before, some times the benefits are dropped from this equation and the analysis focuses on costs only; in this case, the optimization problem is defined as,

$$\min_{\mathbf{p}} \mathbb{E}[C_0(\mathbf{p}) + C_L(\mathbf{p}, t_s) - C_D(t_s)], \tag{9.45}$$

which, although practical, may be misleading since it does not take into consideration the profits; which implies that the project is not necessarily economically feasible.

In some special cases, the cost-benefit problem presented in Eqs. 9.44 and 9.45 can be solved as an unconstrained optimization. However, restrictions may appear depending upon the particular considerations of the problem at hand. For example, if the cost of saving lives is modeled using the LQI (see Sect. 9.6.2) it enters into the optimization as a restriction on the investments in saving lives. Frequently, the numerical solution of Eqs. 9.44 and 9.45 requires some mathematical manipulation. In particular, the optimization becomes complicated when computing the probability becomes an optimization problem itself (see Chap. 2). In these cases, solving Eq. 9.43 becomes a two level optimization; for more details see [19, 81, 82] for a numerical solution. However, for simple and small practical applications, standard software such as Mathcad^{TM} or Mathlab^{TM} can be used to find a numerical solution.

The practicality of finding a unique optimum has been criticized extensively. In practice, it is frequently more useful to define a range of possible solutions from which the designer may choose according to additional considerations that are beyond the physical performance of the system. Within this context, the ALARP region is a concept used in the analysis of critical facilities for risk management purposes. It defines a zone where there is "a level of risk that is tolerable and cannot be reduced further without the expenditure of costs that are disproportionate to the benefit gained or where the solution is impractical to implement" [83]. In other words, it is the region that separates tolerable and unacceptable risk. Finding the ALARP region provides important evidence for engineering decisions.

9.8.2 Illustrative Examples

In the following, we will present several examples that illustrate and integrate the cases presented in this chapter.

Example 9.52 Consider a bridge subjected to extreme events (e.g., earthquakes) that occur according to a Poisson process with rate $\lambda = 1$/year. Every time an event occurs, it may cause the failure of the system with probability $P_f(p)$, where p is the mean resistance of the system. If the system fails, it is immediately repaired and taken to "as good as new" condition and restarts operation immediately. The objective of the example is to define the ALARP ("As Low as Reasonably Practicable") region for the project in terms of the design parameter $\mu_R = p$.

In this example, the probability of failure is defined in terms of a random demand and resistance (see Chap. 2). Both the demand and the resistance are assumed to be lognormally distributed with parameters $\mu_S = 10$ and $COV_S = 0.75$; and $\mu_R = p$ and $COV_R = 0.25$. The financial and cost assumptions are the following: $C_B = \$1.5 \cdot 10^7$ (i.e., base construction cost); $b = 0.085 \cdot C_B$; $\gamma = 0.02$; $C_0(p) = C_B + \$5 \cdot 10^5 \cdot (p/5)^a$, with $a = 2.25$; and $C_L = C_0(p) + 5C_B$ (includes direct and indirect losses).

Based on these assumptions, the objective function can be formulated as follows (see Eqs. 9.32 and 9.37):

$$\begin{aligned}\mathbb{E}[Z(p)] &= \frac{b}{\gamma} - C_0(p) - C_L(p)\frac{\lambda P_f(p)}{\gamma + \lambda P_f(p)}\\ &= \frac{0.085 \cdot C_B}{0.02} - \left(C_B + 5 \cdot 10^5 \left(\frac{p}{5}\right)^{2.25}\right) - \left((C_0(p) + 5C_B)\frac{\lambda P_f(p)}{\gamma + \lambda P_f(p)}\right)\end{aligned} \tag{9.46}$$

In order to find the ALARP region, we need to build the function $\mathbb{E}[Z(p)]$ (Eq. 9.46), whose component elements are shown in Fig. 9.8. Clearly, for the project to be feasible $\mathbb{E}[Z(p)] > 0$, thus, the feasible region can be bounded by $41 \leq p \leq 74$. This region can be divided in two parts; this is, before and after the optimum value $p^* = 56$,

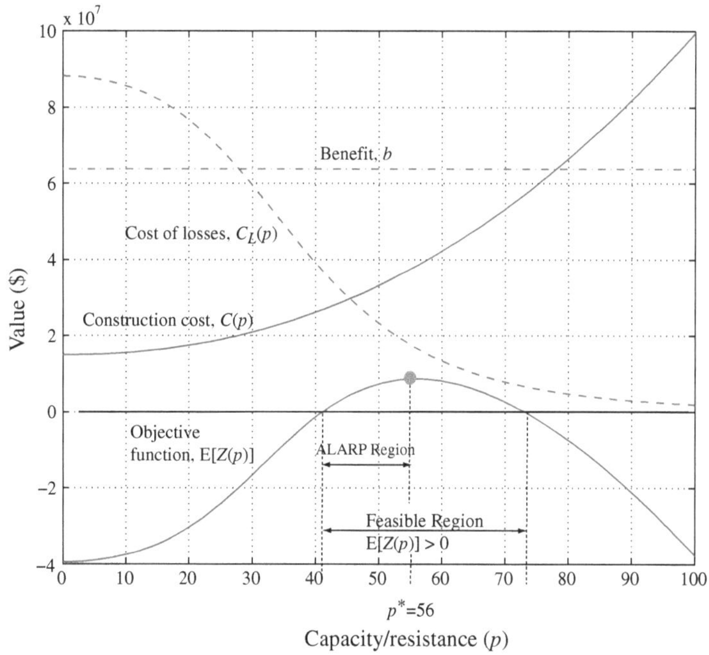

Fig. 9.8 Optimum design parameter and definition of the ALARP region

(for which $\mathbb{E}[Z(p^*)] = 8.71 \cdot 10^6$). Then, in this particular case, the ALARP region corresponds to the range of values of p within the region $41 \leq p \leq (p^* = 56)$ [18]. Note that any value of $p > p^*$ and within the feasible region, implies an unnecessary larger investment to obtain a profit that can be achieved with a smaller p.

Example 9.53 Decisions about investments in a project may be viewed from different perspectives; in particular, the private and public sector have a different approach. This is mainly reflected in two parameters: the expected benefit and the discount rate. The purpose of this example is to compare the objective functions, the optimum design parameters (i.e., p^*), and the feasible region for typical conditions of both a public and a private investors.

Consider a system systematically reconstructed with times between failures that occur with probability density $f(t)$, which is assumed to be exponential with rate $\lambda(p) = 1/p^{1.5}$. The cost assumptions are the following: $C_B = \$5 \cdot 10^7$ (i.e., base construction cost); $b = \beta C_B$; $C_0(p) = C_B + \$7.5 \cdot 10^5 \times (0.1p)^a$, with $a = 1.75$; and $C_L = C_B + 2.1C_0$ (includes all cost of losses).

For the particular case of failure events that follow a Poisson process with rate $\lambda(p)$, the objective function is [18]:

$$
\begin{aligned}
\mathbb{E}[Z(p)] &= \frac{b}{\gamma} - C_0(p) - C_L h^*(\gamma, p) \\
&= \frac{b}{\gamma} - C_0(p) - C_L \frac{\lambda(p)}{\gamma} \\
&= \frac{\beta C_B}{\gamma} - (\$5 \cdot 10^7 + \$7.5 \cdot 10^5 \times (0.1p)^{1.75}) - (\$5 \cdot 10^7 + 2.1C_0)\frac{\lambda(p)}{\gamma}.
\end{aligned}
$$

The form of $h^*(\gamma, p)$ is derived from the fact that $h^*(\gamma, p) = f^*(t, p)/(1 - f^*(t, p))$ and $f^*(t, p) = \lambda(p)/(\gamma + \lambda(p))$. Note that in this formulation, the rate of the process depends on the parameter p.

Frequently, in the public sector both the expected benefits and the discount rates are smaller than in the private sector. Typical values of the discount rate, for the public sector, are $0.02 \leq \gamma \leq 0.05$ and for the private $0.07 \leq \gamma \leq 0.12$. Regarding the benefits, the factor β may vary; for public investments it is within the range $0.03 \leq \beta \leq 0.08$, and for the private sector in the interval $0.07 \leq \beta \leq 0.15$. Based on these ranges, four cases were studied; the objective functions are shown in Fig. 9.9 and the description of the cases and the results in Table 9.9.

The results show that the optimum design criteria for public investments are larger than those for private investments. This is basically due to the fact that public investments operate, in most cases, with smaller discount rates.

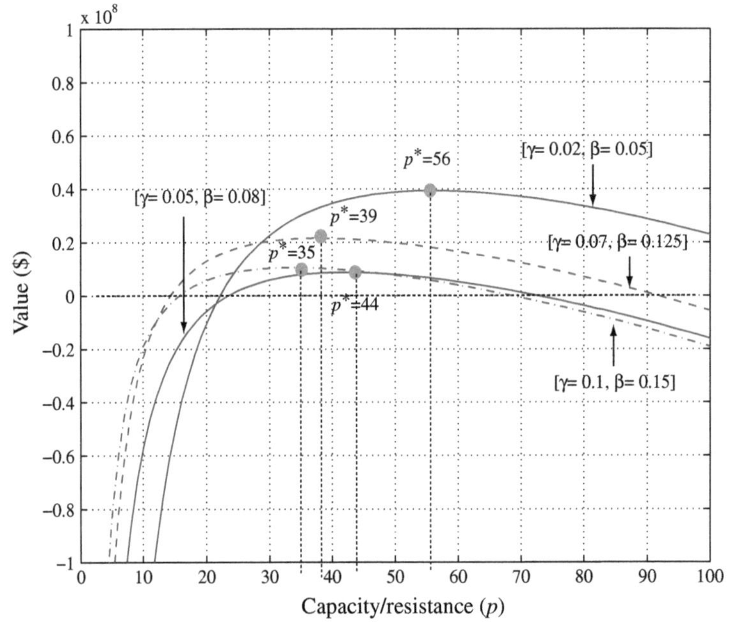

Fig. 9.9 Comparison of typical objective functions for public and private owner conditions

Table 9.9 Comparison of financial criteria for public and private investors

Owner	γ	β	p^*_{opt}	$\lambda(p^*_{opt})$	$\mathbb{E}[Z(p^*_{opt})]$	Feasible region
Public	0.02	0.05	56	$2.4 \cdot 10^{-2}$	$3.94 \cdot 10^7$	[22, 131]
Public	0.05	0.08	44	$3.4 \cdot 10^{-2}$	$8.69 \cdot 10^6$	[24, 73]
Private	0.07	0.125	39	$4.1 \cdot 10^{-2}$	$2.16 \cdot 10^7$	[15, 92]
Private	0.10	0.15	35	$4.8 \cdot 10^{-2}$	$1.05 \cdot 10^7$	[16, 69]

9.9 Summary and Conclusions

The assessment of costs which the owner (or stakeholders) will incur during the life cycle of a project to keep it operating is referred to as the life-cycle cost analysis (LCCA). The LCCA is an economic alternative for project evaluation, in which the decision criteria is the lowest long-term life-cycle cost of a set of projects. This approach can be used as a tool for comparing a set of project alternatives in terms of their long-term cost-effectiveness; or as a modeling strategy for selecting the design and management (e.g., maintenance) requirements. The determination of cost-based optimum parameters constitutes a new design paradigm in engineering. Engineering systems should therefore not be designed simply for requirements specified in codes of practice, but rather designed and operated based on cost optimization criteria. This means that safety and risk control strategies should be defined within a cost-effectiveness framework and not as arbitrary measures based only on the system's physical performance. Several models and analytical solutions to carry out a LCCA are presented in this chapter and illustrated with examples.

References

1. Tellus Institute, CSG/Tellus Packaging Study: inventory of material and energy use and air and water emissions from the production of packaging materials. Technical Report (89-024/2) (prepared for the Council of State Governments and the United States Environmental Protecion Agency). Jellus Institute, Boston, MA, 1992
2. US Environmental Protection Agency (EPA), Life-cycle assessment: principles and practice. US Environmental Protection Agency, EPA/600/R-06/060, Cincinnati, 2006
3. J.C. Bare, P. Hofstetter, D.W. Pennington, H.A. Udo de Haes, Midpoints versus endpoints: the sacrifices and benefits. Int. J. Life-cycle Assess. **5**(6), 319–326 (2000)
4. J.E. Padgett, C. Tapia, Sustainability of natural hazard risk mitigation: a life-cycle analysis of environmental indicators for bridge infrastructure. J. Infrastruct. Syst., ASCE (2013)
5. C. Tapia, J.E. Padgett, Multi-objective optimisation of bridge retrofit and post-event repair selection to enhance sustainability. *Structure and Infrastructure Engineering: Maintenance, Management, Life-Cycle Design and Performance*, page doi:10.1080/15732479.2014.995676 (2015)
6. K.F. Sieglinde, R.P. Stephen, *NIST Handbook 135: Life Cycle Costing Manual for the Federal Energy Management Program* (U.S. Government Printing Office, Washington, 1995)
7. A.J. Dell'Isola, S.J. Kirk, *Life Cycle Cost Data* (McGraw Hill, New York, 1983)

8. American Society for Testing and (ASTM), Materials. *Standard Practice for Measuring Life-cycle Costs of Buildings and Building Systems* (ASTM, Philadelphia, 1994)
9. New South Wales Treasury, *Total Asset Management: Life Cycle Costing Guideline*. TAM-2004; New South Wales Treasury, New South Wales, 2004
10. SAE International, *Reliability, Maintainability, and Supportability Guidebook*, 3rd edn. RMS Committee (SAE International, 1995)
11. SAE International, *Reliability and Maintainability Guideline for Manufacturing Machinery and Equipment*, 3rd edn. SAE (SAE International, 1999)
12. A.S. Goodman, M. Hastak, *Infrastructure Planning Handbook: Planning Engineering and Economics* (ASCE Press, New York, 2006)
13. S.J. Kirk, A.J. Dell'Isola, *Life-Cycle Costing for Design Professionals* (McGraw Hill, New York, 1995)
14. D. Paez-Pérez, M. Sánchez-Silva, A dynamic principal-agent framework for modeling the performance of infrastructure. Eur. J. Oper. Res (2016). In Press
15. D. Paez-Pérez, M. Sánchez-Silva, Modeling the complexity of performance of infrastructure (2016). Under review
16. M. Sánchez-Silva, D. Rosowsky, Risk, reliability and sustainability in the developing world. ICE Struct.: Spec. Issue Struct. Sustain. **161**(4), 189–198 (2008)
17. UN. Brundland Commission, *Our common future*. UN World Commission on Environment and Development (1987)
18. R. Rackwitz, Optimization and risk acceptability based on the life quality index. Struct. Saf. **24**, 297–331 (2002)
19. R. Rackwitz, Optimization—the basis of code making and reliability verification. Struct. Saf. **22**(1), 27–60 (2000)
20. Y.K. Wen, Y.J. Kang, Minimum building lifecycle cost design criteria. i: methodology. J. Struct. Eng., ASC **127**(3), 330–337 (2001)
21. D. Val, M. Stewart, Decision analysis for deteriorating structures. Reliab. Eng. Syst. Saf. **87**, 377–385 (2005)
22. J. Von Neummann, O. Morgenstern, *Theory of Games and Economic Behavior*, 3rd edn. (Princeton University Press, Princeton, 1953)
23. J.S. Nathwani, M.D. Pandey, N.C. Lind, *Engineering Decisions for Life Quality: How Safe is Safe Enough?* (Springer, London, 2009)
24. J. Zhuang, Z. Liang, T. Lin, F. De Guzman, Theory and practice in the choice of social discount rate for cost-benefit analysis: a survey. Asian Development Bank—Series on Economic Working Papers, ERD 94:1–50 (2007)
25. F. Ramsey, A mathematical theory of saving. Econ. J. **38**, 543–549 (1928)
26. L. Young, Determining the discount rate for government projects. Working paper, New Zealand Treasury (2002)
27. A. Harberger, *Project Evaluation: Collected Papers* (The University of Chicago Press, Chicago, 1972)
28. S. Frederick, Valuing future life and future lives: a framework for understanding discounting. J. Econ. Psychol. **27**, 667–680 (2006)
29. R. Rackwitz, A. Lentz, M.H. Faber, Socio-economically sustainable civil engineering infrastructures by optimization. Struct. Saf. **27**, 187–229 (2005)
30. R. Rackwitz, The philosophy behind the Life Quality Index and empirical verification. Joint Committee of Structural Safety (JCSS)-Basic Documents on Risk Assessment in Engineering: Document N4, DTU—Denmark (2008)
31. E. Paté-Cornell, Discounting in risk analysis: capital versus human safety, in *Risk, Structural Engineering and Human Error*, ed. by M. Grigoriu (University of Waterloo Press, Waterloo, 1984)
32. P.O. Johansson, Is there a meaningful definition of the value of statistical life? Health Econ. **20**, 131–139 (2001)
33. S. Bayer, D. Cansier, Intergenerational discounting: a new approach. J. Int. Plan. Lit. **14**(3), 301–325 (1999)

34. R.B. Corotis, Public versus private discounting for life-cycle cost, in *Proceedings of the International Conference on Structural Safety and Reliability ICOSSAR'05*, ed. by G. Augusti, G.I. Schueller, M. Ciampoli. Millress Rotterdam the Netherlands, August (2005)
35. S. Bayer, Intergenerational discounting: a new approach. Tubinger Diskussionsbeitrag **145**, 1–26 (1998)
36. D. Nishijima, K. Straub, M.H. Faber, Inter-generational distribution of the life-cycle cost of an engineering facility. J. Reliab. Struct. Mater. **3**(1), 33–46 (2007)
37. S.E. Chang, M. Shinozuka, Life-cycle cost analysis with natural hazard risk. ASCE-J. Infrastruct. Syst. **2**(3), 118126 (1996)
38. D.M. Neves, L.C. Frangopol, P.J.S. Cruz, Cost of reliability improvement and deterioration delay of maintained structures. Comput. Struct. **82**(13–14), 1077–1089 (2004)
39. L. Ochoa, M. Hendrickson, H.S. Matthews, Economic input-output life-cycle assessment of us residential buildings. J. Infrastruct. Syst. **8**, 132–138 (2002)
40. Y. Itoh, T. Kitagawa, Using co_2 emission quantities in bridge lifecycle analysis. Eng. Struct. **25**, 565–577 (2003)
41. ISO, *Structural Reliability: Statistical Learning Perspectives*. International Organisation of Standardisation, Geneva (2000)
42. IISI, *World Steel Life-cycle Inventory—methodology report*. International Iron and Steel Institute, Committee on Environmental Affairs, Brussels (2002)
43. M. Nisbet, M. Marceau, M. VanGeem, *Environmental Life Cycle Inventory of Portland Cement Concrete* (Portland Cement Association, Stokie, 2002)
44. H. Gervasio, L.S. da Silva, Comparative life-cycle analysis of steel-concrete composite bridges. Struct. Infrastruct. Eng. **4**, 251–269 (2008)
45. E.J. Mishan, Evaluation of life and limb: a theoretical approach. J. Polit. Econ. **79**(4), 687–705 (1971)
46. R. Zeckhauser, Procedures for valuing lives. Public Policy **23**(4), 419–464 (1975)
47. W.B. Arthur, The economics of risk to life. Am. Econ. Rev. **71**(1), 54–64 (1980)
48. M.D. Pandey, J.S. Nathwani, Life quality index for the estimation of societalwillingness-to-pay for safety. Struct. Saf. **26**, 181–199 (2004)
49. A.J. Krupnick, A. Alberini, M. Cropper, N. Simon, B. O'Brien, R. et al. Goeree, Age, health and willingness to pay for mortality risk reduction. Discussion paper, resources for future, DP00-37, Washington (2000)
50. J.K. Hammitt, Valuing changes in mortality risk: lives saved versus life years saved. Rev. Env. Econ. Policy **1**, 228–240 (2007)
51. J.E. Aldy, W.K. Viscusi, Age differences in the value of statistical life: revealed preference evidence. Rev. Environ. Econ. Policy **1**, 241–260 (2001)
52. J.K. Hammitt, Valuing mortality risk: theory and practice. Environ. Sci. Technol. **34**, 1396–1400 (2007)
53. K. Fischer, M. Virguez-Rodriguez, M. Sánchez-Silva, M.H. Faber, On the assessment of marginal life saving costs for risk acceptance criteria. Struct. Saf. **44**, 37–46 (2013)
54. R. Rackwitz, The effect of discounting, different mortality reduction schemes and predictive cohort life tables on risk acceptability criteria. Reliab. Eng. Syst. Saf. **91**, 469–484 (2006)
55. M.D. Pandey, J.S. Nathwani, N.C. Lind, The derivation and calibration of the life quality index (LQI) from economical principles. Struct. Saf. **28**, 341–360 (2006)
56. J. Nathwani, N. Lind, M. Pandey, Affordable safety by choice: the life quality method. Institute for Risk Research. University of Waterloo, Waterloo (1997)
57. T.O. Tengs, M.E. Adams, J.S. Pliskin, D.G. Safran, J.E. Siegel, M.C. Weinstein, Five-hundred life-saving interventions and their cost-effectiveness. Risk Anal. **15**(3), 369–390 (1995)
58. O. Ditlevsen, Life quality index revisited. Struct. Saf. **26**, 443–451 (2004)
59. O. Ditlevsen, P. Friis-Hansen, Life quality allocation indexan equilibrium economy consistent version of the current life quality index. Struct. Saf. **27**, 262–275 (2005)
60. Organisation for Economic Co-operation & Development (OECD). *Statistics database*, OECD. http://www.oecd.org (2011)

61. M.H. Faber, E. Virguez-Rodriguez, Supporting decisions on global health and life safety investments, in *11th International Conference on Applications of Statistics and Probability in Civil Engineering, ICASP11*, Balkema, August (2011)
62. Organisation for Economic Co-operation & Development (OECD). *Employment outlook*, OECD. http://www.oecd.org (2011)
63. N. Keyfitz, *Applied Mathematical Demography* (Springer, New York, 1985)
64. O. Spackova, D. Straub, Cost-benefit analysis for optimization of risk protection under budget constraints. Risk Anal. **35**(5), 941–959 (2015)
65. E. Rosemblueth, E. Mendoza, Optimization in isostatic structures. J. Eng. Mech., ASCE, (EM6):1625–42 (1971)
66. E. Rosemblueth, Optimum design for infrequent disturbances. Structural Division, ASCE, 102-ST9:1807–1825 (1976)
67. A.M. Hasofer, Design for infrequent overloads. Earthq. Eng. Struct. Dyn. **2**(4), 387–388 (1974)
68. J.D. Campbell, A.K.S. Jardine, J. McGlynn, *Asset Management Excellence: Optimizing Equipment Life-cycle Decisions* (CRC Press, Florida, 2011)
69. M. Sánchez-Silva, R. Rackwitz, Implications of the high quality index in the design of optimum structures to withstand earthquakes. J. Struct., ASCE **130**(6), 969–977 (2004)
70. Y.K. Wen, Y.J. Kang, Minimum building lifecycle cost design criteria. II: applications. *J. Struct. Eng., ASCE*, **127**(3), 338–346 (2001)
71. I. Iervolino, M. Giorgio, E. Chioccarelli, Gamma degradation models for earthquake-resistant structures. Struct. Saf. **45**, 48–58 (2013)
72. A. Petcherdchoo, J.S. Kong, D.M. Frangopol, L.C. Neves, NLCADS (New Life-Cycle Analysis of Deteriorating Structures) User's manual; a program to analyze the effects of multiple actions on reliability and condition profiles of groups of deteriorating structures. Engineering and Structural Mechanics Research Series No. CU/SR-04/3, Department of Civil, Environmental, and Architectural Engineering, University of Colorado, Boulder Co (2004)
73. D.M. Frangopol, M.J. Kallen, M. van Noortwijk, Probabilistic models for life-cycle performance of deteriorating structures: review and future directions. Program. Struct. Eng. Mater. **6**(4), 197–212 (2004)
74. D.M. Frangopol, D. Saydam, S. Kim, Maintenance, management, life-cycle design and performance of structures and infrastructures: a brief review. Struct. Infrastruct. Eng. **8**(1), 1–25 (2012)
75. RCP, *COMREL-V8.0*. RCP, http://www.strurel.de/comrel.htm (2012)
76. R.E. Barlow, F. Proschan, *Mathematical Theory of Reliability* (Wiley, New York, 1965)
77. E.E. Lewis, *Introduction to Reliability Engineering* (Wiley, New York, 1994)
78. K.W. Lee, *Handbook on Reliability Engineering* (Springer, London, 2003)
79. D.R. Cox, *Renewal Theory* (Metheun, London, 1962)
80. Y.K. Wen, *Structural Load Modeling and Combination for Performance and Safety Evaluation* (Elsevier Science, New York, 1990)
81. R.E. Melchers, *Structural Reliability-Analysis and Prediction* (Ellis Horwood, Chichester, 1999)
82. A. Haldar, S. Mahadevan, *Probability, Reliability and Statistical Methods in Engineering Design* (Wiley, New York, 2000)
83. U.K. Legislation, *Health and safety at work Act 1974* (1974)

Chapter 10
Maintenance Concepts and Models

10.1 Introduction

One of the main objectives of life-cycle analysis is to provide a framework for the design of an optimal maintenance policy; that is, to define a program of interventions that maximizes the profit derived from the existence of the project while assuring its safety and availability. Maintenance activities are understood to include all physical processes that are intended to increase the useful life of the system. These activities may be initiated because the system is observed to be in a particular system state identified as a fault or failure (generally referred to as *reactive* or *corrective* maintenance), or they may be initiated before such a fault is observed (generally referred to as *preventive* maintenance). This chapter addresses some of the maintenance issues involved in managing infrastructure systems and describes methods for developing optimal maintenance strategies. It also presents a review of current and widely used methods as well as a detailed discussion of two relatively new methods that are highly relevant for managing infrastructure systems.

10.2 Overview of Maintenance Planning

10.2.1 Definition of Maintenance

Maintenance is defined as a set of actions taken in order to keep a system (e.g., machine, building, infrastructure) operating at or above a pre-specified level of service. Maintenance differs from reconstruction in that it is planned and executed during the operational phase of the system, prior to planned complete replacement. The British Standards BS4778-3.1 (1991) or BS3811 (1993) defines maintenance as [1]:

M. Sánchez-Silva and G.-A. Klutke, *Reliability and Life-Cycle Analysis of Deteriorating Systems*, Springer Series in Reliability Engineering,
DOI 10.1007/978-3-319-20946-3_10

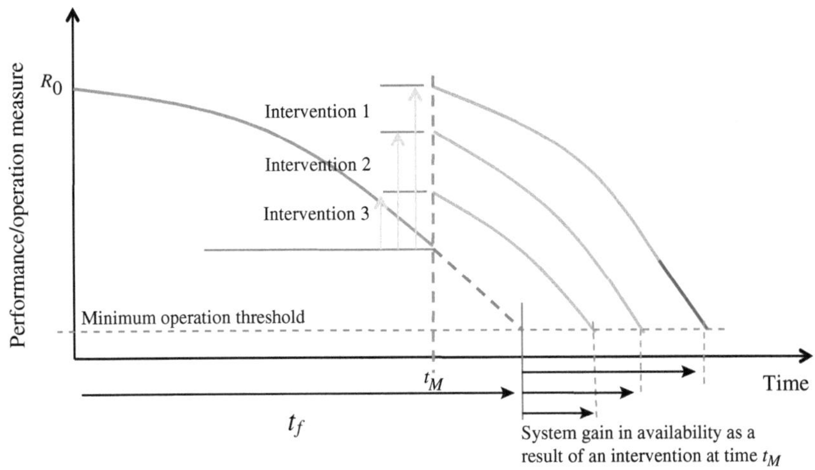

Fig. 10.1 Effect of various intervention measures on the expected time to failure

> "the process of maintaining an item in an operational state by either preventing a transition to a failed state or by restoring it to an operational state following failure".

Maintenance comprises both the technical and associated administrative actions intended to preserve a system at, or restore it to, a level in which it can perform its required function (BS3811, 1984).

The long-term benefits of both preventive and reactive maintenance include improving the availability and extending the lifetime of the system (Fig. 10.1), reducing replacement cost, decreasing system downtime and improving spares inventory management [1]. Maintenance and replacement issues for deteriorating systems have been extensively studied in many engineering fields. In addition to many books on the subject, there is a vast literature of research papers related to maintenance. In particular, various state of the art reviews on maintenance methods have been published during the last decades; see for instance [2–9].

10.2.2 Classification of Maintenance Activities

The standard approach to classifying maintenance activities divides them into *preventive* and *corrective* or *reactive* actions.

Preventive maintenance involves all actions directed toward reducing future costs associated with failure (i.e., the drop in performance indicators below a minimum operational level) while the system is in a satisfactory operating condition. Preventive maintenance is associated with activities such as planned component replacement and structural retrofitting or upgrading, and also includes so-called *essential* maintenance, which are the activities necessary to avoid imminent failure. In many cases, preventive maintenance may require the system be taken out of service for some time, and therefore there may be associated downtimes, but the objective is that these times be minimal and may be performed during non-peak operating times. Preventive maintenance may or may not be based on monitoring the condition of the system while it is operating.

On the other hand, corrective maintenance focuses on the interventions required once a failure has occurred. Corrective maintenance is frequently more expensive than preventive maintenance since the cost may include, in addition to the repair cost, higher downtime costs or replacement of undamaged system components. While preventive maintenance is commonly carried out based on a predefined policy (e.g., fixed time intervals), corrective maintenance is performed at unpredictable time intervals because failure times cannot be known *a priori*.

Maintenance activities may also be classified based on the extent of the intervention; this is, the increase in improvement of the system's performance relative to its original state (Fig. 10.2). Thus, if maintenance is required and executed, four possible strategies may be considered [1]:

- *Perfect maintenance*: the intervention takes the system to its initial condition (*as good as new*).
- *Minimal maintenance*: at a system failure, the intervention takes the system to an operational state but does not materially improve the condition realized just before the failure (*as bad as old*).
- *Imperfect maintenance*: the condition of the system after the intervention is somewhere inbetween *as good as new* and *as bad as old*.
- *Update maintenance*: the system is taken to a performance condition that is better than the initial condition (*better than new*).

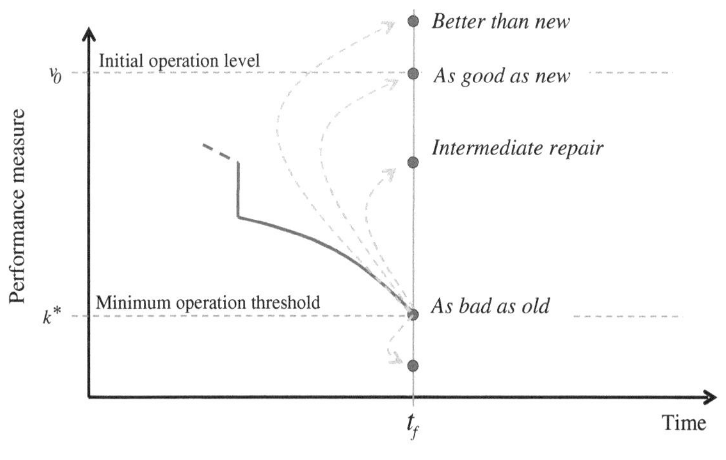

Fig. 10.2 Possible repair strategies

In addition, particularly in preventive maintenance, there is always the possibility that the system condition is degraded instead of improved. This type of maintenance, which is not commonly intentional, is described as "*worse*" maintenance and frequently results in system failure (e.g., equipment breakdown) [1].

10.2.3 Maintenance Management

Maintenance management is a subject of great interest in many engineering areas from manufacturing to engineered structures. In this section we present and discuss briefly some key concepts related to maintenance planning and maintenance policies. Additional information and discussions can be found in [2, 3, 5–7, 10–12].

Maintenance Planning

Maintenance planning is concerned with estimating the time of interventions and the extent of repairs. It is commonly based on general guidelines and engineering judgment; and in most cases, it is prescriptive and does not take into account the structure-specific characteristics or make optimal use of the observed performance data [13].

Scheduling times and extent of a maintenance program is commonly expressed as an optimization problem whose objective is to maximize the system availability at minimum cost. This means maintaining the system operating in acceptable conditions during the maximum length of time. Classical maintenance strategies include:

- *Periodic maintenance*: consists of periodically inspecting, servicing and updating parts of the system to prevent failure; it is also called *time-based* maintenance.
- *Predictive maintenance*: it is carried out based on the results of inspection or diagnosis of the system. Compared to periodic maintenance, predictive maintenance can be interpreted as *condition-based* maintenance.
- *Corrective maintenance*: its is executed only after the system failure, bringing the system back in service.

Maintenance Policies

Many maintenance policies for systems or components have been reported in the literature [14]; they can be grouped into the following (see [1]):

- *Periodic*: maintenance is carried out at fixed time intervals regardless of the failure history.
- *Age-dependent*: maintenance is carried out at some predetermined age or repaired upon failure.
- *Failure limit*: maintenance is performed only when the failure rate (or any performance indicator) reaches a predefined threshold level; the system is also repaired at failures.

- *Sequential*: maintenance is carried out at time intervals, which become shorter with time.
- *Repair limit*: this policy evaluates the system at failure and is divided into: *repair cost limit* and *repair time limit*. In the former the system is repaired if the repair cost is less than a pre-specified value; otherwise, the equipment is replaced. In the latter, the limit is set based on the repair time instead of costs.
- *Repair number counting*: the system is replaced at kth failure; the first $k-1$ failures are addressed as minimal repair. Upon replacement, the process restarts.
- *Warranty-based*: maintenance and replacement are defined according to the conditions specified in warranty policies.

When dealing with groups of components there are some additional policies among which the *Group maintenance* strategy is the most common. This policy can be divided into:

- *T-age group replacement*: the systems or its components are replaced when the system is of age T.
- *M-failure group*: calls for a system inspection, repair or replacement after m failures have been observed.
- *Combined case*: combines *T-age* and *m-failure* policies selecting whichever comes first.

Further information on these policies can be found in [1, 14, 15].

10.2.4 The Role of Inspections in Maintenance Planning

In many complex systems, particularly infrastructure systems, it may not be possible to observe the condition of the system continuously. In such systems, deliberate inspections aimed at determining the condition of the system at a given time play a major role and are an integral part of a maintenance strategy. In many cases, inspections may determine the level of degradation experienced by the system; in other cases, such as in stand-by or protective systems, they may simply determine whether the system is operational or not. In either case, inspections return valuable information to the operator that can be used in scheduling future interventions. However, inspections bear costs that must be considered in maintenance planning. For example, inspections may require that operations be discontinued or curtailed, resulting in a loss or reduction of productive output during the inspection. Inspections may require destructive testing, in which case some replacement or repair cost will be incurred regardless of the state of the system. Inspecting systems in remote locations (e.g., bridges or remote roads) may involve considerable costs for a maintenance crew to access the location. These costs must be included in determining an overall maintenance plan for systems requiring inspection.

The definition of a maintenance strategy is strongly related with the inspection policy. In Fig. 10.3 we present a tree-like structure that describes the relationship

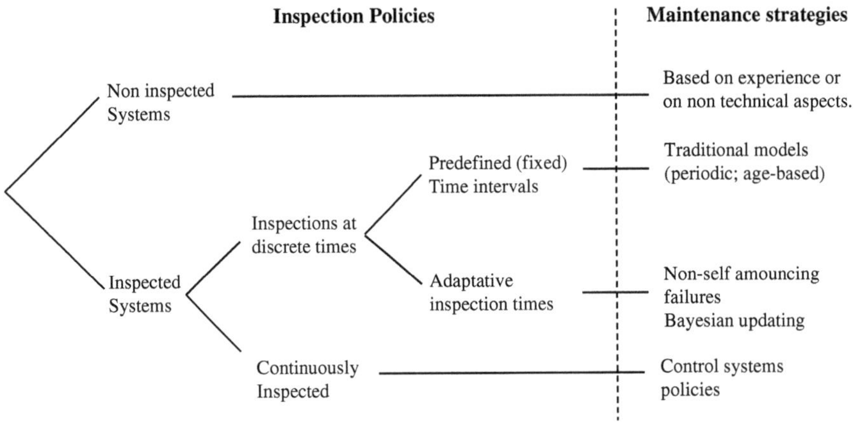

Fig. 10.3 Relationship between inspection and maintenance policies

between inspection and maintenance policies. The figure is not intended to be comprehensive but to make the point that the strategy to evaluate the state (condition) of the system over time is central to an effective maintenance strategy. In many studies the problem of maintenance is addressed independently of the inspection policy; this is equivalent to the upper case in Fig. 10.3. However, an optimal maintenance policy requires balancing the cost/benefit relationship of a particular inspection program. Some factors that influence such decision include direct costs, accessibility, impact on the system availability and criticality of the system, among others.

Bayesian Updating as a Result of Inspections

In systems that can be monitored sporadically via inspections, new data may be acquired that could be used to update performance estimates. For instance, if a bridge structure is damaged after an earthquake, its future performance depends on its condition after the event and not only on the initial state. Thus, if there is information available about the state of the bridge via inspections, it should be incorporated into the analysis to obtain a better estimation of its future performance. In this regard, Bayesian analysis provides a suitable framework to incorporate new information as to how the system evolves with time [16, 17]. Details on Bayesian analysis are provided in the Appendix; here we present an example to illustrate the value of Bayesian updating based on inspections.

Example 10.54 Consider a system whose initial state is $V(0) = v_0 = 100$ (in appropriate units). The system degrades over time as a result of shocks, which occur randomly in time. Based on past records of similar systems, it has been observed that shock sizes are exponentially distributed with parameter $\lambda = 0.1$ with a coefficient of variation COV $= 25\,\%$. The system was inspected after the first two shocks and the results showed that after the first one, the system state went down by 38.25 units and the second event brought it further down 14.25 additional units. Then, we are interested in re-evaluating the parameter λ to better estimate its future performance.

The shock size probability functions can be written as:

$$G_Y(y) = P(Y \leq y) = 1 - \exp(-\lambda y) \quad \text{and} \quad g_Y(y) = \lambda \exp(-\lambda y) \tag{10.1}$$

It is known that if the Poisson rate parameter, λ, is a random variable, it is reasonably to assume a gamma prior distribution [18]; i.e.,

$$g'(\lambda) = \frac{v(v\lambda)^{k-1}}{\Gamma(k)} e^{-v\lambda}; \quad \lambda > 0 \tag{10.2}$$

According to the information available (i.e., mean $\mu_\lambda = 0.1$ with a coefficient of variation COV = 25 %) the parameters for the prior distribution are: $k = 1/COV^2 = 1/0.25^2 = 16$ and $v = k/\mu_\lambda = 160$ [19]; which leads to:

$$g'(\lambda) = \frac{160(160\lambda)^{16-1}}{\Gamma(16)} e^{-160\lambda}; \quad \lambda > 0 \tag{10.3}$$

On the other hand, the sum of n-events exponentially distributed with rate λ can be computed as [18, 19]:

$$f(y_1, y_2, \ldots, y_n|\lambda) = \prod_{i=1}^{n} \lambda e^{-\lambda y_i} = \lambda^n e^{-\lambda S_y} \tag{10.4}$$

where $S_y = \sum_{i=1}^{n} y_i$. Thus, since the new information shows that the total damage caused by the first two shocks is $S_y = y_1 + y_2 = 38.25 + 14.25 = 52.5$, the likelihood function of λ becomes:

$$L(\lambda) = f(y_1, y_2|\lambda) = \lambda^n e^{-\lambda S_y} = \lambda^2 e^{-\lambda \cdot 52.5} \tag{10.5}$$

Then, the posterior distribution is computed by using Eq. A.56:

$$\begin{aligned} f''(\lambda|S_y) &= \frac{1}{K} L(\lambda) f'(\lambda) \\ &= \frac{1}{K} \left(\lambda^n e^{-\lambda S_y}\right) \frac{v(v\lambda)^{k-1}}{\Gamma(k)} e^{-v\lambda} \end{aligned} \tag{10.6}$$

where K is the denominator in Eq. A.56. After some manipulation, the posterior distribution for λ can then be computed as [18]:

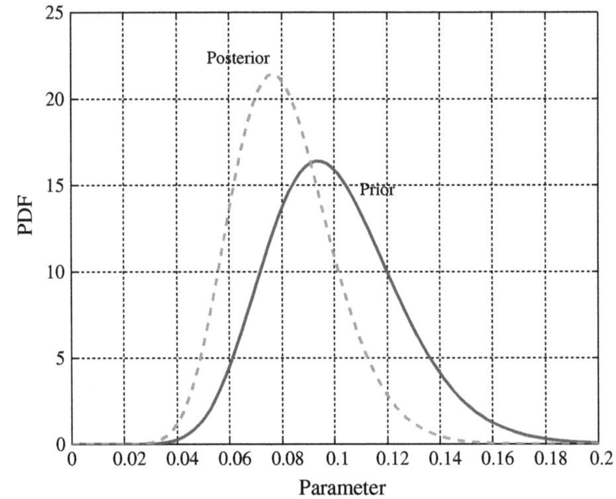

Fig. 10.4 Prior and posterior density of the parameter λ

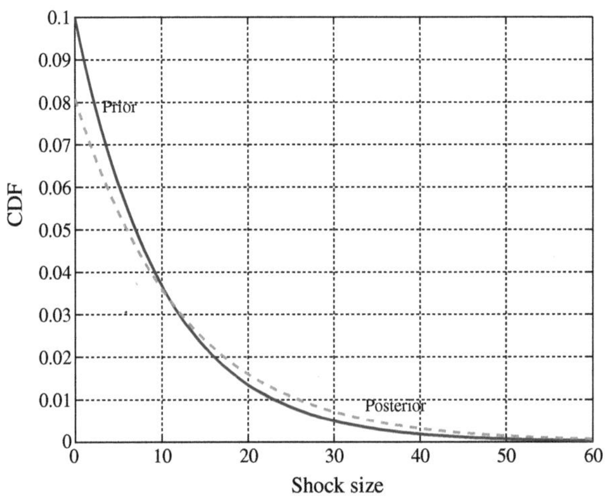

Fig. 10.5 Prior and posterior density of shock sizes

$$
\begin{aligned}
f''(\lambda|S_y) &= \frac{(v+S_y)^{k+n}\lambda^{k+n-1}e^{-\lambda(v+S_y)}}{\Gamma(k+n)} \\
&= \frac{(160+52.5)^{16+2}\lambda^{16+2-1}e^{-\lambda(160+52.5)}}{\Gamma(16+2)} \\
&= \frac{(212.5)^{18}\lambda^{17}e^{-\lambda(212.5)}}{\Gamma(18)}
\end{aligned}
\tag{10.7}
$$

The prior and posterior density function for the parameter λ are shown in Fig. 10.4. Clearly the new observations lead to a difference in the behavior of the parameter. Then, the parameter of the new shock size distribution can be replaced by the estimator of the posterior, computed as in Eq. A.57; this is:

$$\hat{\lambda}'' = \int_{-\infty}^{\infty} \lambda f''(\theta) d\lambda \tag{10.8}$$

Then, the prior and posterior density function of shock sizes will be different as shown in Fig. 10.5. The parameter of the posterior will be $\lambda'' = 0.0809$, which is about 20 % smaller than the rate initially assumed.

The Possibility of Fallible Inspections

The result of inspections is not always accurate; it may fail to identify

- if there is a need for an intervention; and/or
- the extent of the required intervention.

The need for an intervention can be expressed in terms of an indicator function $I(\mathbf{q})$ such that $I(\mathbf{q}) = 1$ indicates that an intervention is required and $I(\mathbf{q}) = 0$ that it is not; where $\mathbf{q}$ are the parameters involved in the inspection process (e.g., methodology, accuracy of evaluation). The indicative function I has been called also *detectability* function [20]. Mori and Ellingwood [20] argue that this function may not be necessarily a step function but a monotonically increasing function that has a second-order effect on the limit state probability.

Consider that the system state at time t is $V(t, \mathbf{p})$, where $\mathbf{p}$ is a random vector parameter that takes into account the system properties (e.g., material, geometry) and s^* is the system's acceptability performance threshold.[1] This means that the system does not comply with the performance standards if $V(t, \mathbf{p}) \leq s^*$. Then the results of an inspection can be classified as:

- *Type A*: the structure is in a *good* state (operating above the minimum threshold level, s^*) but the result of the inspection suggests that it is not and that an intervention is required. This conditional probability can be expressed as:

$$P_A(t) = P(I(\mathbf{q}) = 1 | V(t, \mathbf{p}) > s^*) \tag{10.9}$$

  The probability that the result of the inspection is correct (i.e., an intervention is not required) is then:

$$\begin{aligned} P_{\bar{A}}(t) &= 1 - P_A(t) \\ &= 1 - P(I(\mathbf{q}) = 1 | V(t, \mathbf{p}) > s^*) \\ &= P(I(\mathbf{q}) = 0 | V(t, \mathbf{p}) > s^*). \end{aligned} \tag{10.10}$$

[1] The value of s^* may be k^* as described in previous chapters, or any other value of interest for that matter.

- *Type B*: the structure is in a *bad* state but the result of the inspection is that it is in good state and it should not be repaired. Similarly, this conditional probability can be computed as:

$$P_B(t) = (I(\mathbf{q}) = 0|V(t, \mathbf{p}) \leq s^*) \tag{10.11}$$

Then, the probability that the inspection is correct (i.e., an intervention is required) in this case is:

$$\begin{aligned} P_{\bar{B}}(t) &= 1 - P_B(t) \\ &= 1 - P(I(\mathbf{q}) = 0|V(t, \mathbf{p}) \leq s^*) \\ &= P(I(\mathbf{q}) = 1|V(t, \mathbf{p}) \leq s^*). \end{aligned} \tag{10.12}$$

In most cases, as a result of deterioration, the probability that an intervention is required increases with time t. Streicher et al. [21] state that since, frequently, the performance indicator function $V(t, \mathbf{p})$ has a similar form as the failure function, failure and repair events become dependent events. Several inspection-based models for maintenance management will be presented and discussed in the following sections. Many models for optimizing inspection policies have been proposed in the literature; a good review is presented in [5] and a detail discussion is presented in [15].

10.3 Performance Measures for Maintained Systems

A typical sample path of a repairable engineered system is presented in Fig. 10.6. Basically system operation alternates between operation, i.e., "on" state; and failure, i.e., "off" state. The times during which the system is operating are called *uptimes*, and those during which the system is not operating are called *downtimes*.

Availability is the most common measure used to describe the performance of repairable systems. Intuitively, availability is used to measure the relationship between the length of time that the system operates appropriately to the length of time it does not (either due to failures or during times of repair). For repairable systems, depending on the particular assumptions made on the system and type of repairs, there are often several equivalent ways to compute system availability. Lie [22] gives a comprehensive classification of existing definitions of availability within different contexts; herein we will present only a few cases of particular interest.

Pointwise or *instantaneous* availability, $A(t)$, is defined as the probability that the system (component) performs satisfactorily (i.e., within the tolerances), at a given instant of time t [15, 23]. Point availability is defined as:

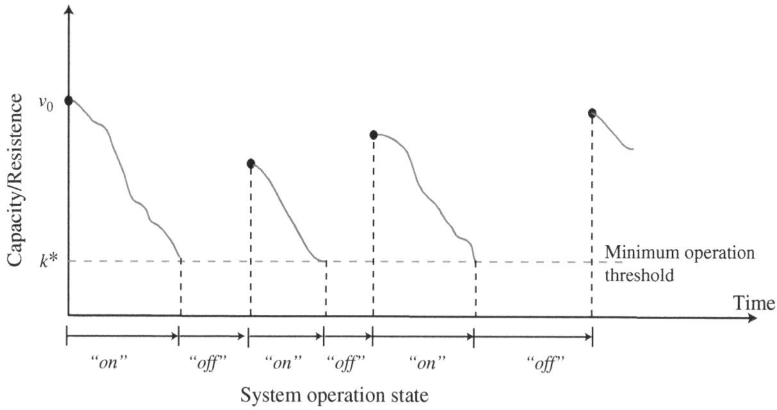

Fig. 10.6 Sample path of "on" and "off" states of reparable systems

$$A(t) = P\{\text{system is working at time } t\} \tag{10.13}$$

If the mission has a fixed length, say T, then the *mission* availability is given by

$$A(T) = \frac{1}{T}\int_0^T A(\tau)d\tau \tag{10.14}$$

and equals the expected fraction of time during the mission length T that the system is *up* (i.e., operating satisfactorily).

If the system is maintained indefinitely, the *steady-state*, *asymptotic* or *limiting interval* availability is defined as [23]:

$$A = \lim_{t\to\infty} \frac{1}{t}\int_0^t A(\tau)d\tau. \tag{10.15}$$

Other definitions of availability and a detailed discussion can be found in [24, 25]. In particular, the problem of availability for the case of multi-component systems is of great importance and has been discussed elsewhere [15, 23, 26].

Less common performance measure used to describe repairable systems included the mean time between failures (MTBF) and Mean Time To Repair (MTTR), which are, respectively, the expected length of a "typical" *on* phase in a cycle and the expected length of a "typical" *off* phase of a cycle (see Fig. 10.6); these measures are used only when the *on* and *off* phases each constitute an i.i.d. sequence.

In many models of maintained systems, it is assumed that repairs or replacements are instantaneous. In this situation, availability is not an appropriate performance measure, and typical performance measures involve total maintenance cost. In these models, as we will see in the next section, different costs are associated with repairs or replacements. If we define $C(t)$ to be the total cost of a maintenance policy in the interval (0,t], then $\mathbb{E}[C(t)]$ represents the expected total cost over that period (reflecting the random nature of the failure process). For a fixed mission length T, the relevant cost-based performance measure is $\mathbb{E}[C(T)]$, and if the planning horizon is infinite, the expected cost rate

$$K \equiv \lim_{t\to\infty} \frac{\mathbb{E}[C(t)]}{t} \tag{10.16}$$

(long-run expected cost per unit time) is used as the performance measure.

10.4 Simple Preventive Maintenance Models

Maintenance strategies have been widely studied in the literature; see [5] and references therein for an extensive survey of preventive maintenance models. In this section, we present two simple maintenance strategies that include both preventive maintenance (repair or replacement before failure) and reactive or corrective maintenance (repair or replacement at failure). In both of these strategies, we assume that actual deterioration is not observable, but the lifetime distribution of a new system is known. In the first strategy, termed *age replacement*, the system is replaced at failures or whenever its lifetime exceeds a fixed age. In the second strategy, termed *periodic replacement*, the system is preventively replaced at fixed, predetermined times, and is repaired or replaced at failures in between replacement epochs. In subsequent sections, we present two more sophisticated models that are particularly useful for infrastructure systems; these include models for systems that can be continuously monitored, and models for systems with non-self-announcing failures.

10.4.1 Age Replacement Models

In the standard age replacement model, the system is replaced upon failure or when it reaches a predetermined critical age α (Fig. 10.7). New systems, whether replaced at failure or preventively, are assumed to have statistically independent and identical lives. Age-replacement models are used in cases where the risk of failure increases with age and failures have very serious consequences, as might be the case with infrastructure systems (preventive maintenance is generally suboptimal for non-aging components [27]). Age replacement policies have been studied extensively with applications in various engineering fields; see for instance [15, 28–34]. Among replacement policies with *i.i.d.* lifetimes of new systems, stationary, non-randomized age replacement policies have been shown [35, 36] to be optimal among all "reasonable" policies (those that consider the entire replacement history).

Suppose that whenever the system is replaced preventively, a cost C_1 is incurred, and when the system is replaced at a failure, a cost C_2 is incurred, with $C_2 > C_1$.

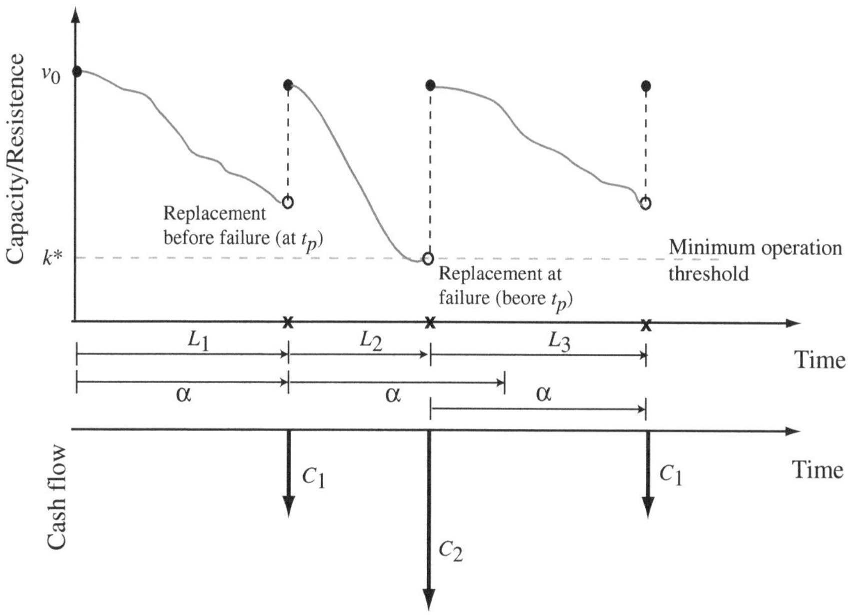

Fig. 10.7 Age replacement policy

Further, let the lifetime of a new system have distribution function F with mean $\mu < \infty$, and suppose that replacements are instantaneous. Then, the sequence of replacement times (either planned or unplanned) constitutes a renewal process, and the times between renewals has distribution

$$G(t; \alpha) = \begin{cases} F(t) & \text{for} \quad t < \alpha \\ 1 & \text{for} \quad t \geq \alpha. \end{cases} \tag{10.17}$$

(here we explicitly note the dependence of the distribution on the critical age α). Now the cost incurred in the interval $(0, t]$ is given by

$$C(t; \alpha) = C_1 N_1(t; \alpha) + C_2 N_2(t; \alpha), \tag{10.18}$$

where $N_1(t; \alpha)$ and $N_2(t; \alpha)$ are, respectively, the number of preventive and corrective replacements by time t when the policy uses the critical age α. Note that we ignore the cost of the initial system, as it has no bearing on the optimal age-replacement strategy. When the planning horizon is infinite, our objective is to find the critical age α that minimizes the long run expected cost per unit time (or expected cost rate), i.e.

$$K(\alpha) = \lim_{t \to \infty} \frac{\mathbb{E}[C(t; \alpha)]}{t} = \lim_{t \to \infty} \frac{C_1 \mathbb{E}[N_1(t; \alpha)] + C_2 \mathbb{E}[N_2(t; \alpha)]}{t} \tag{10.19}$$

Let us say that a cycle begins with a replacement and ends with the next replacement. Because cycles are independent and statistically identical, we can use results from renewal theory to express $K(\alpha)$ as

$$K(\alpha) = \frac{\text{Expected cost in a cycle}}{\text{Expected length of a cycle}} \tag{10.20}$$

Since the cycle ends with a preventive replacement if the system lifetime exceeds α and with a corrective replacement otherwise, the expected cost of a cycle is given by

$$C_1\bar{F}(\alpha) + C_2F(\alpha), \tag{10.21}$$

and the expected length of a cycle is given by

$$\int_0^\alpha u dF(u) + \alpha\bar{F}(\alpha) = \int_0^\alpha \bar{F}(u)du. \tag{10.22}$$

Putting these expressions into Eq. 10.20, we have ([15])

$$K(\alpha) = \frac{C_1\bar{F}(\alpha) + C_2F(\alpha)}{\int_0^\alpha \bar{F}(u)du} \tag{10.23}$$

Note that when $\alpha = \infty$, this policy describes the case of replacements only at failure. In this case the long run expected cost rate becomes

$$K(\infty) = \lim_{\alpha\to\infty} K(\alpha) = \frac{C_2}{\mu} \tag{10.24}$$

Optimal Maintenance Policy

The optimal maintenance policy can be determined by finding α^* that minimizes the right hand side of Eq. 10.20. If we assume that the lifetime distribution F has density f, an optimal policy can be derived based on the nature of the failure rate $h(t) = f(t)/\bar{F}(t)$ [15, 29, 31]. If $h(t)$ is continuous and strictly increasing, then

- if $h(\infty) = \lim_{t\to\infty} h(t) > \frac{C_2}{\mu(C_2-C_1)}$, there exists a finite and unique α^* that minimizes (10.20), and α^* satisfies

$$h(\alpha^*)\int_0^{\alpha^*} \bar{F}(u)du - F(\alpha^*) = \frac{C_1}{C_2 - C_1}, \tag{10.25}$$

  and the corresponding optimal expected cost rate is

$$K(\alpha^*) = (C_2 - C_1)h(\alpha^*); \tag{10.26}$$

- if $h(\infty) \leq \frac{C_2}{\mu(C_2-C_1)}$, then $\alpha^* = \infty$ and the system is replaced only at failures. In this case, the expected cost rate is given by Eq. 10.24.

As noted earlier, if $h(t)$ is non-increasing, it is never advantageous to replace preventively, and the optimal replacement age is $\alpha^* = \infty$.

Example 10.55 Consider a system with exponential lifetimes with mean μ, that is $F(t) = 1 - \exp(-t/\mu),\ t \geq 0$. The expected long-run cost per unit time for age replacement can be calculated from Eq. 10.23 as

$$\begin{aligned} K(\alpha) &= \frac{C_1 \exp(-\alpha/\mu) + C_2(1 - \exp(-\alpha/\mu))}{\int_0^\alpha (\exp(-u/\mu))du} \\ &= \frac{C_1 \exp(-\alpha/\mu) + C_2(1 - \exp(-\alpha/\mu))}{\mu(1 - \exp(-\alpha/\mu))} \\ &= \frac{1}{\mu}\left(\frac{C_1 \exp(-\alpha/\mu)}{1 - \exp(-\alpha/\mu)} + C_2\right) \end{aligned} \qquad (10.27)$$

Here, the right hand side is strictly decreasing with α, so that $\alpha^* = \infty$. This result is consistent with the optimal maintenance policy described above, since

$$h(\infty) = \lim_{t\to\infty} \frac{\frac{1}{\mu}\exp(-(t/\mu))}{\exp(-(t/\mu))} = \frac{1}{\mu} \leq \frac{C_2}{\mu(C_2 - C_1)}. \qquad (10.28)$$

Intuitively, preventive replacement is not justified when lifetimes are memoryless (exponential), as we are as likely to replace a long-lived system with a short-lived one as vice-versa.

Example 10.56 Consider a system with lognormal failure times with mean $\mu = 25$ and two possible coefficients of variation $COV = 0.2$ and $COV = 0.4$. Suppose the cost of scheduled maintenances is $C_1 = \$100$ and the cost of replacement in case of failure is $C_2 = \$300$. Compute the expected cost rate as a function of the maintenance times and find the optimal solution.

The cost rate may be computed based on Eq. 10.20, which can be easily evaluated numerically. The results are shown in Fig. 10.8. Note that as the coefficient of variation increases, the cost rate gets closer to the limiting solution ($\alpha \to \infty$); i.e., no preventive replacements. The optimal age at replacement and the corresponding cost rates are:

$$\begin{aligned} &\text{for}\quad \text{COV} = 0.4 \quad \alpha_1^* = 15.15 \quad \text{and} \quad K(\alpha_1^*) = \$8.56/\text{year}, \quad \text{and} \\ &\text{for}\quad \text{COV} = 0.2 \quad \alpha_2^* = 17.7 \quad \text{and} \quad K(\alpha_2^*) = \$6.24/\text{year}. \end{aligned}$$

The limiting solution can be computed analytically as (Eq. 10.24):

$$K(\infty) = \lim_{\alpha\to\infty} K(\alpha^*) = \frac{\$300}{25} = \$12/\text{year}.$$

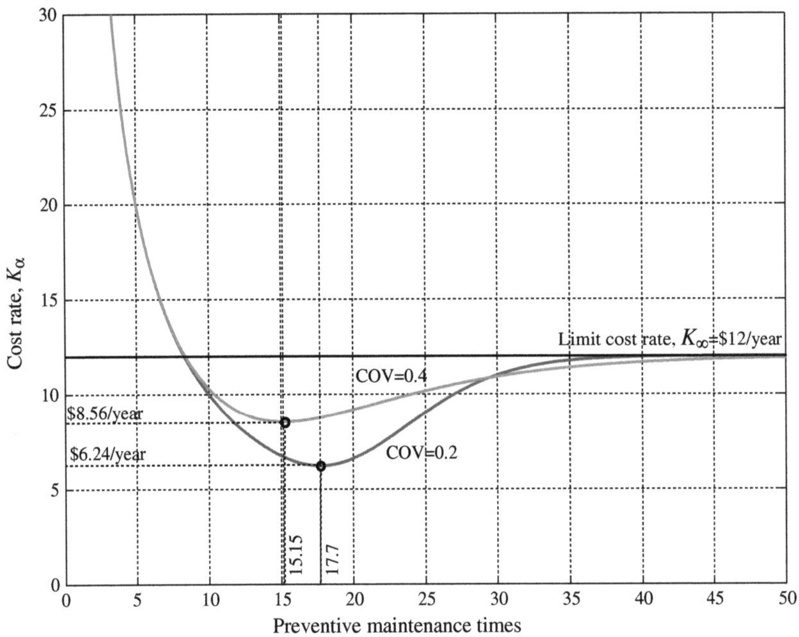

Fig. 10.8 Age replacement policy; maintenance time intervals and limiting solution

Example 10.57 Consider the case and the data used in the previous example (Example 11.57) to compute analytically the optimal solution.

First we need to evaluate the failure rate $h(t) = f(t)/\bar{F}(t)$, which clearly approaches to infinity as $t \to \infty$ and it is continuous and strictly increasing. Then, it is also clear that

$$h(\infty) > \left[\frac{C_2}{\mu_t(C_2 - C_1)}\right] = \left[\frac{300}{25(300 - 100)}\right] = 0.06,$$

which implies that the optimal times for preventive maintenance can be computed using Eq. 10.25. The derivation of the minimum according to Eq. 10.25 is shown graphically in Fig. 10.9. The corresponding minimum cost rates are then computed using Eq. 10.26,

$$K(\alpha_1^*) = (C_2 - C_1)h(\alpha_1^*) = 200 \cdot h(15.15) = \$8.56/\text{year}$$
$$K(\alpha_2^*) = 200 \cdot h(17.7) = \$6.24/\text{year}$$

Age Replacement with Discounting

As discussed in Chap. 9, life-cycle cost analysis requires that decisions are made at time 0 for costs that are incurred after time 0, and thus future costs must be discounted.

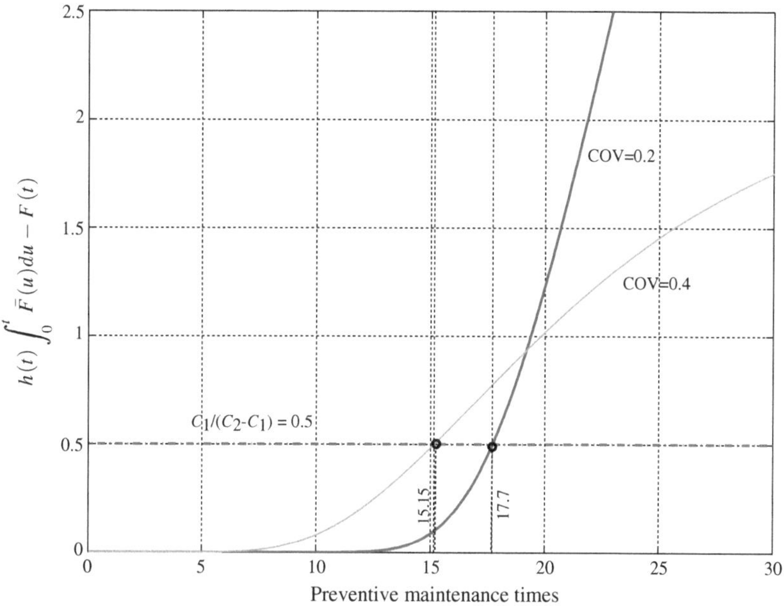

Fig. 10.9 Selection of the optimal intervention time

Assuming continuous discounting with rate $\gamma > 0$, the present value (time 0) cost of a cycle that begins at time t can be written as [15]:

$$C_1 e^{-\gamma(t+\alpha)} 1_{L>\alpha} + C_2 e^{-\gamma(t+L)} 1_{L\leq\alpha}, \tag{10.29}$$

where L is a random variable with distribution function F (representing the lifetime of the system in the cycle). Equation 10.23 can be modified to include the discount rate as follows:

$$K(\alpha) = \frac{C_1 e^{-\gamma\alpha} \bar{F}(\alpha) + C_2 \int_0^{\alpha} e^{-\gamma u} dF(u)}{\gamma \int_0^{\alpha} e^{-\gamma u} \bar{F}(u) du}. \tag{10.30}$$

In the limiting case where $\alpha \to \infty$ (no preventive replacements), we have

$$K(\infty) = \frac{C_2 F^*(\gamma)}{1 - F^*(\gamma)}, \tag{10.31}$$

where F^* is the Laplace-Stieltjes transform of F. Similarly to the case without discounting, optimal solutions for the age replacement parameter α can be derived for some special cases [15, 29]. With

$$Z = \frac{C_1[1 - F^*(\gamma)] + C_2 F^*(\gamma)}{(C_2 - C_1)[1 - F^*(\gamma)]/\gamma}, \tag{10.32}$$

we have that if $h(t)$ is continuous and strictly increasing,

- $h(\infty) > Z$ implies that there exists a finite and unique α^* that satisfies:

$$h(\alpha)\int_0^{\alpha} e^{-\gamma u}\bar{F}(u)du - \int_0^{\alpha} e^{-\gamma u}dF(u) = \frac{C_1}{C_2 - C_1} \tag{10.33}$$

  and the expected cost rate is:

$$\mathbb{E}[C(\alpha^*)] = \frac{1}{\gamma}(C_2 - C_1)h(\alpha^*) - C_1; \tag{10.34}$$

- $h(\infty) \leq Z$ implies that $\alpha^* = \infty$; this means that the component is only replaced at failures an the expected cost rate is computed as in Eq. 10.31.

10.4.2 Periodic Replacement Models

An alternative to age replacement involves preventively replacing the system at scheduled times $\tau_1, \tau_2, \ldots$, where $\tau_n = n\tau, n = 1, 2, \ldots$ (thus the time between planned replacement is fixed at τ). If the system fails between planned replacements, it is repaired (to some level) and made operational. Unlike the age replacement policy, in the periodic maintenance policy, replacements always occur τ time units after the last *planned* replacement; there is no age limit on any of the systems in operation. We let the cost of each planned replacement be C_1; planned replacements always take the system to a "good as new" condition. The cost of repairing the system at a failure (i.e. between replacements) is given by C_2 (see Fig. 10.10).

This paradigm allows for a variety of different maintenance strategies between replacements, such as complete repair ("good as new") or minimal repair ("bad as old") at failures. In these models, we again assume that replacement and repair both take a negligible amount of time to perform. Typically, complete repairs at failure might be used when the cost of a failure is much higher than the cost of a planned replacements (perhaps because of the level of disruption to the system), while minimal repairs at failure might be used when the cost of repair is less than that of a planned (complete) replacements.

We again assume that lifetimes of new systems are independent and have distribution function F and mean μ. Because planned replacements always bring the system to a "good as new" state, the times of these replacements constitute a renewal process. Let us define a cycle as the time between successive planned replacements; we can again approach the problem of minimizing long-run expected cost per unit time by analyzing the cost on each (statistically identical) cycle. Let N_i denote the number of failures (and therefore the number of corrective interventions) during the ith cycle (i.e., during the interval $[(i-1)\tau, i\tau]$. Then the expected total cost incurred in the ith cycle is

$$\mathbb{E}[C_i(\tau)] = C_1 + C_2\mathbb{E}[N_i], \tag{10.35}$$

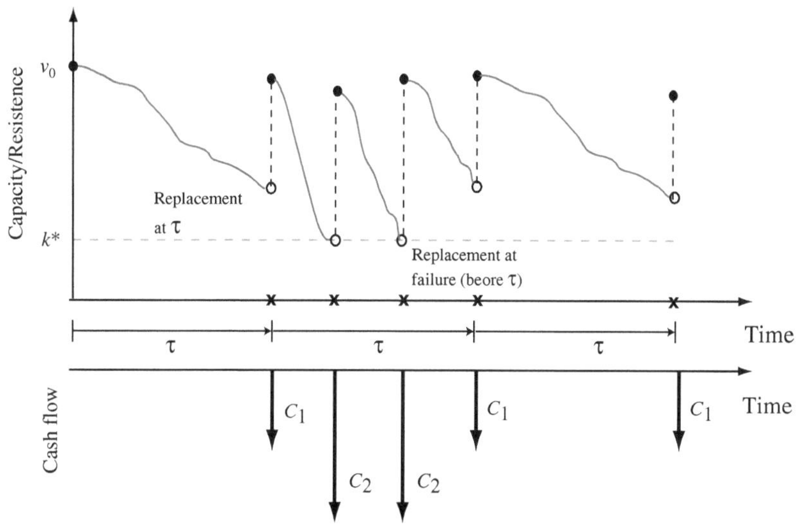

Fig. 10.10 Sample path of replacement at a fixed time interval or at failure

as each cycle comprises one planned replacement and a random number of replacements at failures. Note that the expected cycle length is simply τ.

For periodic replacements, the analysis of an optimal policy revolves around the expression for $\mathbb{E}[N_i]$, the expected number of repairs between successive planned replacements. In what follows, we consider two different types of repairs with periodic replacement.

10.4.3 Periodic Replacement with Complete Repair at Failures

In the case illustrated in Fig. 10.10, repairs between planned replacements bring the system to a good-as-new state, and thus times between repairs also form a renewal process. Therefore $\mathbb{E}[N_i]$ in Eq. 10.35 is simply the renewal function $M(t)$ associated with F, evaluated at τ:

$$\mathbb{E}[C_i(\tau)] = C_1 + C_2 M(\tau). \tag{10.36}$$

Here

$$M(t) = \sum_{n=1}^{\infty} F_n(t)$$

where F_n is the nth Stieltjes convolution of F with itself (see Chap. 3). Alternatively, $M(t)$ may be evaluated using the expression

$$M(t) = \int_0^t h(u)du, \tag{10.37}$$

where $h(u)$ is the failure rate associated with F.

Again employing a renewal argument, the cost rate for this maintenance policy is given by the ratio of the mean cost on a cycle to the mean cycle length:

$$K(\tau) = \frac{C_1 + C_2 M(\tau)}{\tau} \tag{10.38}$$

In the limiting case where $\tau \to \infty$ (interventions are carried out only at failures), we have, using the elementary renewal theorem (Chap. 3, Theorem 29),

$$K(\infty) = \lim_{\tau\to\infty} K(\tau) = \lim_{\tau\to\infty} \frac{C_1 + C_2 M(\tau)}{\tau} = \frac{C_2}{\mu}, \tag{10.39}$$

which is just the cost of replacement at failure times the rate of failures.

Optimal Policy

The objective is to find the optimal planned replacement interval τ^* that minimizes the cost rate $K(\tau)$ (Eq. 10.38). Differentiating $K(\tau)$ with respect to τ and setting the expression equal to zero we obtain

$$\tau m(\tau) - M(\tau) = \frac{C_1}{C_2}, \tag{10.40}$$

where $m(t) \equiv dM(t)/dt$ is the renewal density. In practice, minimization of the cost function requires evaluating the renewal function, which often must be done numerically. Some asymptotic expansions and numerical model are available in the literature [31].

Once τ^* has been obtained from Eq. 10.40, the optimal cost rate is given by

$$K(\tau^*) = C_2 m(\tau^*). \tag{10.41}$$

Again, planned replacements only make sense if the lifetime distribution of the component fulfills some aging condition such as IFR, NBU or NBUE [31].

Example 10.58 Consider a system where components have Gamma distributed lifetimes with parameters $n = 2$ and $\lambda > 0$. For this special case of the Gamma distribution, the renewal function has the following expression [31]

$$M(t) = \frac{\lambda t}{2} - \frac{1 - \exp\{-2\lambda t\}}{4}.$$

The cost rate using a planned replacement interval τ is then

$$K(\tau) = \frac{C_1 + C_2 M(\tau)}{\tau}$$

then, the optimal maintenance interval τ^*, can be obtained by making $dK(\tau)/d\tau = 0$; and therefore solving

$$\frac{d}{d\tau} M(\tau) = \frac{M(\tau)}{\tau} + \frac{C_1}{C_2 \tau}$$

A finite solution for τ^* can be found if $C_1/C_2 < 1/4$; in other words, failure replacements are at least four times more expensive than preventive replacements [31].

Example 10.59 Consider a system where the cost of planned replacements is $C_1 =$ \$50 and the cost of replacement at failure is $C_2 =$ \$300. Let us consider two different time to failure time distributions, both with mean $\mu = 50$ years. The first has uniform density

$$f_1(t) = \begin{cases} \frac{1}{100} & 0 \leq t < 100 \\ 0 & otherwise \end{cases}$$

and the second has a lognormal density with COV$= 0.25$. Then, for the first case, we have

$$\begin{aligned} M(\tau) = \int_0^{t_p} h(u)du &= \int_0^{\tau} \frac{f_1(u)}{1 - F_1(u)} du \\ &= \int_0^{\tau} \frac{1/100}{1 - u/100} du \end{aligned}$$

and the cost rate can be evaluated as in Eq. 10.38

$$K(\tau) = \frac{C_1 + C_2 M(\tau)}{\tau} = \frac{C_1 + C_2 \int_0^{\tau} \frac{1/100}{1 - u/100} du}{\tau}$$

This expression is minimized at $\tau^* = 41$ years at a cost of $K(\tau^*) =$ \$5.08/year. In the second case, a closed form expression for the cost rate is difficult to obtain, but it can be minimized numerically using software such as Matlab^{TM}. In this case, the optimal planned replacement interval is $\tau^* = 29$ years with $K(\tau^*) =$ \$1.92/year. Figure 10.11 plots the cost rate as a function of the replacement interval τ for both cases and shows the optimal values.

Complete Repair with Discounting

Again, using a continuous discounting function $\exp(-\gamma t)$ with $\gamma > 0$, the discounted total expected cost on a cycle for a planned replacement interval τ is [15]:

$$\mathbb{E}[C_i(\tau)] = C_1 \exp(-\gamma\tau) + C_2 \int_0^{\tau} m(t) \exp(-\gamma t)dt, \tag{10.42}$$

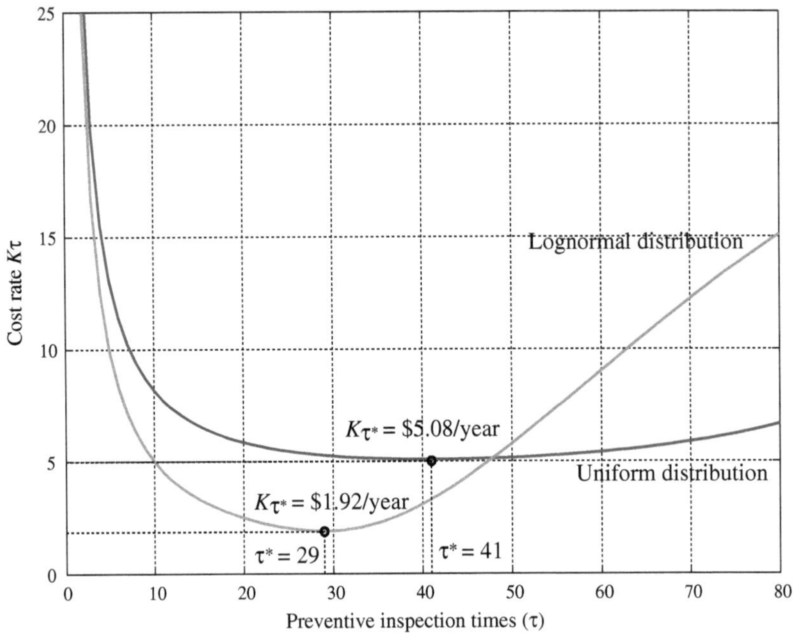

Fig. 10.11 Cost rate as function of the replacement times for two probability distribution functions

and therefore, the discounted cost rate can be computed as

$$K(\tau;\gamma) = \frac{C_1 \exp(-\gamma\tau) + C_2 \int_0^{\tau} \exp(-\gamma t) m(t) dt}{1 - \exp(-\gamma\tau)} \tag{10.43}$$

Following the same reasoning structure as in the previous section; i.e., differentiating $K(\tau;\gamma)$ (Eq. 10.43) with respect to τ and setting the expression equal to zero, we have

$$\frac{1 - \exp(-\gamma\tau^*)}{\gamma} m(\tau^*) - \int_0^{\tau^*} \exp(-\gamma t) m(t) dt = \frac{C_1}{C_2} \tag{10.44}$$

Then, the optimal time interval is obtained by solving for τ^* in Eq. 10.44; the optimal cost rate is:

$$K(\tau^*;\gamma) = \frac{C_2}{\gamma} m(\tau^*) - C_1 \tag{10.45}$$

Example 10.60 Based on the data used in Example 10.59 and considering that the time between failures follows a lognormal distribution with mean $\mu = 50$ and COV$= 0.25$, we are interested in evaluating the discounted cost rate. For comparative purposes, the effect of three discount rates on the cost rate were evaluated; they are: $\gamma = \{0.03,\ 0.05,\ 0.1\}$.

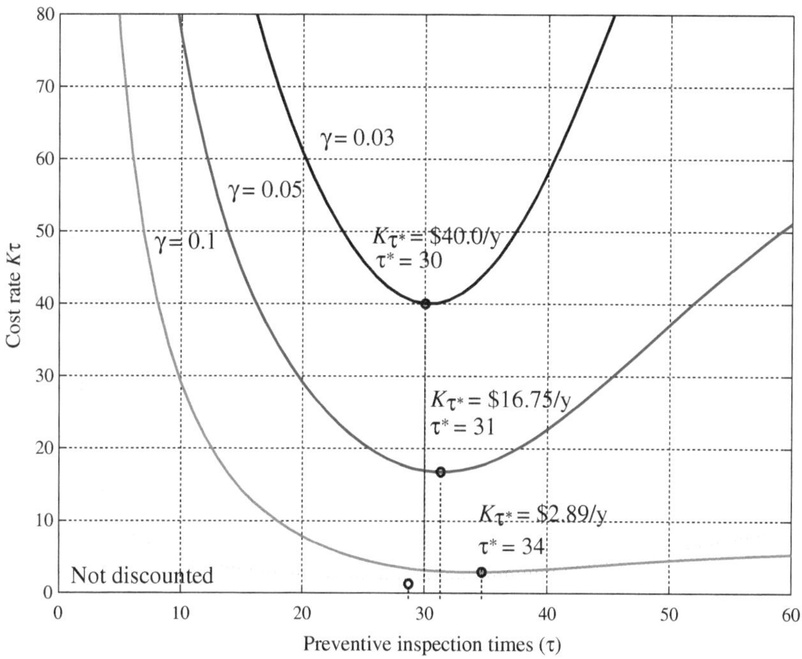

Fig. 10.12 Discounted cost rate for periodic replacements

The cost rate in every case was computed according to Eq. 10.43. The results are shown in Fig. 10.12. It can be observed that larger discount rates lead to smaller values of the discounted cost rate $K_{\tau,\gamma}$. Although thee is not much difference between the optimal times; i.e., $\tau^* = \{29,\ 30,\ 31,\ 34\}$, the values of the cost rate do change significantly, $K_{\tau,\gamma} = \{1.92,\ 40,\ 16.75,\ 2.89\}$; these values are indicated in the figure. The optimal cost rate results can be validated using Eq. 10.45 where $m(\tau^*)$ needs to be evaluated numerically.

No Replacement at Failure

Consider a particular case in which the system is maintained at time τ; but if it fails before τ it is not repaired and remains without operating until the time τ, where it is repaired (Fig. 10.13). This type of problem is common in cases when inspections to detect the condition of the system can only be carried out at fixed time intervals.

The mean time from failure to failure detection is:

$$\int_0^{\tau} (\tau - t)dF(t) = \int_0^{\tau} F(t)dt \tag{10.46}$$

where $F(t)$ is the probability distribution of the time until failure with mean μ. If C_1 is the cost of planned replacement and C_3 the downtime cost per time unit (Fig. 10.13), the expected cost rate becomes [15]

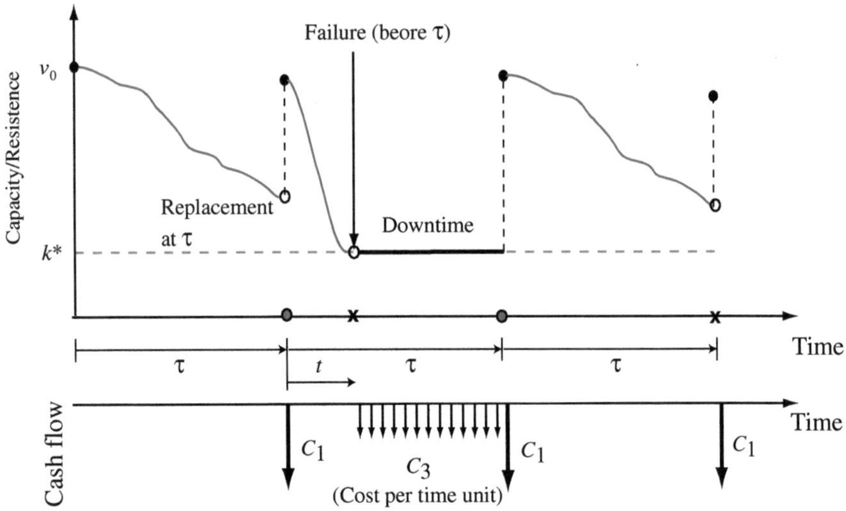

Fig. 10.13 Sample path of replacement at a fixed time intervals only

$$K(\tau) = \frac{1}{\tau}\left[C_3 \int_0^{\tau} F(t)dt + C_1\right] \tag{10.47}$$

Differentiating Eq. 10.47 with respect to τ and equating to 0,

$$\tau F(\tau) - \int_0^{\tau} F(t)dt = \frac{C_1}{C_3}; \qquad \text{or} \qquad \int_0^{\tau} tdF(t) = \frac{C_1}{C_3} \tag{10.48}$$

If $\mu > C_1/C_3$ there exists an optimal time τ^* that uniquely satisfies Eq. 10.48; and the corresponding optimal cost rate becomes [15],

$$K(\tau^*) = C_3 F(\tau^*) \tag{10.49}$$

10.4.4 Minimal Repair at Failures

For large, complex systems, it is often too expensive to completely replace the system at failures, so we may consider a maintenance strategy that does only what is necessary to make the system operational if it fails between planned replacements. This might be the case for a system consisting of many components, where we prefer to replace a failed component rather than the entire system. In this case the repair after failure renders the system operational with the same failure rate as before failure. This approach has been used extensively in electrical and mechanical systems [37]; and some modifications for special problems, mainly related to cost

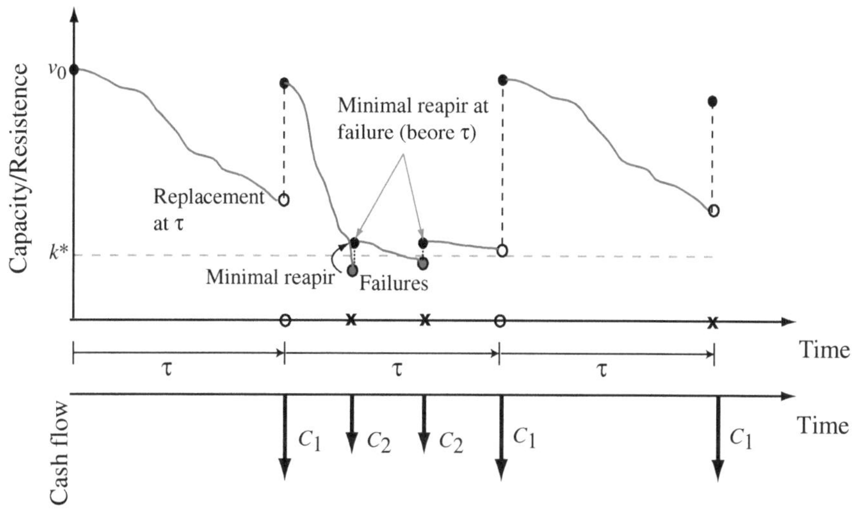

Fig. 10.14 Minimal repair replacement policy

optimization, have been proposed in [38–42]. Figure 10.14 shows a sample path of periodic replacement with minimal repair.

Again, we let F denote the distribution of the lifetime of a new system, and suppose that each time the system fails, it undergoes *minimal repair*. By minimal repair, we mean that, if the successive times between failures of a minimally repaired system are denoted by $X_1, X_2, X_3, \ldots$, then

$$\Pr(X_n \le t | X_1 + X_2 + \cdots + X_{n-1} = t) = \frac{F(t+x) - F(t)}{\bar{F}(t)}, n = 2, 3, \ldots, x > 0, t \ge 0; \tag{10.50}$$

that is, a system that fails at time t and is minimally repaired operates from t onward as if had operated continuously for t time units. Of course, the right hand side of Eq. 10.50 can also be written as

$$\int_t^{t+x} h(u)du, \tag{10.51}$$

where h is the failure rate associated with F, so minimal repair implies that the failure rate of the system in service is unchanged just after the repair.

For a new system that begins operating at time 0 and is subsequently minimally repaired, it can be shown [15] that the number of failures $N(t)$ in $[0, t)$ has distribution

$$\Pr(N(t) = n) = \frac{[H(t)]^n}{n!} e^{-H(t)}, \quad n = 0, 1, 2, \ldots, \tag{10.52}$$

where $H(t) = \int_0^t h(u)du$ is the *cumulative hazard function*. That is, the number of failures in $[0, t)$ for a minimally repaired system has a Poisson distribution with mean $H(t)$. Moreover, if $h(t)$ is increasing, then $\lim_{t\to\infty} h(\infty)$ exists (it may be ∞), and the expected times between successive failures is a decreasing sequence whose limiting value is $1/h(\infty)$.

Recalling Eq. 10.35, the expected cost during a planned replacement cycle of length τ of a minimally repaired system becomes

$$\mathbb{E}[C_i(\tau)] = C_1 + C_2 H(\tau), \tag{10.53}$$

and the long-run expected cost per unit time (the cost rate) is

$$K(\tau) = \frac{C_1 + C_2 H(\tau)}{\tau} \tag{10.54}$$

For the case of no planned replacements (minimal repairs only), we have

$$K(\infty) = \lim_{\tau\to\infty} K(\tau) = \lim_{\tau\to\infty} C_2 \frac{H(\tau)}{\tau} = C_2 h(\infty), \tag{10.55}$$

provided $h(\infty)$ exists (it may be infinite).

Optimal Policy

As in the other models described in this chapter, the objective of an optimal policy is to determine the replacement interval τ^* that minimizes the cost rate. Differentiating the right hand side of Eq. 10.54 with respect to τ and setting it equal to 0, we obtain

$$\tau h(\tau) - H(\tau) = \frac{C_1}{C_2}; \quad \text{or} \quad \int_0^\tau u dh(u) = \frac{C_1}{C_2}. \tag{10.56}$$

If $h(t)$ is continuous and strictly increasing, and if additionally $\int_0^\infty u dh(u) > C_2/C_1$, then there exists a unique solution for τ^* and the corresponding cost rate is

$$K(\tau^*) = C_2 h(\tau^*) \tag{10.57}$$

Replacement with Discounting

As discussed in Chap. 9 and in previous sections, life-cycle cost analysis requires that decisions are made at $t = 0$ and, therefore, costs after time 0 should be discounted. Again assuming continuous discounting with rate $\gamma > 0$, the discounted cost rate can be written as

$$K(\tau; \gamma) = \frac{C_1 e^{-\gamma\tau} + C_2 \int_0^\tau e^{-\gamma u} h(u)du}{1 - \exp(-\gamma\tau)} \tag{10.58}$$

The optimal replacement interval τ^* incorporating the discount rate then satisfies

$$\frac{1-e^{-\gamma\tau}}{\gamma}h(\tau) - \int_0^{\tau} e^{-\gamma u}h(u)du = \frac{C_1}{C_2} \tag{10.59}$$

with the corresponding optimal cost rate,

$$K(\tau^*) = \frac{C_2}{\gamma}h(\tau^*) - C_1 \tag{10.60}$$

There are many generalizations to the basic minimal repair model, incorporating, for example, age-dependent repair costs, a limited number of minimal repairs before complete replacement and imperfect minimal repairs (see [15] or [5] for extensive references).

10.4.5 Summary of Periodic Replacements

The periodic replacement models presented in this section share some basic structure in their formulation. In each of these models, the cost rate has the form

$$K(\tau) = \frac{C_1 + C_2\Phi(\tau)}{\tau} \tag{10.61}$$

where Φ may represent M in Eq. 10.38; or H in Eq. 10.54 depending upon the case considered.

Similarly, for the periodic replacement with discounting,

$$K(\tau, \gamma) = \frac{C_1 e^{-\gamma\tau} + C_2\int_0^{\tau} e^{-\gamma u}\phi(u)du}{1-e^{-\gamma\tau}} \tag{10.62}$$

where $\phi(t) = \Phi'(t)$ in Eqs. 10.43 (i.e., $m(t)$) and 10.58 (i.e., $h(t)$). The optimal solution, i.e., optimal preventive maintenance time $\tau = \tau^*$, can be obtained by derivation with respect to τ and equating to 0. Note that for the case of age replacement, the corresponding equations are slightly different: these are: Eq. 10.23 for the cost rate and Eq. 10.30 for the discounted cost rate.

The main expressions for each model are summarized in Table 10.1. The cases of combined replacement models; i.e., age, periodic and block replacements; as well as those related to imperfect maintenance are discussed in [15, 31].

Table 10.1 Summary of the main quantities for different maintenance policies

Quantity	Expression*	Equation
Age-replacement models:		
Cost rate	$K(\alpha) = \frac{C_1\bar{F}(\alpha)+C_2F(\alpha)}{\int_0^{\alpha}\bar{F}(u)du}$ and $K(\infty) = \frac{C_2}{\mu}$	10.20–10.24
Optimum	$h(\alpha^*)\int_0^{\alpha^*}\bar{F}(u)du - F(\alpha^*) \geq \frac{C_1}{C_2-C_1}$	10.25
Discounted	$K(\alpha,\gamma) = \frac{C_1e^{-\gamma\alpha}\bar{F}(\alpha)+C_2\int_0^{\alpha}e^{-\gamma u}dF(u)}{\gamma\int_0^{\alpha}\bar{F}(u)du}$	10.30
Periodic replacement: complete repair:		
Cost rate	$K(\tau) = \frac{C_1+C_2M(\tau)}{\tau}$ and $K(\infty) = \frac{C_2}{\mu}$	10.38–10.39
Optimum	$\tau^*m(\tau^*) - M(\tau^*) = \frac{C_1}{C_2}$	10.40
Discounted	$K(\tau,\gamma) = \frac{C_1e^{-\gamma\tau}+C_2\int_0^{\tau}e^{-\gamma u}m(u)du}{1-e^{-\gamma\tau}}$	10.43
Minimal repair at failures:		
Cost rate	$K(\tau) = \frac{C_1+C_2H(\tau)}{\tau}$ and $K(\infty) = C_2h(\infty)$	10.54–10.55
Optimum	$\tau^*h(\tau^*) - H(\tau^*) = \frac{C_1}{C_2}$	10.56
Discounted	$K(\tau,\gamma) = \frac{C_1e^{-\gamma\tau}+C_2\int_0^{\tau}e^{-\gamma u}h(u)du}{1-e^{(-\gamma\tau)}}$	10.58

*Go to the appropriate section for the restrictions in the applicability of these equations

10.5 Maintenance Models for Infrastructure Systems

Most large infrastructure systems have particular characteristics that distinguish their maintenance activities from, for example, those associated with vehicles, consumer products, or electronic devices. The first distinction concerns the long design lifetimes of infrastructure elements, which are typically measured in decades rather than in months or years. Because of this fact, infrastructure maintenance planning acknowledges that significant technological advances may take place between replacement or major refurbishment intervals, and future life cycle planning may need to be revised accordingly between large subcomponent rehabilitations. Thus periodic replacement with statistically identical subcomponents is generally not an appropriate assumption for infrastructure systems. Moreover, because of their intended long design lives, usage of infrastructure components is often difficult to predict with accuracy; it may increase significantly during its initial life before decreasing significantly during its later life, when newer alternatives may eventually make it obsolete. Clearly, degradation is highly influenced by usage, so that usage must explicitly be taken into account in planning maintenance activities.

Second, vehicles, consumer products and electronic devices are often comprised of off-the-shelf components whose failure characteristics have been well studied and documented. In contrast, infrastructure systems are often designed for particular applications, and although they may use well-studied materials, design and usage may be closer to one-off products, and failure characteristics are much less certain.

Third, although sensor technology is rapidly improving, it is still generally very difficult to continuously monitor the state of infrastructure degradation. For example, it may be difficult to monitor crack degradation in large concrete subcomponents. Moreover, it may not be possible to identify imminent system failures (i.e., system degradation has exceeded a safety threshold, the system is still operating, but failure may be close at hand).

As discussed at the beginning of the chapter, an important aspect of maintenance planning for infrastructure systems involves inspections, whose purpose is to assess system condition. Because infrastructure typically remains in place and may be in remote locations, inspections are generally costly and time consuming. Unlike pulling aircraft into a maintenance facility to inspect for fuselage or wing cracks, for example, inspectors must be sent to the field to check bridges for cracks visually. Inspections also typically involve removing the system from use for a significant period of time, which again is costly; while a company can plan capacity to remove aircraft from service for inspection and repair, this is typically not the case for infrastructure systems. To help mitigate the cost of inspections, more and more systems are designed now with embedded sensors that can provide real-time information on system state. However, there are difficulties that arise in fusing data from various sensors and sensor types, and decision making will likely involve sophisticated modeling of sensor information. In addition, sensor can fail and may need to be maintained/replaced as well. For these reasons, typical maintenance models that have appeared over the course of the last decades may not be appropriate for infrastructure management.

In summary, maintenance of infrastructure systems is in constant evolution and therefore must be supported by both physical advancements and developments in modeling and decision support. In the following two sections, we present two approaches for maintenance modeling that are particularly relevant to infrastructure maintenance. One approach addresses systems that can be continuously monitored (e.g. by sensors), and the second approach addresses systems that must be inspected to determine if they are above operating thresholds or not.

10.6 Maintenance of Permanently Monitored Systems

In this section we present a maintenance strategy based on *impulse control* models in which the time at which maintenance is carried out and the extent of interventions are optimized simultaneously to maximize the cost-benefit relationship. In the model the optimal timing and size of interventions are executed according to the system state, which is obtained from permanent monitoring. The model assumes that an infrastructure maintenance policy is mainly dominated by its mechanical performance. Impulse control models have been applied in diverse areas such as finance, to optimize a portfolio of risky assets with transaction costs, or to find the best strategy to execute a position in a risky asset [43, 44]; inventory control, to find the optimal size and timing of order placement [45]; and insurance, to find the optimal dividend payment for an insurance company [46]. Recently, this approach has been used in the context of optimal maintenance policies. This section is adapted from [47, 48].

10.6.1 Impulse Control Model for Maintenance

We assume that a system (e.g., structure, bridge) is subject to degradation caused by shocks that occur according to a compound Poisson process. Each shock causes a random amount of damage according to the function g as described in Sect. 4.10. We define the system capacity process $\mathscr{V} = \{V(t), t \geq 0\}$ by

$$V(t) = v_0 - \sum_{i=1}^{N(t)} g(Y_i, V(T_i^-)) \tag{10.63}$$

where $N(t)$ is a Poisson random variable with parameter $\lambda t > 0$, $\{T_i\}_{i \in \mathbb{N}}$ are the times at which shocks occur, $\{Y_i\}_{i \in \mathbb{N}}$ are independent, identically distributed, non-negative shock sizes with distribution function F, and the initial system capacity is $V(0^-) = v_0$ (Fig. 10.15). As mentioned in previous chapters, the damage inflicted by a shock may depend on both the shock size and the system capacity at the time of the shock.

We define an impulse control policy as follows.

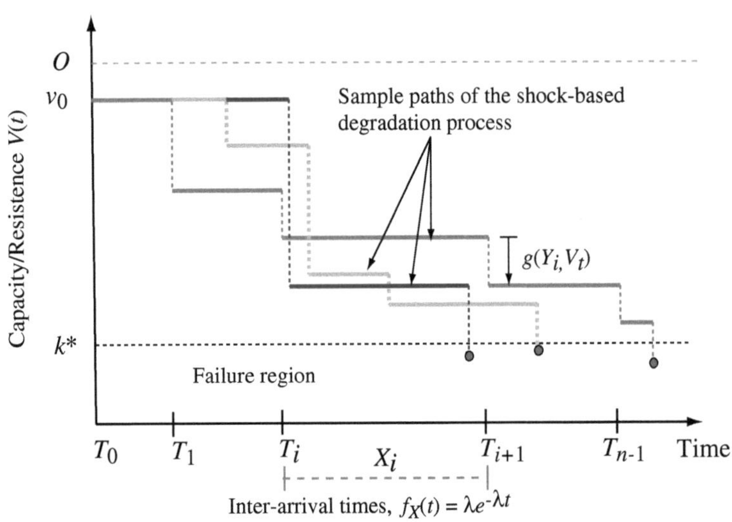

Fig. 10.15 Sample path of a shock-based degradation model

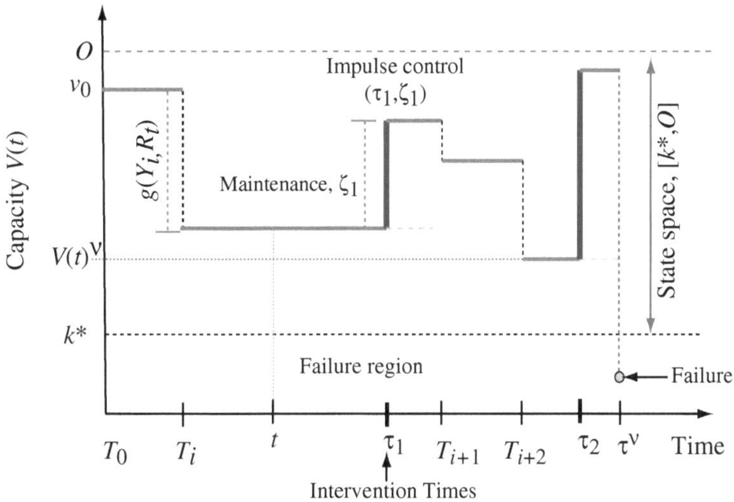

Fig. 10.16 General description of the impulse control model

Definition 48 A *maintenance policy* for the system is a double sequence $\nu = \{(\tau_i, \zeta_i)\}_{i \in \mathbb{N}}$ comprising maintenance times τ_i at which the performance is improved by an amount ζ_i. The policy is an *impulse control* if satisfies the following conditions:

1. $0 \leq \tau_i \leq \tau_{i+1}$ for all $i \in \mathbb{N}$,
2. τ_i is a stopping time with respect to the filtration $\mathscr{F}_t = \sigma\{V(s^-)|s \leq t\}$ for $t \geq 0$,
3. ζ_i is an $\mathscr{F}_{\tau_i}$-measurable random variable,

In the definition above, the second condition requires that we be able to determine whether the ith maintenance has been performed by time t or not by observing the history of the process up until time t, and the third condition requires that the improvement made at the ith maintenance be determined by the history of the process up until time τ_i. The class of impulse control policies is very general and includes periodic maintenance policies.

Given an impulse control ν, we define the controlled process $V^\nu(t)$ by

$$V^\nu(t) = v_0 - \sum_{i=1}^{N(t)} g(Y_i, V(T_i-)) + \sum_{\tau_i \leq t} \zeta_i. \tag{10.64}$$

Figure 10.16 shows a sample path of a controlled process.

Since we are interested in keeping the system capacity above a pre-defined threshold $k^* \geq 0$, we assume that the system fails when the capacity falls to or below this level. At this time the process is stopped; i.e., the system is abandoned after first failure (see Chap. 5). The time of failure of the controlled process is denoted by

$$\tau^\nu = \inf\{t > 0 | V^\nu(t) \leq k^*\}. \tag{10.65}$$

We denote by τ the time of failure of the uncontrolled process $\mathscr{V}$. For simplicity, in what follows we take $k^* = 0$. While k^* denotes a lower limit for the process, we also assume that there is a maximum (i.e., optimal) performance level O that cannot be improved. Therefore, any maintenance activity at time τ_i must satisfy that $\zeta_i \in [0, O - V^\nu(\tau_i-)]$, where $V(\tau_i-)$ is the state of the system just before the maintenance. In this case we say that the policy is *admissible*.

If we denote $\mathbb{E}_{v_0}[\cdot] := \mathbb{E}[\cdot|V^\nu(0^-) = v_0]$, for a given admissible ν and initial component state $v_0 \in [0, O]$, then the expected benefits minus costs is given by

$$J(v_0, \nu) = \mathbb{E}_{v_0}\left[\int_0^{\tau^\nu} e^{-\delta s} G(V^\nu(s))ds - \sum_{\tau_i < \tau^\nu} e^{-\delta \tau_i} C(V^\nu(\tau_i-), \zeta_i)\right], \tag{10.66}$$

where G is a non-negative continuous, increasing and concave function on $[0, O]$ with $G(0) = 0$, C is a continuous function, increasing in both variables, and δ is the discount factor (see Chap. 9). Note that the first term inside the expectation in Eq. (10.66) represents the discounted benefits, where the function G can be interpreted as a utility function. The second term represents the discounted costs of interventions with $C(v, \zeta)$ being the cost of bringing the system from level v to level $v + \zeta$.

The objective of the problem is to find the policy that maximizes the expected benefits minus costs among all admissible impulse controls, that is

$$Z(v_0) = \sup_\nu J(v_0, \nu) \tag{10.67}$$

for a given level $v_0 \in [0, O]$. It is generally very difficult to calculate $Z(v_0)$ directly from Eq. 10.67. Instead of finding $Z(v_0)$ directly, we will solve the problem for all $v \in [0, O]$ at once, that is, we will find the *value function*

$$Z(v) = \sup_\nu J(v, \nu) \tag{10.68}$$

and evaluate this function at v_0. Although apparently this is a harder problem, we will characterize Z as the unique solution of a certain equation and solve this equation numerically. From the definition of the value function, we can easily see that $Z \geq 0$, since we can always choose to do nothing. Also, $Z(0) = 0$ and V is bounded. We will use these properties in the derivations below to characterize the function Z.

10.6.2 Determining the Optimal Maintenance Policy

In this section we present the fundamental theoretical results that allow us to determine Z in Eq. (10.68). We state these results without proof; all proofs can be found in [47].

Lemma 49 *Let T be a stopping time with respect to the filtration $\mathscr{F}_t$. Then for all $v \in [0, O]$*

$$Z(v) \geq \mathbb{E}_v \left[\int_0^{T \wedge \tau} e^{-\delta s} G(V(s))ds + e^{-\delta T} Z(V(T)) I_{\{T<\tau\}} \right]. \tag{10.69}$$

Furthermore, we have equality in (10.69) *if it is not optimal to perform any maintenance on the system before time T.*

In order to characterize the value function Z in Eq. (10.68) we need to define two important operators. The first one is the *intervention operator* $\mathscr{M}$ defined as

$$\mathscr{M}f(v) = \sup_{0 \leq \zeta \leq O-v} f(v+\zeta) - C(v,\zeta) \tag{10.70}$$

for a given function f defined on $[0, O]$ and v in the same interval. Note that we take the supremum over the interval $[0, O - v]$ in order to consider only admissible policies. We are interested in applying $\mathscr{M}$ to the function Z. If we consider any policy ν such that $\tau_1 = 0$ and write $\nu = (0, \zeta) \cup \{(\tau_i, \zeta_i)\}_{i \geq 2} = (0, \zeta) \cup \hat{\nu}$, then by Eqs. 10.68 and 10.66

$$Z(v) \geq J(v, \nu) = -C(v, \zeta) + J(v + \zeta, \hat{\nu}). \tag{10.71}$$

Since $\hat{\nu}$ is arbitrary we can take the supremum over all controls ν and obtain

$$Z(v) \geq Z(v + \zeta) - C(v, \zeta). \tag{10.72}$$

Now, taking the supremum over all admissible ζ, we obtain

$$Z(v) \geq \sup_{0 \leq \zeta \leq O-v} Z(v+\zeta) - C(v,\zeta) = \mathscr{M}Z(v). \tag{10.73}$$

We will use this inequality in the characterization of the function Z. The second operator that we will use is the *infinitesimal generator* $\mathscr{A}$ of the uncontrolled Markov process $\mathscr{V}$, that is:

$$\mathscr{A}f(v) = \lambda \left(\int_0^\infty f(v - g(y, v))dF(y) - f(v) \right) \tag{10.74}$$

for f and v as in Eq. 10.70. The infinitesimal generator has the property that, for bounded f, the process

$$e^{-\delta t} f(V(t)) - f(v) + \int_0^t e^{-\delta s} \left(\delta f(V(s)) - \mathscr{A}f(V(s)) \right) ds \tag{10.75}$$

is a martingale with respect to $\mathscr{F}_t$ (see [46, 49]). Taking expectations in Eq. 10.75 and using Optional Sampling Theorem [49] we obtain the so-called Dynkin's Formula; i.e., given $T_1 \leq T_2$ almost sure (a.s.) finite stopping times, then

$$\mathbb{E}[e^{-\delta T_2} f(R(T_2)) - e^{-\delta T_1} f(V(T_1))] \\ = \mathbb{E}\left[\int_{T_1}^{T_2} e^{-\delta s}\left(\mathscr{A} f(V(s)) - \delta f(V(s))\right) ds\right]. \tag{10.76}$$

We will use this formula with f replaced by Z to completely describe the value function.

Since the process $\mathscr{V}$ is Markovian, in order to obtain an optimal policy it is necessary to consider only at present state of the system, and not how the system arrived at the present state. So, given a state v we want to know if an intervention is required or not. We use the intervention operator $\mathscr{M}$ to answer this question. From Eq. 10.73 $Z \geq \mathscr{M}Z$, and we can divide the state space $[0, O]$ into the subsets:

$$A = \{v \in [0, O] : Z(v) = \mathscr{M}Z(v)\} \tag{10.77}$$

and

$$B = \{v \in [0, O] : Z(v) > \mathscr{M}Z(v)\}. \tag{10.78}$$

For $v \in A$ we maintain the system immediately and improve the performance by ζ^*, where

$$\mathscr{M}Z(v) = Z(v + \zeta^*) - C(v, \zeta^*) = \sup_{0 \leq \zeta \leq O - v} Z(v + \zeta) - C(v, \zeta). \tag{10.79}$$

Therefore, we call the set A the *maintenance region*. For the other states, i.e. those in B, we do nothing and let the system evolve. We call the set B the *no maintenance region* (Fig. 10.17). It is important to stress that because of the Markov property, this classification of states will always be the same and does not depend on time.

Now, for $v \in B$ it is optimal to leave the system alone, therefore, we obtain equality in (10.69), and using Dynkin's Formula we have that $\delta Z(v) - \mathscr{A}Z(v) = G(v)$. We formalize the existence and uniqueness results in the following theorems (for proofs see [47]).

Theorem 50 *The value function Z solves the equation*

$$\min\{\delta Z(v) - \mathscr{A}Z(v) - G(v), Z(v) - \mathscr{M}Z(v)\} = 0, \tag{10.80}$$

for all $v \in [0, O]$.

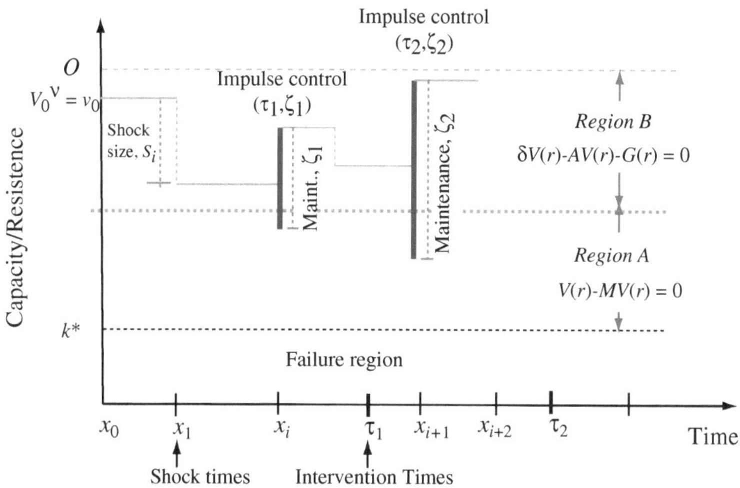

Fig. 10.17 Description of the impulse control model

Theorem 51 *Let f be a non-negative bounded function on $[0, O]$ that solves Eq. 10.80 such that $f(0) = 0$. Then $f = Z$.*

With respect to existence of optimal controls, it is possible that ζ^* is not attainable in equation (10.79). In this case there is no attainable optimal policy, but we can find controls with expected profit as close as possible to the value function V.

To obtain the optimal policy is necessary to find the value function Z by solving Eq. 10.80. Once we have Z, we can compute the maintenance and no maintenance regions. Typically, we use numerical methods (for example, the standard Jacobi iteration method described in [50]) to find the optimal policy.

Example 10.61 (Adapted from [47]) Consider an infrastructure component whose performance may be continuously monitored. We suppose that performance is normalized and evaluated within the interval $[0, 1]$, with $v = 1$ indicating that the system is in "as good as new" condition and $v = k^* = 0$ indicating that the component has failed. Furthermore, let's assume that the structure is located in a seismic region where earthquake occurrence times follow a Poisson process with rate $\lambda = 0.5$.

As a result of each earthquake (shock), the structure may be damaged. The structural damage inflicted by each shock is the result of both the earthquake motion characteristics (shock size) and the structural capacity. For the purpose of this example, we assume that shock sizes Y_i are *iid* log-normally distributed random variables with mean $\mu = 0.2$ and $COV = 0.25$, and that structural damage is a monotonically increasing process with function $g(y, v)$ given by

$$g(y, v) = \beta \frac{y}{v} \tag{10.81}$$

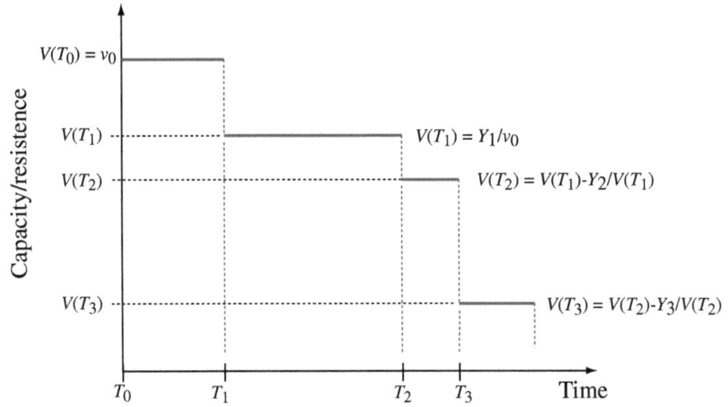

Fig. 10.18 Shock-based degradation model conditioned on damage state

where β is an arbitrary constant; in this example we take $\beta = 1$. An illustrative sample path of this process is shown in Fig. 10.18. Note also that it is not easy to find an analytical expression for the the probability distribution of $\frac{Y_i}{V(T_i^-)}$ and the corresponding convolution after a given number of shocks.

The objective of the study is to determine an optimal maintenance policy for this continuously monitored structure. In practice, it may still be necessary to determine the system capacity through an inspection, but we assume that inspections can be performed at any time at no cost. To determine the optimal maintenance policy requires first the assessment of benefits and costs needed to evaluate the function J (i.e., cost benefit relationship; Eq. 10.66). For this example, let the benefit derived form the existence of the project be given by

$$G(v) = (\phi\ C_0) \cdot \frac{1}{\alpha}(1 - e^{-\alpha v}), \tag{10.82}$$

where $C_0 = 100$, $\phi = 0.275$ and $\alpha = 0.5$. Note that this curve has the form of an exponential utility function. Furthermore, consider that the costs associated with an intervention are given by the following expression:

$$C(v, \zeta) = C_0 \zeta^{\frac{1}{2}} + kC_0(1 - v), \tag{10.83}$$

where the constant $k = 0.1$ reflects the fixed costs of any intervention. Note that the intervention costs are proportional to the current state of the system and grow with the square root of the size of the intervention. For both benefit and cost, these values are discounted to the time of the decision by using a discount factor $\delta = 0.05$.

The analysis consists of two steps. First, we determine the impulse-control policy; i.e., for every structural state v, we find the intervention intensity ζ that maximizes the expected profit (Eq. 10.66). This step requires partitioning the state space into a region where no maintenance should be performed and a region for which maintenance

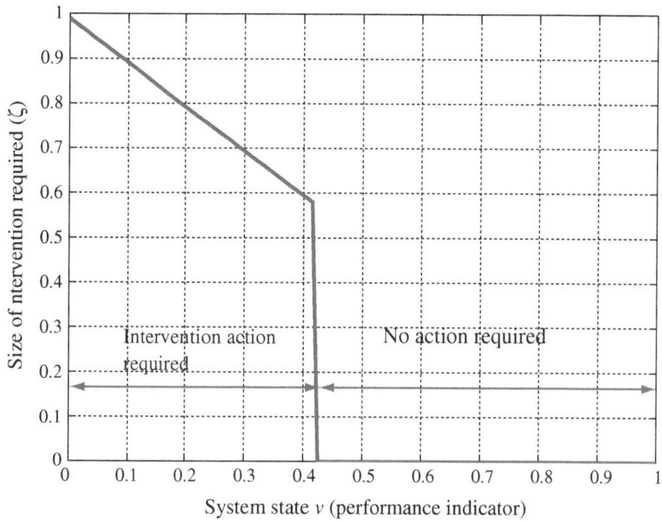

Fig. 10.19 Optimal impulse-control strategy (Adapted from [47])

necessary. Second, we determine the value function Z that provides the maximum expected profit if the intervention program is implemented.

Using the numerical approach described in [47, 48], we obtain the the impulse-control policy given in Fig. 10.19. Note that so long as the system capacity exceeds 0.42 ($v > 0.42$), no maintenance should be performed. However, if the capacity falls to or below 0.42, maintenance is required, at a level shown in Fig. 10.19. For instance, if an inspection shows the capacity to be $v = 0.3$, maintenance effort of $\zeta = 0.7$ is optimal, which will bring the system to a good-as-new condition.

If maintenance is carried out under this policy, the maximum expected profit can be obtained in Fig. 10.20, where the x-axis corresponds to the initial state of the system, i.e., v_0 and the y-axis shows maximum profit Z for the intervention program shown in Fig. 10.19.

The sensitivity of the maintenance policy with respect to the discount rate is shown in Fig. 10.21. For comparison purposes, two different deterioration functions g (Eq. 10.63) were considered. In Fig. 10.21a, the function g was selected as defined in Eq. 10.81; while in Fig. 10.21b the analysis was carried out for $g(v, y) = y$, which means that shock sizes, are *iid* and the damage accumulation does not depend on the previous state of the system.

It should be first noted that, for both functions, as the discount rate becomes larger, the range of structural states for which an intervention is required becomes smaller. This is justified by the fact that interventions are only required if the system state is closer to failure; then, although interventions are more expensive, they are discounted with a higher rate. In addition, it can be observed also that if the effect of damage

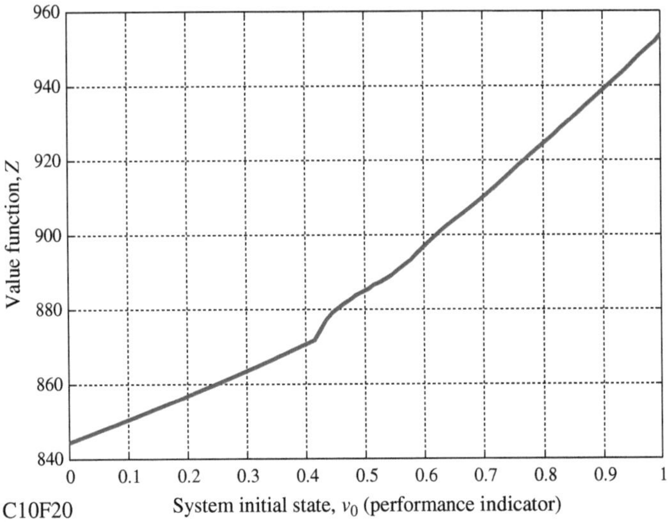

Fig. 10.20 Value function for the optimal impulse-control strategy (Adapted from [47])

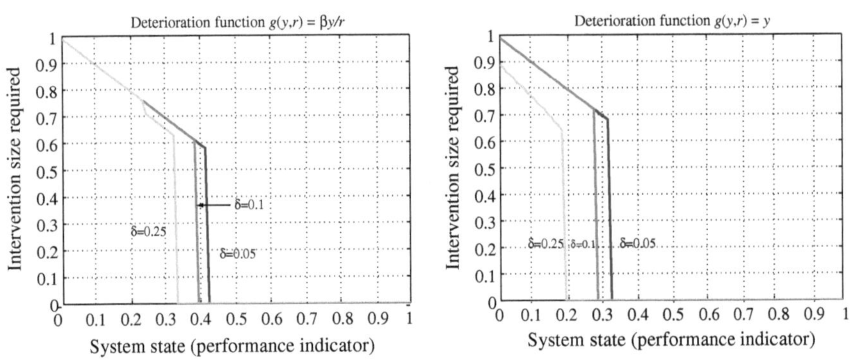

Fig. 10.21 Effect of the discounting rate on the intervention program for two deterioration functions (i.e., g) (adapted from [47])

accumulation is taken into account, the region of system states where an intervention is required is larger than the region for the case of no damage accumulation.

Finally, the effect of the shock sizes on the maintenance policy for the case in which damage accumulation is taken into consideration is presented in Fig. 10.22. For given mean shock size it is clear that larger coefficients of variation (COV) imply larger failure probabilities and, therefore, the region where interventions are required becomes also larger. In addition, the effect of the mean, for a fixed COV, is similar than in the previous case. However, intervention space is larger in this case than in the first case.

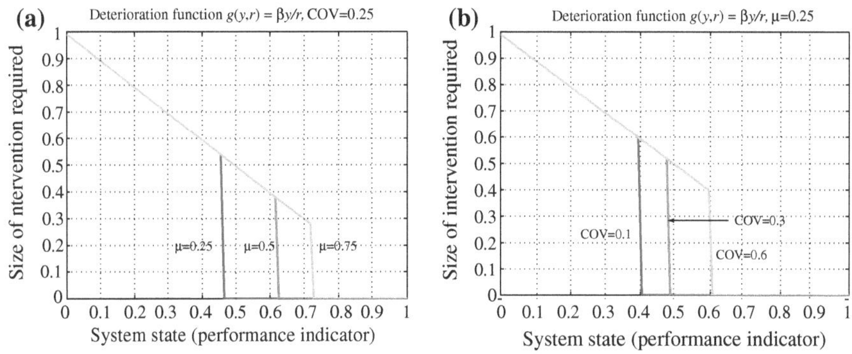

Fig. 10.22 Effect of the mean and covariance of shock sizes on the intervention program (adapted from [47])

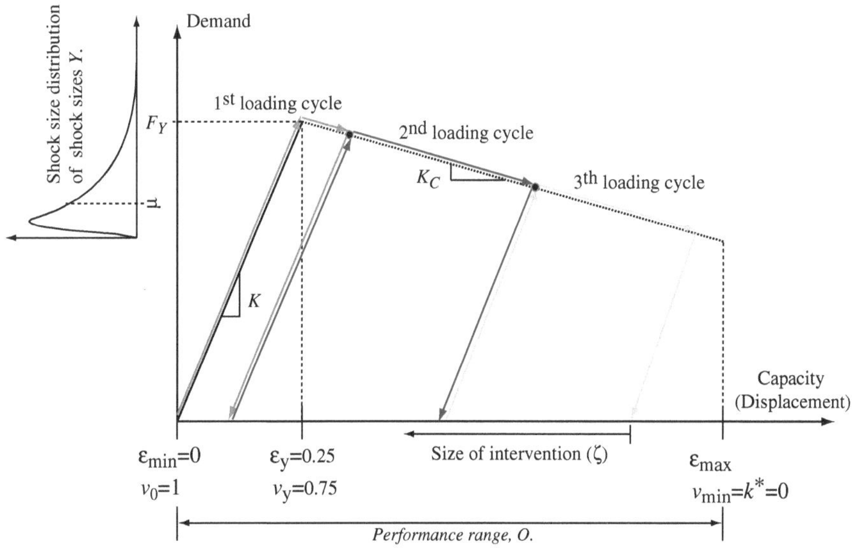

Fig. 10.23 Sample path of a structural deterioration process described by a bilinear constitutive model

Example 10.62 (Adapted from [48]) Consider now the case of a structure whose performance is described by a bilinear constitutive model as shown in Fig. 10.23; where $K = 2$, $K_C = 0.2$ and $\varepsilon_Y = 0.25$.

The structure is subject to successive extreme events. If the demand (shock) is not large enough to take the structure out of the elastic range, no damage will be reported. The excursions into the inelastic range will define the degradation process by redefining the initial displacement state and the extension of the elastic range for next

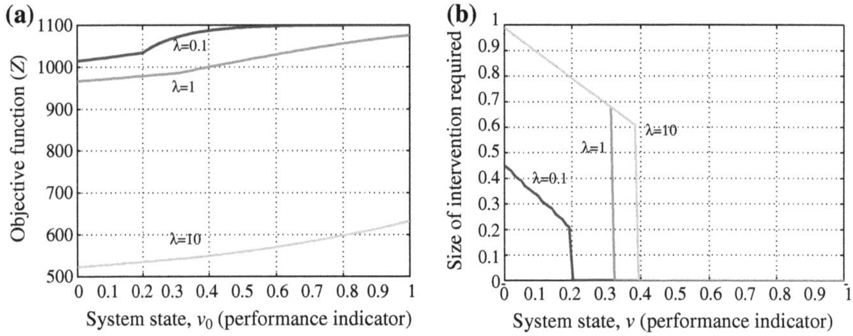

Fig. 10.24 Results from the optimization: **a** Objective function; **b** optimal maintenance policy

iteration. Damage in this case will be measured in terms of the residual displacement; then, after a shock of size y, the change in the residual displacement v can be computed as:

$$g(y, v) = \begin{cases} 0 & \text{if } y \leq K\varepsilon_Y - K_C(1-v)\dfrac{K}{K+K_C} \\ \left(\dfrac{y}{K} + 1 - v - \varepsilon_Y\right)\left(1 + \dfrac{K_C}{K}\right) - 1 + v & \text{if } y > K\varepsilon_Y - K_C(1-v)\dfrac{K}{K+K_C}. \end{cases} \tag{10.84}$$

where ε_Y is as indicated in Fig. 10.23. Note that if an intervention ζ is carried out, it will be directed to reduce the initial displacement, for the subsequent loading cycle, by retrofitting the structure.

The purpose of this example is to identify the optimal maintenance policy. Both the utility and cost of intervention functions have the same form as in the previous example; i.e., equations (10.82) and (10.83) with the following parameters: $C = 100$, $k = 0.1$, $\delta = 0.05$. Shocks sizes are assumed to be lognormally distributed with $\mu = 0.4$ and COV=0.35. For comparison purposes, the analysis was carried out for three different event occurrence rates $\lambda = 0.1$, $\lambda = 1$ and $\lambda = 10$.

The optimal maintenance strategy and the cost-benefit relationship are shown in Fig. 10.24. The maximum expected benefit is shown in Fig. 10.24a; while the the optimal maintenance policy for all three cases considered is presented Fig. 10.24b. The results show that the effect of the shock rate λ on the total profit is as expected; lower rates lead to larger profits and to a smaller intervention region. Note that when the rate becomes very small, the value of the objective function reaches a maximum value of \$1100. On the other hand, the intervention policies also change depending upon de occurrence rate. In this case, the state space for which maintenance actions are required is larger for higher rates (see Fig. 10.24b). In this case, it is interesting to observe that for $\lambda = 0.1$ interventions do not require to take the structure to its original condition (i.e., "as good as new") but to a lower level. For instance, for $\lambda = 0.1$, if the condition of the system is $v = 0.1$ the size of the intervention would be $\zeta = 0.3$ and the final state of the system would be $v = 0.1 + 0.3 = 0.4$. The main reason for this is that since events are highly spaced in time, the structure can operate for a long period of time without failure.

10.7 Maintenance of Systems with Non Self-announcing Failures

Many systems degrade over time in a manner that is not outwardly visible. At some point, symptoms of serious degradation may become apparent, signaling that imminent failure is likely. If this occurs, the system is immediately shut down and repaired or replaced. For example, a bridge may appear to be operational even when internal damage may exceed desirable levels. Before degradation is outwardly apparent, however, it may be possible to inspect the system to determine whether the system is operating within acceptable limits. It may be the case that inspections can determine whether the system is operating above the acceptable threshold, but may not be able to determine the exact level of degradation. For example, the inspection may involve a simple load test that is either passed or failed. Of course, the system may fail catastrophically between inspections before we can identify the imminent failure state; thus the objective of inspections is to find the system below the operating threshold but before catastrophic failure occurs. We say that such a system has *non-self-announcing* failures. Typically, inspections involve significant expense and/or system downtime, and thus they are treated as a resource that must be used wisely. Current maintenance strategies for non-self-announcing failures have generally considered periodic inspections with fairly restrictive assumptions on the deterioration process; e.g. [51, 52]. More recent work [53, 54] has identified opportunities to improve on periodic inspection schemes by taking system lifetime information into account. This section will investigate some inspection strategies for these systems.

10.7.1 A General Modeling Framework

Consider a system in operation that is subject to deterioration and possible failure. As long as the system capacity is above a threshold level k^*, we say the system is operational (system is "up"), but when the system capacity falls below the threshold level, we say that system has failed (system is "down").

Let's suppose that inspections can determine whether or not the device in use is operational (i.e., it is operating above the threshold level k^*), but cannot determine the level of degradation. If the device is found to be failed at an inspection, a complete

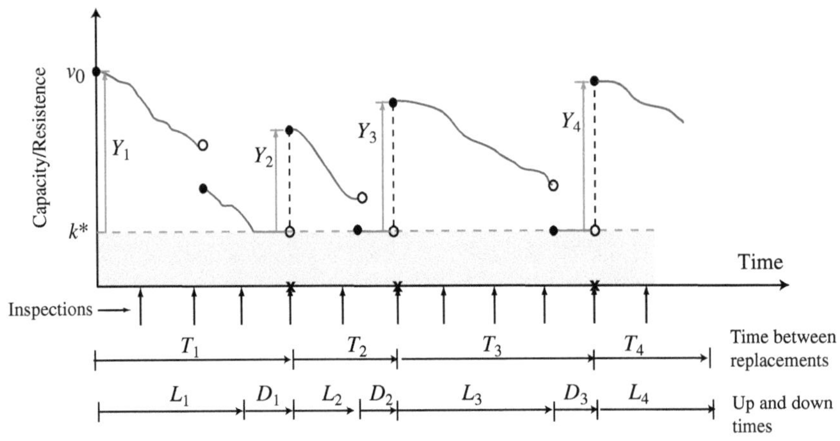

Fig. 10.25 Sample path for system with non-self-announcing failures

replacement is made with a statistically identical new system. If the device is found to be operational, the system is left undisturbed. A typical sample path for this type of system is shown in Fig. 10.25; note that when the device fails, the system will remain out of service until the next inspection time. Let us define $\{L_1, L_2, \ldots\}$ to be the sequence of lifetimes in which the system is operational, and $\{D_1, D_2, \ldots\}$ to be the sequence of times during which the system operates below the threshold level. We will call the former "up" times and the latter "down" times (Fig. 10.25).

Beginning with a new system at time 0, inspections are scheduled at predetermined times $\tau_1, \tau_2, \ldots$. Furthermore, let $\{T_1, T_2 \ldots\}$ be the times between replacements ("cycle times"). After the system is maintained, inspections are again scheduled at times $\tau_1, \tau_2, \ldots$, and the process repeats itself. We assume that inspections and replacements take negligible time. In this way, the system operates through a sequence of maintenance cycles that begin with a new system and end at the first inspection that finds the system failed, as illustrated in Fig. 10.25.

For this model, the objective is to determine a sequence of inspection times to appropriately balance the inspection capacity (rate of inspections) with the system downtime; that is, to find an inspection strategy that most effectively minimizes system downtime. The performance measures we use are the *limiting average availability*, defined as

$$A_{av} := \frac{\int_0^t P(V(s) > k^*)ds}{t}, \tag{10.85}$$

where $V(s)$ is the remaining life (i.e., capacity/resistance) of the system in service at time s, and the *long run inspection rate*

$$\beta := \lim_{t\to\infty} \frac{\mathbb{E}[N_t]}{t}, \tag{10.86}$$

where N_t is the number of inspections made up to time t [55].

We assume that successive lifetimes are independent, identically distributed random variables with cumulative distribution function F. In this case, the system regenerates at the time of an inspection that finds the system failed (resulting in a replacement), and the limiting average availability A_{av} has a particularly simple expression as the ratio of mean system lifetime to mean cycle time; i.e.

$$A_{av} = \frac{\mathbb{E}[L]}{\mathbb{E}[T]} \tag{10.87}$$

(note that since all cycles are independent and statistically identical, for ease of notation we have dropped the subscript that denotes the cycle).

The long-run inspection rate β is given by the ratio of the expected number of inspections in the cycle to the expected cycle length; i.e.

$$\beta = \frac{\mathbb{E}[N]}{\mathbb{E}[T]}, \tag{10.88}$$

where N denotes the number of inspections in a cycle (starting with a new system, the number of inspections until the system is first found failed).

Equations (10.87) and (10.88) follow from basic regenerative process theory [56]. Note that these performance measures are *competing* in the sense that the cost of improving A_{av} is generally that β also increases. The main interest in this section is to find an efficient inspection strategy that maximizes availability for a given inspection rate.

10.7.2 Periodic Inspections

As described in previous sections, the most widely used inspection strategy for deteriorating equipment is to schedule inspections periodically; that is, inspections are made at multiples of a fixed inter-inspection time τ. This system is easy to implement and relatively straightforward to analyze. Recall that F represents the lifetime of a new system, and suppose initially that F is known in advance (in subsequent sections, we will determine F based on some assumed properties of the deterioration process). To compute $\mathbb{E}[T_1]$ in Eqs. 10.87 and 10.88, note that a cycle ends at (random) time $N\tau$, where N is (as above) the number of inspections in a cycle. Therefore

$$\mathbb{E}[T] = \tau\, \mathbb{E}[N] = \tau \sum_{m=0}^{\infty} P(N > m)$$
$$= \tau \sum_{m=0}^{\infty} P(L > m\tau) = \tau \sum_{m=0}^{\infty} \overline{F}(m\tau)$$

Thus, from Eq. 10.87, the limiting average availability for periodic inspections is given by

$$A_{av} = \frac{\int_0^\infty \overline{F}(u)du}{\tau \sum_{m=0}^{\infty} \overline{F}(m\tau)} \tag{10.89}$$

The inspection rate for periodic inspections is simply the reciprocal of the inter-inspections time, that is

$$\beta = \tau^{-1} \tag{10.90}$$

In the expressions above, we have assumed that the failure distribution F is known. In many cases, it may be estimated using observed failure times. In some special cases, we may be able to compute it directly using assumptions on both the nominal life distribution and the characteristics of degradation process. Recall that the nominal life (see Chap. 4) of a system represents a physical attribute of a new system that degrades due to usage. The following examples show how availability can be determined in these special cases. The results in these examples are extracted from [55, 57–59].

Determining Availability Under Periodic Inspections

Let's assume that the system deteriorates due to shocks that occur according to a compound Poisson process. Let the nominal lives of new systems be independent and identically distributed random variables $X_1, X_2, \ldots$ with common distribution function A. Further, let λ be the rate of the Poisson shock process and B the distribution of sizes of successive shocks (shock sizes are assumed to be independent and identically distributed and are denoted by $Y_1, Y_2, \ldots$).

To determine availability, we must compute $\mathbb{E}[L]$ and $\mathbb{E}[T]$ in Eq. 10.87.

We first examine the numerator of the expression. For $t \geq 0$, let $D(t)$ be the accumulated damage by time t; that is, if $M(t)$ denotes the number of shocks by time t,

$$D(t) = \begin{cases} \sum_{i=1}^{M(t)} Y_i, & M(t) > 0 \\ 0, & M(t) = 0 \end{cases}, \tag{10.91}$$

and let $H(z, t) = P(D(t) \leq z)$ be the distribution function of $D(t)$. Then we have

$$P(L > t) = P(D(t) < X_1) = \int_0^\infty \int_0^z H(dy, t)A(dx) = \int_0^\infty H(z, t)A(dz), \tag{10.92}$$

Conditioning on $M(t)$, it follows that

$$H(z,t) = \sum_{n=0}^{\infty} P(D(t) \le z | M(t) = n) P(M(t) = n) = \sum_{n=0}^{\infty} B^{(n)}(z) e^{-\lambda t} \frac{(\lambda t)^n}{n!}, \tag{10.93}$$

where $B^{(n)}$ denotes the n-fold convolution of B with itself; i.e., the distribution of the sum of n shocks. Plugging in to the expression for $P(L > t)$ above, we have

$$\begin{aligned} P(L > t) &= \int_0^{\infty} \sum_{n=0}^{\infty} B^{(n)}(z) e^{-\lambda t} \frac{(\lambda t)^n}{n!} A(dz) \\ &= \sum_{n=0}^{\infty} e^{-\lambda t} \frac{(\lambda t)^n}{n!} \int_0^{\infty} B^{(n)}(z) A(dz). \end{aligned} \tag{10.94}$$

So we have

$$\begin{aligned} \mathbb{E}[L] = \int_0^{\infty} P(L > t) dt &= \int_0^{\infty} \sum_{n=0}^{\infty} e^{-\lambda t} \frac{(\lambda t)^n}{n!} \int_0^{\infty} B^{(n)}(z) A(dz) dt \\ &= \int_0^{\infty} \sum_{n=0}^{\infty} B^{(n)}(z) A(dz) \int_0^{\infty} e^{-\lambda t} \frac{(\lambda t)^n}{n!} dt \\ &= \frac{1}{\lambda} \int_0^{\infty} \sum_{n=0}^{\infty} B^{(n)}(z) A(dz). \end{aligned} \tag{10.95}$$

If we let $R(z) = \sum_{n=1}^{\infty} B^{(n)}(z)$, then, $R(z)$ can be interpreted as the mean number of shocks required to reach a cumulative shock magnitude of at least z. This gives

$$E[L] = \frac{1}{\lambda} \int_0^{\infty} (R(z) + 1) A(dz) = \frac{1}{\lambda} \Big[\int_0^{\infty} R(z) A(dz) + 1 \Big]. \tag{10.96}$$

The term R plays the role of a renewal function indexed on the cumulative shock magnitude. In general, closed-form expressions for R are difficult to obtain, but there are fairly efficient techniques available to compute these terms numerically; see [60, 61].

Unlike the numerator, the denominator of the availability expression depends on the inspection policy used. Assuming periodic inspections every τ units, let $I(t)$ count the number of inspections by time t; i.e.,

$$I(t) = \sup\{n : n\tau \le t\} \tag{10.97}$$

Then the number of inspections required to find the system failed is $I(L) + 1$, and

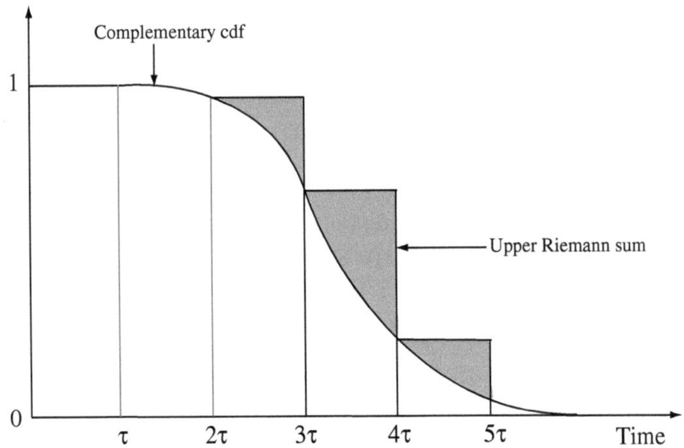

Fig. 10.26 Complementary cdf and upper Riemann sum for periodic inspections

$$\begin{aligned}\mathbb{E}[T] &= \tau\ \mathbb{E}[I(L)+1] \\ &= \tau\left(\sum_{n=1}^{\infty} P(I(L)\geq n)+1\right) \\ &= \tau \sum_{n=0}^{\infty} P(L>n\tau),\end{aligned} \tag{10.98}$$

where $P(L > t)$ appears above in the expression for $\mathbb{E}[L]$ (Eq. 10.95).

An expression for the limiting average availability for periodic inspections can then be obtained putting together the expressions for $\mathbb{E}[L]$ and $\mathbb{E}[T]$ in Eq. 10.87 [58],

$$A_{av} = \frac{\int_0^{\infty} R(z)A(dz)+1}{\lambda\tau \sum_{n=0}^{\infty} P(L>n\tau)}. \tag{10.99}$$

This expression involves computing a renewal-type function, which is in general difficult. However, the denominator of the expression for availability leads to a very nice graphical illustration of the relationship between mean life time, mean down time, and mean cycle time. Note that the denominator expresses mean cycle time as the upper Riemann sum of the complementary distribution function of the lifetime, where the partition is determined by the inspection times. This relationship is illustrated in Fig. 10.26.

Because the area under the complementary distribution function of lifetime is $\mathbb{E}[L]$, and the area under the upper Riemann sum is $\mathbb{E}[T]$, the shaded area represents the mean down time. Figure 10.26 suggests that we might use the inspection resources more effectively if we move the inspection times around to match the shape of the distribution of L. For example, a better inspection scheme can be obtained if

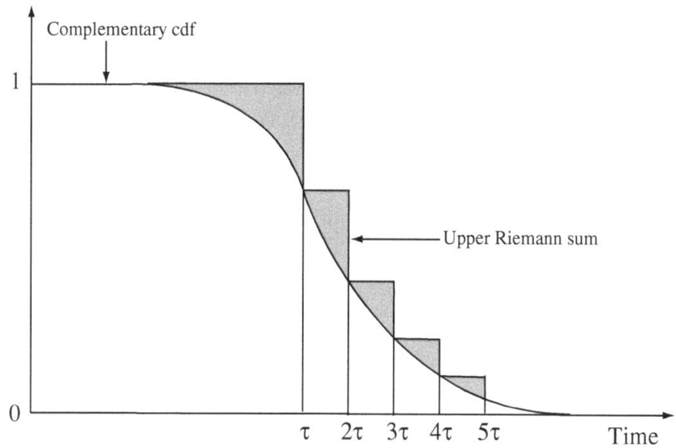

Fig. 10.27 A potentially improved inspection scheme—unequally spaced inspections

inspection times are selected as shown in Fig. 10.27 (notice that it has less shaded area, so less downtime). This idea will be pursued in the next section.

The results in this section can be generalized slightly to consider degradation as the superposition of a compound Poisson shock process and a deterministic graceful degradation process (see [58]); in this case, all the results shown above hold with very minor modifications.

10.7.3 Availability for Periodic Inspections (Markovian Deterioration)

A somewhat more complicated situation arises when we consider a Markovian degradation process. Here renewal arguments cannot be used because cycles are no longer independent and identically distributed. Nevertheless, it is possible to derive an expression for the limiting average availability under periodic inspections. This section is abstracted from [59].

For this model, let the state of the operating environment be governed by a continuous time Markov chain $\mathscr{W} = \{W(t), t \geq 0\}$ with finite state space $E = \{1, 2, \ldots, N\}$, infinitesimal generator $\mathbf{Q} = [q_{ij}]$, and stationary distribution $\boldsymbol{\pi}$. When the environment is in state j, the system deteriorates at rate γ_j, and without loss of generality, we will assume the states are ordered such that $0 < \gamma_1 < \cdots < \gamma_N$.

Again we assume that the nominal lives $X_1, X_2, \ldots$ are *iid*. and independent of the Markov chain. As in the sections above, we let $\{L_1, L_2, \ldots\}$ be the sequence of lifetimes. Finally, define $\{R_n, n = 0, 1, 2 \ldots\}$ to be the sequence of replacement times (with $R_0 := 0$).

Note that if the initial distribution of the Markov chain $\mathscr{W}$ is $\boldsymbol{\pi}$ (i.e., the environment begins in steady state), the sequence of device lifetimes $\{L_n, n = 1, 2, 3, \ldots\}$ is *not* a sequence of independent and identically distributed random variables, because the distribution of L_{n+1} depends on W_n, and W_n depends on L_n. Thus we must characterize the probability structure of the state of the environment embedded at replacement times. To this end, let $W_n = W(R_n)$. Then $\hat{\mathscr{W}} = \{W_n, n = 0, 1, 2, \ldots\}$ is an irreducible Markov chain with transition probability matrix $\hat{\mathbf{P}}$ and stationary distribution $\boldsymbol{\nu}$.

Theorem 52 *The paired process* $(\hat{\mathscr{W}}, \mathscr{R}) = \{(W_n, R_n), n = 0, 1, 2, \ldots\}$ *is a Markov renewal process.*

Proof The proof is somewhat technical and appears in [59].

Note that this result says that each new device begins in an environmental state that depends on the state of the environment in which the previous device failed. Thus, we cannot employ the usual renewal-theoretic arguments to arrive at an expression for A_{av}. We can, however, employ some slightly more sophisticated theory based on the notion of *semi-regenerative* processes. Semi-regenerative processes are processes that possess a type of conditional independence; in this case we state (again without proof) some properties of the system state process $\{Z(t); t \geq 0\}$.

Theorem 53 *The process* $\{Z(t); t \geq 0\}$ *has the following properties*

(i) $\{Z(t); t \geq R_n\}$ *is conditionally independent of* $\{Z(u); u \leq R_n\}$ *and* $\{(W(R_k), R_k), k = 0, 1, \ldots, n\}$ *given* R_n;

(ii) the distribution of $\{Z(t); t \geq R_n\}$ *given* $W(R_n) = j$ *equals that of* $\{Z(t); t \geq 0\}$ *given* $Z(0) = j$.

That is, $\{Z(t); t \geq 0\}$ *is a semi-regenerative process with respect to the Markov renewal process* $(\hat{\mathscr{W}}, \mathscr{R})$.

The results of this theorem allow us to express the limiting average availability as a ratio of mean time to first failure (mean lifetime) to mean time to first replacement, where the expectations are taken with respect to the stationary distribution $\boldsymbol{\nu}$. Then, limiting average availability is given by [59]

$$A_{av} = \frac{\sum_{i=1}^{N} \nu_i \mathbb{E}_{\sqsupset}[L_1]}{\sum_{i=1}^{N} \nu_i \mathbb{E}_{\sqsupset}[R_1]}, \tag{10.100}$$

where $\mathbb{E}_{\sqsupset}[\,\cdot\,] = \mathbb{E}[\,\cdot \mid W_0 = i]$. The term ν describes the stationary distribution of the environment embedded at maintenance times, and $\mathbb{E}_i$ denotes the conditional expectation given the initial state of the environment is i. Intuitively, the Markov chain that describes the environment is not distributed according to the stationary distribution π at maintenance times, but rather according to a "biased" distribution ν.

While these results are quite elegant, they do not lend themselves easily to computation. However, they do provide some structural understanding about degradation processes in a random environment and illustrate how easy it might be to apply renewal-theoretic results incorrectly, which in this case, might significantly overestimate availability. Additional details on the derivation and the scope of this approach can be seen in [59].

10.7.4 An Improved Inspection Policy: Quantile-Based Inspections

Note that, at an inspection, periodic inspections use no information about the time since the last cycle began (i.e. the *age* of the system in use) to schedule the next inspection. Since system lifetimes are not generally memoryless, periodic inspections may tend to "overinspect" at times where failures are less likely to occur, and "underinspect" at times where failures are more likely to occur, as Figs. 10.26 and 10.27 suggest.

An alternative to periodic inspections uses the distributional information of the lifetime to schedule inspections more advantageously; that is, to achieve the same availability with a smaller inspection rate. Consider a policy whereby we select a fixed quantile $0 < \alpha < 1$ in advance, and then determine inspection times as follows [53]:

$$\tau_1 = \sup\{t > 0 : P(L > t) \geq \alpha\}, \tag{10.101}$$

$$\tau_n = \sup\{t > 0 : P(L > t | L > \tau_{n-1}) \geq \alpha\}, \quad n \geq 2. \tag{10.102}$$

If F is continuous and strictly increasing, then

$$\tau_n = \overline{F}^{-1}(\alpha^n) \quad n = 1, 2 \ldots \tag{10.103}$$

We call this policy "Quantile-Based Inspections" (QBI) with quantile α and denoted it by QBI(α). This policy has the following property.

Theorem 54 *If the lifetime distribution of L is IFR (DFR),*[2] *then the inter-inspection times of QBI(α) are non increasing (nondecreasing).*

Proof We prove the result for the IFR case; the DFR case follows similarly. If F is IFR, then $\overline{F}(x+t)/\overline{F}(t)$ is non increasing in t for all $x > 0$. Therefore, for all n,

$$\frac{P(L > (\tau_{n+1} - \tau_n) + \tau_n)}{P(L > \tau_n)} \leq \frac{P(L > (\tau_{n+1} - \tau_n) + \tau_n)}{P(L > \tau_{n+1})} \tag{10.104}$$

that is

[2] IFR—Increasing Failure Rate; DFR—Decreasing Failure Rate.

$$P(L > \tau_{n+1}|L > \tau_n) \leq P(L > (\tau_{n+1} - \tau_n) + \tau_{n-1}|L > \tau_{n-1}). \tag{10.105}$$

Now by the definition of the τ_n's

$$P(L > (\tau_{n+1} - \tau_n) + \tau_{n-1}|L > \tau_{n-1}) \geq \alpha \tag{10.106}$$

and

$$(\tau_{n+1} - \tau_n) + \tau_{n-1} \leq \tau_n, \tag{10.107}$$

and therefore

$$\tau_{n+1} - \tau_n \leq \tau_n - \tau_{n-1}, \tag{10.108}$$

and the interinspection times form a nonincreasing sequence. □

Therefore, for deteriorating systems (F is IFR), the longer the system has been operating, under QBI, the shorter it will be between successive inspections. Note that the only time QBI(α) and periodic inspections produce the same sequence of inspection times is when lifetimes have the exponential distribution.

To evaluate the availability of QBI(α), we first compute the expected cycle length $\mathbb{E}[T]$:

$$\begin{aligned}
\mathbb{E}[T] &= \sum_{n=1}^{\infty} \tau_n \, P(\tau_{n-1} < L \leq \tau_n) = \sum_{n=1}^{\infty} \tau_n \left(\overline{F}(\tau_{n-1}) - \overline{F}(\tau_n)\right) \\
&= \sum_{n=1}^{\infty} \tau_n \, (\alpha^{n-1} - \alpha^n) \\
&= (1-\alpha) \sum_{n=1}^{\infty} \overline{F}^{-1}(\alpha^n) \, \alpha^{n-1},
\end{aligned} \tag{10.109}$$

and therefore, the limiting availability of QBI(α) becomes

$$A_{av} = \frac{\int_0^\infty \bar{F}(u)du}{(1-\alpha)\sum_{n=1}^{\infty} \overline{F}^{-1}(\alpha^n)\,\alpha^{n-1}} \tag{10.110}$$

To compute the limiting inspection rate, note that quantile-based inspections are designed so that the conditional probability that an inspection finds the system failed, provided the system was working at the last inspection, is a constant, namely $1-\alpha$. Thus, the number of inspections required to find a failure on each cycle has a geometric distribution, with long-run inspection rate is given by

$$\begin{aligned}
\beta &= \frac{1/(1-\alpha)}{(1-\alpha)\sum_{n=1}^{\infty} \tau_n \alpha^{n-1}} \\
&= \frac{1}{(1-\alpha)^2 \sum_{n=1}^{\infty} \tau_n \alpha^{n-1}}
\end{aligned} \tag{10.111}$$

Table 10.2 Availability and inspection rate for different inspection schemes

		Weibull(2, 10)		Weibull(4, 10)	
		PI	QBI	PI	QBI
$\alpha = 0.5$	A_{av}	0.760	0.790	0.776	0.866
	β	0.178	0.178	0.191	0.191
$\alpha = 0.6$	A_{av}	0.806	0.833	0.817	0.891
	β	0.235	0.235	0.246	0.246
$\alpha = 0.8$	A_{av}	0.901	0.915	0.904	0.942
	β	0.516	0.516	0.520	0.520
$\alpha = 0.9$	A_{av}	0.950	0.956	0.951	0.968
	β	1.079	1.079	1.068	1.068
$\alpha = 0.95$	A_{av}	0.975	0.977	0.975	0.983
	β	2.205	2.205	2.169	2.169

This expressions are challenging to compute analytically, but they can be investigated numerically (see example). Further details about this approach can be found in [53].

Example 10.63 Compare the periodic and quantile-based inspection policies assuming that random lifetimes that follow the Weibull distribution (Adapted from [54]).

Because the quantile-based inspection strategy involves the evaluation of quantile functions, it is difficult to compare analytically with periodic inspections. However, the superiority of quantile-based inspection schemes can be shown numerically. Recall that the Weibull distribution has cumulative distribution function

$$F(t) = 1 - \exp\left\{-\left(\frac{t}{\theta}\right)^{\zeta}\right\}, \quad t \geq 0, \text{ and } \theta, \zeta > 0. \tag{10.112}$$

The Table 10.2 compares inspection rate and limiting average availability for two Weibull distributions with parameters $\theta = 2, \zeta = 10$ and $\theta = 4, \zeta = 10$. The entries in the table are obtained by fixing β for both periodic (PI) (Eq. 10.90) and quantile-based (QBI) (Eq. 10.111) inspections, and then computing the resulting limiting average availability from Eqs. 10.89 and 10.110, respectively. Note that for a given inspection rate α, quantile-based inspections have higher availability than periodic inspections. As expected, as the inspection rate increases, both availabilities tend toward 1.

10.8 Summary

This chapter summarizes both basic maintenance concepts and a set of relevant models for planning infrastructure management and operation. In the first part of the chapter we focus on relevant definitions and a classification of different maintenance types and policies. In the second part of the chapter three basic and widely used maintenance strategies are presented: maintenance at regular time intervals; age-replacement models; and periodic replacement policies (Table 10.1). In the last part, this chapter describes two new and specific inspection and maintenance models which provide more realistic solutions to actual infrastructure applications. The first of these new models can be used for optimizing the maintenance for systems that are permanently monitored. This approach is based on impulse control models and allows to define the size of interventions that maximizes the profit. The second model addresses the case of scheduling inspections of systems with non-self-announcing failures. Here we consider periodic inspections at regular time intervals and compare this strategy to quantile-based inspections. A model for the case of shock-based deterioration is presented in which the effectiveness of the inspections is evaluated as the difference between the areas under the complementary cumulative distribution function and the upper Riemann sum.

References

1. K.B. Misra, *Handbook of Performability Engineering* (Springer, London, 2008)
2. W.P. Pierskalla, J.A. Voelker, A survey of maintenance models: the control and surveillance of deteriorating systems. Nav. Res. Logist. Q. **23**, 353–388 (1976)
3. Y.S. Sherif, M.L. Smith, Optimal maintenance models for systems subject to failure -a review. Nay. Res. Log. Quart. **28**, 47–74 (1981)
4. K. Bosch, U. Jensen, Maintenance models: a survey: parts 1 and 2 (in german). OR Spektrum **5**(105–118), 129–148 (1983)
5. C. Valdez-Flores, R.M. Feldman, A survey of preventive maintenance models for stochastically deteriorating single unit systems. Nav. Res. Logist. Q. **36**, 419–446 (1989)
6. D. Cho, M. Parlar, A survey of maintenance models for multilayer systems. Eur. J. Oper. Res. **51**, 123 (1991)
7. R. Dekker, Applications of maintenance optimization models: a review and analysis. Reliab. Eng. Syst. Saf. **51**, 229–240 (1996)
8. D. Sherwin, A review of overall models for maintenance management. J. Qual. Maint. Eng. **6**(3), 138–164 (2000)
9. D.M. Frangopol, D. Saydam, S. Kim, Maintenance, management, life-cycle design and performance of structures and infrastructures: a brief review. Struct. Infrastruct. Eng. **8**(1), 1–25 (2012)
10. I.B. Gerstbakh, *Models of Preventive Maintenance* (North Holland, New York, 1977)
11. J.D. Campbell, A.K.S. Jardine, J. McGlynn, *Asset Management Excellence: Optimizing Equipment Life-Cycle Decisions* (CRC Press, Florida, 2011)
12. A. Van Horenbeek, P. Pintelon, L. Muchiri, Maintenance optimization models and criteria. White paper (2011), https://lirias.kuleuven.be/bitstream/123456789/270349/1/
13. M.D. Pandey, Probabilistic models for condition assessment of oil and gas pipelines. Int. J. Non-Destr. Test. Eval. **31**(5), 349–358 (1998)
14. H. Wang, H. Pham, *Reliability and Optimal Maintenance* (Springer, London, 2006)
15. T. Nakagawa, *Maintenance Theory of Reliability* (Springer, London, 2005)
16. A. Gelman, J.B. Carlin, H.S. Stern, D.B. Rubin, *Bayesian Data Analysis* (Chapman & Hall/CRC, New York, 2000)
17. N. Fenton, M. Neil, *Risk Assessment and Decision Analysis with Bayesian Networks* (CRC Press, Boca Raton, 2012)

18. N.T. Kottegoda, R. Rosso, *Probability, Statistics and Reliability for Civil and Environmental Engineers* (McGraw Hill, New York, 1997)
19. A.H.-S. Ang, W.H. Tang, *Probability Concepts in Engineering: Emphasis on Applications to Civil and Environmental Engineering* (Wiley, New York, 2007)
20. Y. Mori, B. Ellingwood, Maintaining reliability of concrete structures. i: role of inspection/repair. J. Struct. ASCE **120**(3), 824–835 (1994)
21. H. Streicher, A. Joanni, R. Rackwitz, Cost-benefit optimization and risk acceptability for existing, aging but maintained structures. Struct. Saf. **30**, 375–393 (2008)
22. C.H. Lie, C.L. Hwang, F.A. Tillman, Availability of maintained systems: a state-of-the-art survey. AIIE Trans. **9**, 247–259 (1977)
23. E.E. Lewis, *Introduction to Reliability Engineering* (Wiley, New York, 1994)
24. S. Ozikichi (ed.), *Reliability and Maintenance of Complex Systems* (Springer, New York, 1996)
25. K.W. Lee, *Handbook on Reliability Engineering* (Springer, London, 2003)
26. S. Ross, *Introduction of Probability Models* (Academic Press, San Diego, 2007)
27. R. Rackwitz, A. Joanni, Risk acceptance and maintenance optimization of aging civil engineering infrastructures. Struct. Saf. **31**, 251–259 (2009)
28. D.R. Cox, *Renewal Theory* (Metheun, London, 1962)
29. R.E. Barlow, F. Proschan, *Mathematical Theory of Reliability* (Wiley, New York, 1965)
30. R. Cleroux, S. Dubuc, C. Tilquin, The age replacement problem with minimal repair and random repair costs. Oper. Res. **27**, 1158–1167 (1979)
31. T.J. Aven, U. Jensen, *Stochastic Models in Reliability*, Series in Applications of Mathematics: Stochastic Modeling and Applied Probability (41) (Springer, New York, 1999)
32. T. Dohi, N. Kaio, S. Osaki, Basic Preventive Maintenance Policies and Their Variations, in *Maintenance Modeling and Optimization*, ed. by M. Ben-Daya, S.O. Duffuaa, A. Raouf (Kluwer Academic Press, Boston, 2000), pp. 155–183
33. S.H. Sheu, W.S. Griffith, Optimal age-replacement policy with age dependent minimal-repair and random leadtime. IEEE Trans. Reliab. **50**, 302–309 (2001)
34. W. Kuo, M.J. Zuo, *Optimal Reliability Modeling* (Wiley, Hoboken, 2003)
35. M. Berg, A proof of optimality for age replacement policies. J. Appl. Probab. **13**, 751–759 (1976)
36. B. Bergman, On the optimality of stationary replacement strategies. J. Appl. Probab. **17**, 178–186 (1980)
37. C.W. Holland, R.A. McLean, Applications of replacement theory. AIIE Trans. **7**, 42–47 (1975)
38. C. Tilquin, R. Cleroux, Periodic replacement with minimal repair at failure and adjustment costs. Nav. Res. Logis. Q. **22**, 243–254 (1975)
39. P.J. Boland, Periodic replacement when minimal repair costs vary with time. Nav. Res. Logis. Q. **29**, 541–546 (1982)
40. T. Aven, Optimal replacement under a minimal repair strategy: a general failure model. Adv. Appl. Probab. **15**, 198–211 (1983)
41. I. Bagai, K. Jain, Improvement, deterioration and optimal replacement under age-replacement with minimal repair. IEEE Trans. Reliab. **43**, 156–162 (1994)
42. M. Chen, R.M. Feldman, Optimal replacement policies with minimal repair and age dependent costs. Eur. J. Oper. Res. **98**, 75–84 (1997)
43. R. Korn, Some applications of impulse control in mathematical finance. Math. Methods Oper. Res. **50**, 493–518 (1999)
44. M. Junca, Optimal execution strategy in the presence of permanent price impact and fixed transaction cost. Optim. Control Appl. Methods **33**(6), 713–738 (2012)
45. A. Bensoussan, R.H. Liu, S.P. Sethi, Optimality of an (s, s) policy with compound poisson and diffusion demands: a quasi-variational inequalities approach. SIAM, J. Control Optim. **44**(5), 1650–1676 (2005)
46. S. Thonhauser, H. Albrecher, Optimal dividend strategies for a compound poisson process under transaction costs and power utility. Stoch. Models **27**, 120–140 (2011)
47. M. Junca, M. Sánchez-Silva, Optimal maintenance policy for a compound poisson shock model. IEEE - Trans. Reliab. **62**(1), 66–72 (2012)

48. M. Junca, M. Sánchez-Silva, Optimal maintenance policy for permanently monitored infrastructure subjected to extreme events. Probab. Eng. Mech. **33**(1), 1–8 (2013)
49. L.C.G. Rogers, D. Williams, *Diffusions, Markov Processes and Martingales*, vol. 1 (Cambridge Mathematical Library, Cambridge University Press, Cambridge, 2000)
50. H. Kushner, P. Dupuis, *Numerical Methods for Stochastic Control Problems in Continuous Time* (Springer, New York, 1992)
51. R.E. Barlow, L.C. Hunter, F. Proschan, Optimum checking procedures. SIAM J. **4**, 1078–1095 (1963)
52. T. Nakagawa, Optimum inspection policies for a standby unit. J. Oper. Soc. Jpn. **23**, 13–26 (1980)
53. Y. Yang, G.-A. Klutke, Improved inspections schemes for deteriorating equipment. Probab. Eng. Inf. Sci. **14**, 445–460 (2000)
54. Y. Yang, G.-A. Klutke, A distribution-free lower bound for availability of quantile-based inspection schemes. IEEE Trans. Reliab. **50**(4), 419–421 (2001)
55. G-A. Klutke, M. Sánchez-Silva, J. Riascos-Ochoa, Long-term maintenance of deteriorating infrastructure: inspection strategies for incipient failures. in *Proceedings of the Third International Symposium on Life-Cycle Civil Engineering, IALCCE'12*, Vienna, Austria, 3-6 October 2012
56. S.M. Ross, *Stochastic Processes*, 2nd edn. (Wiley, New York, 1996)
57. M.A. Wortman, G.-A. Klutke, H. Ayhan, A maintenance strategy for systems subjected to deterioration governed by random shocks. IEEE Trans. Reliab. **43**(3), 439–445 (1994)
58. G.-A. Klutke, Y. Yang, The availability of inspected systems subject to shocks and graceful deterioration. IEEE Trans. Reliab. **51**(3), 371–374 (2002)
59. P. Kiessler, G.-A. Klutke, Y. Yang, Availability of periodically inspected systems subject to markovian degradation. J. Appl. Probab. **39**, 700–711 (2002)
60. H. Ayhan, J. Limon-Robles, M.A. Wortman, An approach for computing tight numerical bounds on renewal functions. IEEE Trans. Reliab. **48**, 182–188 (1999)
61. D.A. Elkins, M.A. Wortman, On numerical solution of the markov renewal equation: tight upper and lower kernel bounds. Methodol. Comput. Appl. Probab. **3**, 239–253 (2001)

Appendix A
Review of Probability Theory

A.1 Introduction: What Is Probability?

What's in a word? The words "probably" and "probability" are used commonly in everyday speech. We all know how to interpret expressions such as "It will probably rain tomorrow," or "Careless smoking probably caused that fire," although the meanings are not particularly precise. The common usage of "probability" has to do with how closely a given statement resembles *truth*. Note that in common usage, it may be impossible to verify whether the statement is *true* or not; that is, the *truth* may not be knowable. Informally, we use the terms "probable" and "probability" to express a *likelihood* or *chance* of truth.

While these common usages of the term "probability" are effective in communicating ideas, from a mathematical point of view, they lack the precision and standardization of terminology to be particularly functional. Thus scientists and mathematicians have developed various *theories* of probability to address the needs of scientific analysis and decision making. We will use a particular theory that has its origins in the early twentieth century and is now (by far) the most widely used theory of probability. This theory provides a formal structure (entities, definitions, axioms, etc.) that allows us to use other well-developed mathematical concepts (limits, sums, averages, etc.) in a way that remains consistent with our understanding of physical principals. All theories have limitations. Our theory of probability, for instance, will not help us answer questions like, "What is the probability that individual X is guilty of a crime?" or "What is the probability that pigs will fly?" Fortunately, a well-developed theory has well-defined limitations, and we should be able to identify when we have overstepped the bounds of scientific validity.

As we discuss these concepts, keep in mind that it is "probably" inevitable that we will at times encounter conflicts between the colloquial meanings of words and their formal mathematical definitions. These conflicts are natural and are no cause for alarm!

M. Sánchez-Silva and G.-A. Klutke, *Reliability and Life-Cycle Analysis of Deteriorating Systems*, Springer Series in Reliability Engineering,
DOI 10.1007/978-3-319-20946-3

A.2 Random Experiments and Probability Spaces: The Building Blocks of Probability

Our theory of probability begins with the concept of a *random experiment*. The idea is that we intend to perform an experiment that results in (precisely) one of a group of outcomes. We use the term *random* experiment because we cannot be certain in advance about the outcome. That is, we can identify all *possible* outcomes of the experiment, but we do not know in advance which particular outcome will occur. The experiment is assumed to be *repeatable*, in the sense that we could recreate the exact conditions of the experiment. If we repeat the experiment, however, we are not guaranteed that the same outcome will occur. To effectively describe the random experiment, we must be able to: (i) identify its outcomes, (ii) characterize the information available to us about the outcome of the experiment, and (iii) quantify the likelihood that the experiment results in a particular incident. In mathematical terminology, a random experiment will be identified with (actually, is equivalent to) a *probability space*. A probability space consists of three entities: a sample space (we will call it Ω), an event space (we'll call it $\mathscr{F}$), and a probability measure (we'll call it P). Let us discuss each of these entities in turn.

A.2.1 Sample Space

Formally, we define the *sample space* Ω to be the collection of all possible outcomes. Elements of the sample space are distinct and exhaustive (i.e., on any given performance of the experiment, one and only one outcome occurs), and we can think of the sample space as a set of distinct points. The sample space may be discrete (countable or denumerable) or continuous (uncountable or nondenumerable), likewise, it may be finite or infinite.

Example A.1 The experiment consists of tossing a coin three times consecutively. Assuming that we do not allow the possibility of a coin landing on its side (H heads or T Tails), the sample space can be identified as $\{(HHH), (HHT), (HTH), (THH), (HTT), (THT), (TTH), (TTT)\}$. The sample space is discrete and finite.

Example A.2 The experiment consists of two players (A and B) playing hands of poker for \$1 per hand. Each player begins with \$5, and the game continues until one of the players is bankrupt. Here the sample space can be identified as all sequences of the elements A and B such that the number of one letter does not exceed the number of the other letter by more than 5. The sample space is discrete and infinite.

Example A.3 The experiment consists of measuring the diameter of every 5th steel cylinder that leaves a manufacturing line. The sample space consists of sequences of real numbers; it is continuous and infinite.

To reiterate, a sample space is a set of outcomes; it obeys the typical rules that obtain with sets (unions, intersections, complements, differences, etc.).

A.2.2 Event Space

The second element of a probability space is a collection of *so-called* events $\mathscr{F}$. Events themselves consist of particular groups of outcomes. Thus the set of events is a collection of subsets of the sample space. Events can be thought of as characteristics of outcomes that can be identified once the experiment has been performed; that is, they are the "information scale" at which we can view the results of an experiment. In many, but not all, experiments, we can identify individual outcomes of an experiment; in some experiments we can identify only certain characteristics of individual outcomes. Thus the event space characterizes the information that we have available to us about the outcomes of a random experiment; it is the mesh or filter through which we can view the outcomes. Some terminology: we say that "an event has occurred" if the outcome that occurred is contained in that event.

The specification of the event space is not completely arbitrary; in order to maintain consistency, we need to instill some structure (rules) on the event space. The structure makes perfect intuitive sense. First, if we are able to observe that a particular group of outcomes occurred, we should be able to observe that the same group of outcomes did not occur. This means that if a set of outcomes F is in the event space, then the set of outcomes $\overline{F}$ (the complement of F) is also in the event space. Secondly, if we are able to determine if a set of outcomes F_1 occurred, and we are able to determine if a group of outcomes F_2 occurred, then we should be able to determine if either F_1 or F_2 occurred. That is, if F_1 and F_2 are in the event space, then $F_1 \cup F_2$ must be in the event space. Finally, we must be able to observe that *some* outcome occurred; that is, Ω itself must be an event. Note that since Ω is in the event space, so is ϕ, the empty set (also called the impossible event). With these rules for the event space, the smallest event space that we can work with is $\mathscr{F} = \{\Omega, \phi\}$.

Example A.4 Suppose the random experiment is as in Example A.1, and suppose that we are able to observe the outcome of each individual coin toss. Then the event space consists of *all* subsets of the sample space (the power set of the sample space).

Example A.5 Now suppose the random experiment is as in Example A.1, except that we are able to observe only the outcome of the last toss. Then the event space consists of Ω, ϕ, and the sets $\{(HHH), (HTH), (THH), (TTH)\}$ and $\{(HHT), (HTT), (THT), (TTT)\}$.

Note that an event can be determined either by listing its elements or by stating a condition that its elements must satisfy; e.g., if the sample space of our experiment is as in Example A.1, the set $\{(HHT), (HTH), (THH), (HTT)\}$ and the statement "two heads occurred" determine the same event.

A.2.3 Probability Measure

The final element of our probability space is an assignment of probabilities for each event in the event space. Such an assignment is described by a function P that assigns a value to each event. This value represents our belief in the likelihood that the experiment will result in an event's occurrence. The choice of this function quantifies our knowledge of the randomness of the experiment. It is important to remember that a probability measure lives on (assigns values to) *events* rather than *outcomes*, but remember, also, that there are certain situations where individual outcomes can also be events; such events are called *atomic events*.

Definition 55 A sample space Ω of a random experiment is the set of all possible outcomes of the experiment.

Definition 56 An event space $\mathscr{F}$ of a random experiment is a collection of subsets of the sample space that satisfy

- Ω is in $\mathscr{F}$
- If F is in $\mathscr{F}$, then $\overline{F}$ is in $\mathscr{F}$
- If F_1 and F_2 are in $\mathscr{F}$, then $F_1 \cup F_2$ is in $\mathscr{F}$.

Definition 57 A probability measure P for a random experiment is a function that assigns a numerical value to each event in an event space such that

- If F is an event, $0 \leq P(F) \leq 1$.
- $P(\Omega) = 1$.
- If $F_1, F_2, \ldots$ are mutually exclusive events, then

$$P(F_1 \cup F_2 \cup \cdots) = \sum_i P(F_i).$$

These rules guarantee that a probability measure is meaningful and workable and are often referred to as the "Axioms of Probability." Beyond these rules, how we determine which probability measure to use for a given random experiment is a *modeling* issue rather than a mathematical one. Many different choices for probability measures are possible, depending on how we believe the probabilistic mechanism producing the outcomes works.

To summarize, we have fully described any random experiment if we have specified a *probability space* $\{\Omega, \mathscr{F}, P\}$ consisting of a sample space Ω, an event space $\mathscr{F}$, and a probability measure P.

Example A.6 Consider again the random experiment described in Example A.1 and the event space described in Example A.4. If we believe that the coin we are using is *fair* (unbiased), it should follow that each of the atomic events should have the same probability (i.e., be equally likely). If on any given toss, a head is twice as likely as a tail, the probability of the event $\{(HHH)\}$ should be eight times the probability of the event $\{(TTT)\}$, and the events $\{(HTH)\}$ and $\{(THH)\}$ should have the same probability.

The probability axioms lead to several elementary properties of probability. These properties follow easily by considering simple set operations.

Property 1 *For any event* F, $P(F) = 1 - P(\overline{F})$.

Proof F and $\overline{F}$ are mutually exclusive events, and $\Omega = F \cup \overline{F}$. Hence by Axioms 2 and 3,

$$1 = P(\Omega) = P(F \cup \overline{F}) = P(F) + P(\overline{F}), \tag{A.1}$$

and hence $P(F) = 1 - P(F)$.

Property 2 *If* F_1 *and* F_2 *are any events (not necessarily mutually exclusive), then*

$$P(F_1 \cup F_2) = P(F_1) + P(F_2) - P(F_1 \cap F_2). \tag{A.2}$$

Proof By simple set properties,

$$F_1 \cup F_2 = F_1 \cup (\overline{F_1} \cap F_2) \quad \text{and} \quad F_2 = (F_1 \cap F_2) \cup (\overline{F_1} \cap F_2). \tag{A.3}$$

The unions on the right-hand side of each equation are of mutually exclusive events, so by Axiom 3,

$$\begin{aligned} P(F_1 \cup F_2) =& P(F_1) + P(\overline{F_1} \cap F_2) \\ P(F_2) =& P(F_1 \cap F_2) + P(\overline{F_1} \cap F_2) \end{aligned}$$

Solving both equations for $P(\overline{F_1} \cap F_2)$ gives the desired result.

Property 3 *If* $F_1, F_2, \ldots, F_k$ *are any events,*

$$\begin{aligned} P(F_1 \cup F_2 \cup \cdots \cup F_k) = &\sum_i P(F_i) - \sum_{i<j} P(F_i \cap F_j) \\ &+ \cdots + (-1)^{k+1} P(F_1 \cap F_2 \cdots \cap F_k). \end{aligned}$$

Proof Follows from Property 2 by mathematical induction.

A.2.4 Conditional Probability and the Law of Total Probability

The probability measure ensures that we have assigned a probability to every event in the event space of our random experiment. In many situations, we may be able to observe *partial* information about the outcome of an experiment in terms of the occurrence of an event. We would like to have a consistent way of "updating" the probabilities of other events based on this information. To this end, we give an elementary definition of *conditional probability*.

Definition 58 Given events F_1 and F_2, the conditional probability of F_2 given that the F_1 occurs is given by

$$P(F_2|F_1) = \frac{P(F_1 \cap F_2)}{P(F_1)}. \tag{A.4}$$

Of course, this definition only makes sense if $P(F_1) > 0$. For now, we leave the conditional probability undefined if $P(F_1) = 0$, but there are other ways to consistently define the conditional probability in this case.

Now consider a set of events $F_1, F_2, \ldots$ that form a *partition* of the sample space Ω; that is, the events are mutually exclusive ($F_i \cap F_j = \emptyset, i \neq j$) and exhaustive ($\cup_j F_j = \Omega$). The number of events in the partition may be finite or infinite. For any event A, by the properties of the partition, we can write

$$A = [A \cap F_1] \cup [A \cap F_1] \cup \cdots, \tag{A.5}$$

and since the $[A \cap F_1]$'s are mutually exclusive, we have

$$P(A) = P(A \cap F_1) + P(A \cap F_2) + \cdots, \tag{A.6}$$

and using the definition of conditional probability,

$$P(A) = P(A|F_1)P(F_1) + P(A|F_2)P(F_2) + \cdots = \sum_i P(A|F_i)P(F_i). \tag{A.7}$$

This result is known as the *Law of Total Probability* and is very useful.

A.3 Random Variables

A.3.1 Definition

Once we have a probability space that describes our random experiment, there are many things that we can "measure" about each outcome in the sample space. These measurable properties, which depend on the actual outcome realized by the experiment, are termed *random variables*.

Definition 59 A random variable X is a function that assigns a real number $X(\omega)$ to each element ω of the sample space such that for any collection of real numbers C,

$$X^{-1}(C) = \{\omega : X(\omega) \in C\} \tag{A.8}$$

is an event (i.e., is in $\mathscr{F}$).

Mathematically, an assignment of a numerical value to an element of the sample space is a mapping (function) of the sample space to the real line. Such a mapping is called a random variable *provided* we can "trace back" values of the function to events. Formally, a random variable is a function whose *domain* is the sample space and whose *range* is some subset of (or possibly the whole) the real line; that is, a random variable assigns a real number to each element of the sample space. A random variable must have the property that, if we take a particular range of numerical values, the collection of outcomes that gets assigned a value in that range is an event. This last property is called *measurability* and ensures that our probability space is "rich enough" to support the random variable.

Example A.7 Suppose our experiment consists of selecting an individual at random from a classroom with n students. A reasonable choice for a probability space for this experiment might be to choose Ω to be the list of students' id numbers (to make sure each student is uniquely identified), $\mathscr{F}$ to be the power set of Ω, and to choose P such that it assigns value $1/n$ to each atomic event. Now to each outcome in the sample space (each student), assign a numerical value equal to the students' height, weight, cumulative GPA, and score on the last exam.

Example A.8 Consider the random experiment of Example A.1, and suppose we define a function X to be the number of heads in all three tosses. Then $X((HHH)) = 3$, $X((HHT)) = X((HTH)) = X((THH)) = 2$, $X((HTT)) = X((THT)) = X((TTH)) = 1$, $X((TTT)) = 0$. X is a random variable for the event space described in Example A.4 but not for the event space described in Example A.5.

Random variables are termed *discrete* if the possible values they can take on is a discrete set and *continuous* if it is a continuous set.

Example A.9 A manufacturing facility contains a sophisticated CNC pipe bending station. In-process jobs arrive at the bending station from an upstream cutting station, and after processing at the bending station, are placed on a conveyor that takes them to a drilling station. Let X be the number of jobs waiting for processing at the machine at the beginning of a particular day. X is a discrete random variable. Let Y be the amount of time between the first two departures from the machine on a given day. Y is a continuous random variable.

A.3.2 Events Defined by Random Variables

A probability measure is part of the description of a random experiment. A probability measure "lives on" the event space that we have chosen for our random experiment. How do we make a connection between probability and random variables? The answer lies in constructing appropriate events using random variables.

Let X be a random variable defined on a probability space $(\Omega, \mathscr{F}, P)$. For simplicity, suppose X is discrete. Take any real number x, and consider the set

$$F_x = \{\omega \in \Omega : X(\omega) = x\}. \tag{A.9}$$

F_x is an event, and therefore it makes sense to talk about $P(F_x)$. That is, for any real number x, we can use the random variable X to construct an event by considering all sample points whose X-value is x. Such an event is called an event *generated* by the random variable X.

We will use the notation $\{X = x\}$ to indicate the event $\{\omega \in \Omega : X(\omega) = x\}$, and we will write $P(X = x)$ to mean $P(\{\omega \in \Omega : X(\omega) = x\})$. Similarly, we can define events such as $\{X < x\}$, $\{X \geq x\}$, and even such events as $\{X \leq y, X \geq x\}$ and $\{y \leq X \leq x\}$. As long as we associate statements about random variables with events in the event space and use the rules for probability measure, we have no difficulty in assigning the proper probabilities to any event generated by a random variable.

A.3.3 Distribution Function

Suppose we have defined a random variable X on a probability space. For a given x, we know how to interpret the event $\{X \leq x\}$, and how to evaluate its probability. As x varies over the real line, $P(X \leq x)$ defines a function of x; this function is called the *cumulative distribution function* (*distribution function* or *cdf* for short) and it plays a very important role in probability theory.

Definition 60 The distribution function of a random variable X is defined by

$$F(x) = P(X \leq x), \qquad -\infty < x < \infty. \tag{A.10}$$

Note that knowing the cdf of a random variable is equivalent to knowing the probability of each and every event generated by that random variable.

The cdf of any random variable has a number of important properties.

- The cdf is right continuous.
- The cdf is nondecreasing.
- $F(-\infty) = 0$, $F(\infty) = 1$.

The cdf of a discrete random variable is a step function; the cdf of a continuous random variable is a continuous function.

Example A.10 Let X be the number of heads in three consecutive tosses of a fair coin. Then

$$X(\omega) = \begin{cases} 0 & \text{if } \omega = (TTT); \\ 1 & \text{if } \omega \in \{(TTH), (THT), (HTT)\}; \\ 2 & \text{if } \omega \in \{(HHT), (HTH), (THH)\}; \\ 3 & \text{if } \omega = (HHH). \end{cases}$$

Since the coin is fair, the probability measure assigns the following values to the events $\{X = x\}$:

$$P(X = x) = \begin{cases} \frac{1}{8} & \text{if } x = 0; \\ \frac{3}{8} & \text{if } x = 1; \\ \frac{3}{8} & \text{if } x = 2; \\ \frac{1}{8} & \text{if } x = 3. \end{cases}$$

and therefore, the distribution function of X is

$$F(x) = \begin{cases} 0 & \text{if } x < 0; \\ \frac{1}{8} & \text{if } 0 \leq x < 1; \\ \frac{1}{2} & \text{if } 1 \leq x < 2; \\ \frac{7}{8} & \text{if } 2 \leq x < 3; \\ 1 & \text{if } x \geq 3. \end{cases}$$

Example A.11 Let X be an exponentially distributed random variable. Then

$$P(X \leq x) = F(x) = 1 - e^{-\lambda x}, \qquad x > 0. \tag{A.11}$$

A.3.4 Expectation and Moments

We have seen that its distribution function completely specifies the probabilistic structure of a random variable. Only the distribution function is capable of giving us the probability that the random variable takes on values in a particular range. We may, however, be interested in other, less detailed, information about the structure of the random variable. For instance, we might want to know the 95th percentile (value α such that $P(X \leq \alpha) = 0.95$), the median (value β such that $P(X \leq \beta) = P(X \geq \beta)$), or the mean (probabilistic average) of the random variable. Each of these entities is a *number* (rather than a function) and contains some useful information about the random variable. In this section, we will define a probabilistic average that will be of great use to us in characterizing random variables.

The *expectation* operator $\mathbb{E}$ of a random variable X is defined as

$$\mathbb{E}[X] = \int_\Omega X(\omega) P(d\omega), \tag{A.12}$$

or in terms of the distribution function

$$\mathbb{E}[X] = \int_{-\infty}^{\infty} x dF(x). \tag{A.13}$$

Expectation is an *averaging* operation; as you can see from the right-hand side of the definition, it "weights" values assigned by the random variable by their "likelihood" as assigned by the probability measure. We can define the expectation for *functions* of random variables similarly:

$$\mathbb{E}[\phi(X)] = \int_{\Omega} \phi(X(\omega))P(d\omega) = \int_{-\infty}^{\infty} \phi(x)dF(x). \tag{A.14}$$

We refer to $\mathbb{E}[X]$ as the *mean* of X, and we often denote it by μ. If we choose $\phi(X) = X^k$, we have

$$\mathbb{E}[X^k] = \int_{-\infty}^{\infty} x^k dF(x). \tag{A.15}$$

where $\mathbb{E}[X^k]$ is called the kth *moment about zero* of the random variable X. If we choose $\phi(X) = (X - \mu)^k$, we have

$$\mathbb{E}[(X - \mu)^k] = \int_{-\infty}^{\infty} (x - \mu)^k dF(x). \tag{A.16}$$

where $\mathbb{E}[(X-\mu)^k]$ is called the kth *moment about the mean* of the random variable X.

A.3.5 Discrete Random Variables

If X is a discrete random variable, then $F(x)$ is a step function, and $dF(x)$ is computed as a difference $F(x) - F(x^-)$. Note that this difference will be zero except at jump points (steps) of $F(x)$. In this case, $dF(x)$ is known as the *mass* function $p(x)$ and is defined for each jump point x of $F(x)$. Notice that

$$p(x) = dF(x) = F(x) - F(x^-) = P(X \le x) - P(X < x) = P(X = x). \tag{A.17}$$

Thus for a discrete random variable X, $\mathbb{E}[X]$ is calculated as

$$\mathbb{E}[X] = \sum_x xp(x). \tag{A.18}$$

Example A.12 Consider the random variable X in Example A.10. Here

$$dF(x) = p(x) = \begin{cases} \frac{1}{8} & \text{if } x = 0; \\ \frac{3}{8} & \text{if } x = 1; \\ \frac{3}{8} & \text{if } x = 2; \\ \frac{1}{8} & \text{if } x = 3. \end{cases}$$

Then

$$\mathbb{E}(X) = \sum_x xp(x) = 0 \cdot \frac{1}{8} + 1 \cdot \frac{3}{8} + 2 \cdot \frac{3}{8} + 3 \cdot \frac{1}{8} = 1\frac{1}{2} \tag{A.19}$$

and

$$\mathbb{E}(X^2) = \sum_x x^2 p(x) = 0 \cdot \frac{1}{8} + 1 \cdot \frac{3}{8} + 4 \cdot \frac{3}{8} + 9 \cdot \frac{1}{8} = 3. \tag{A.20}$$

A.3.6 Continuous Random Variables

If X is a continuous random variable, then $F(x)$ is a continuous function. Thus it has a derivative $f(x)$; i.e.,

$$dF(x) = f(x)dx \tag{A.21}$$

The derivative $f(x) = \frac{d}{dx}F(x)$ is called the density function of X. Thus for a continuous random variable X. Then, $\mathbb{E}[X]$ is calculated by

$$\mathbb{E}[X] = \int_{-\infty}^{\infty} xf(x)dx. \tag{A.22}$$

Example A.13 Consider the random variable X in Example A.11. Here

$$f(x) = \frac{dF(x)}{dx} = \lambda e^{-\lambda x}. \tag{A.23}$$

This gives

$$\mathbb{E}[X] = \int_{0}^{\infty} x\lambda e^{-\lambda x} dx = \frac{1}{\lambda} \tag{A.24}$$

and

$$\mathbb{E}[X^2] = \int_{0}^{\infty} x^2\lambda e^{-\lambda x} dx = \frac{2}{\lambda^2}. \tag{A.25}$$

A.3.7 Variance and Coefficient of Variation

The second moment about the mean, $\mathbb{E}[(X - \mu)^2]$, is known as the variance of the random variable X and is of great importance in both probability and statistics. It provides a simple measure of the *dispersion* of X around the mean. The variance of

X is written as $Var(X)$ and is often denoted by σ^2. Variance can be computed in terms of the second moment of X by

$$Var(X) = \mathbb{E}[X^2] - (\mathbb{E}[X])^2. \tag{A.26}$$

The square root of the variance is known as the standard deviation, $StDev(X)$, and is denoted by σ.

Also of great importance is the ratio of standard deviation to mean of the random variable, known as the coefficient of variation of X:

$$COV = \frac{StDev(X)}{\mathbb{E}[X]} = \frac{\sigma}{\mu}. \tag{A.27}$$

A.4 Multiple Random Variables: Joint and Conditional Distributions

In many applications, we will be interested in studying two or more random variables defined on the same probability space. For instance, in a manufacturing environment, we might be interested in studying the number of jobs waiting to be processed (the *work-in-process* inventory, or *wip*) at n machines at a given point in time. We are interested in this section in describing the properties of several random variables simultaneously. We will discuss the joint distribution of two random variables, but our discussion extends naturally to several random variables or an entire sequence of random variables.

A.4.1 Events Generated by Pairs of Random Variables

When two random variables X and Y are considered simultaneously, the events generated by X and Y take the form

$$\{X \in E_X \text{ and } Y \in E_Y\} = \{\omega \in \Omega : X(\omega) \in E_X \text{ and } Y(\omega) \in E_Y\}, \tag{A.28}$$

where E_X and E_Y are, respectively, subsets of the range space of X and the range space of Y. Events generated by X and Y are such sets as $\{X < x_1 \text{ and } y_1 < Y \leq y_2\}$ or $\{X \geq x_1 \text{ and } Y \geq y_1\}$, or even $\{X < x_1\}$, which is really the event $\{X < x_1 \text{ and } Y \leq \infty\}$.

To compute probabilities of events generated by pairs of random variables, we need only to find the subset $F \in \mathscr{F}$ of the sample space that the event represents, and then to find the assignment $P(F)$ made by the probability measure to that subset.

A.4.2 Joint Distributions

In the previous section, we defined the cdf of a random variable X to be

$$F(x) = P(X \leq x), \qquad -\infty < x < \infty. \tag{A.29}$$

We can similarly define a joint distribution of two random variables X and Y as

$$F(x, y) = P(X \leq x \text{ and } Y \leq y), \qquad -\infty < x < \infty, -\infty < y < \infty. \tag{A.30}$$

with respect to the *joint* distribution of X and Y, we refer to the cdf of X alone, or of Y alone, as a *marginal* distribution. $F(x, y)$ has the following properties, which correspond to the properties of the marginal distribution functions we have encountered earlier.

- $0 \leq F(x, y) \leq 1$ for $-\infty < x < \infty, -\infty < y < \infty$.
- $\lim_{x \to a^+} F(x, y) = F(a, y)$ and $\lim_{y \to b^+} F(x, y) = F(x, b)$.
- If $x_1 \leq x_2$ and $y_1 \leq y_2$, then $F(x_1, y_1) \leq F(x_2, y_2)$.
- $\lim_{x \to -\infty} F(x, y) = 0$, $\lim_{y \to -\infty} F(x, y) = 0$, $\lim_{x \to \infty, y \to \infty} F(x, y) = 1$.
- Whenever $a \leq b$ and $c \leq d$, then $F(a, c) - F(a, d) - F(b, c) + F(b, d) \geq 0$.

Notice that we can always recover the marginal cdfs from the joint cdf:

$$\lim_{y \to \infty} F(x, y) = F(x, \infty) = F_X(x)$$

$$\lim_{x \to \infty} F(x, y) = F(\infty, y) = F_Y(y)$$

Example A.14 Let the joint distribution of X and Y be given by

$$F(x, y) = \begin{cases} 1 - e^{-x} - e^{-y} + e^{-(x+y)}, & 0 \leq x < \infty, 0 \leq y < \infty; \\ 0 & \text{otherwise.} \end{cases}$$

Then the marginal cdfs of X and Y are, respectively,

$$F_X(x) = \lim_{y \to \infty} F(x, y) = \begin{cases} 1 - e^{-x} & 0 \leq x < \infty; \\ 0 & \text{otherwise.} \end{cases}$$

$$F_Y(y) = \lim_{x \to \infty} F(x, y) = \begin{cases} 1 - e^{-y} & 0 \leq y < \infty; \\ 0 & \text{otherwise.} \end{cases}$$

A.4.3 Determining Probabilities from the Joint Distribution Function

Just as in the one-dimensional case, the joint distribution function of X and Y allows us to compute the probability of *any* event generated by the random variables X and Y. Any event of the form $\{X \leq x \text{ and } Y \leq y\}$ has probability $F(x, y)$. For more complicated events, it is often useful to sketch the event as a region in the (x, y) plane. Doing so, we observe that

$$P(x_1 < X \leq x_2 \text{ and } Y \leq y) = F(x_2, y) - F(x_1, y), \tag{A.31}$$

and

$$P(x_1 < X \leq x_2 \text{ and } y_1 < Y \leq y_2) = F(x_2, y_2) - F(x_1, y_2) - F(x_2, y_1) + F(x_1, y_1). \tag{A.32}$$

Another way to understand the last equality is to examine set relationships. Let

$$\begin{aligned} A &= \{x_1 < X \leq x_2 \text{ and } y_1 < Y \leq y_2\} \\ B &= \{X \leq x_2 \text{ and } Y \leq y_2\} \\ C &= \{X \leq x_1 \text{ and } Y \leq y_2\} \\ D &= \{X \leq x_2 \text{ and } Y \leq y_1\} \end{aligned}$$

We are interested in computing $P(A)$. Notice that any point of the set B that does not lie in A must lie in C or D; i.e.,

$$B = A \cup (C \cup D). \tag{A.33}$$

Moreover, the sets A and $C \cup D$ are mutually exclusive, so that

$$P(B) = P(A) + P(C \cup D). \tag{A.34}$$

Therefore,

$$\begin{aligned} P(A) &= P(B) - P(C \cup D) \\ &= P(B) - \big(P(C) + P(D) - P(C \cap D)\big) \qquad \text{(Property 2, Sect. 2.3)} \\ &= P(B) - P(C) - P(D) + P(C \cap D), \end{aligned}$$

which is what we needed to show.

A.4.4 Joint Mass and Density Functions

As for a single random variable, we can define a joint mass function (for discrete random variables) or density function (for continuous random variables) for a pair of random variables. We may also have one discrete and one continuous random variable, in which case we have a mixture of a mass function and a density function.

When random variables X and Y are both discrete, we define the joint mass function

$$p(i, j) = P(X = i \text{ and } Y = j), \quad \forall i \text{ in the range of } X, j \text{ in the range of } Y \quad \text{(A.35)}$$

The joint mass function has the following properties:

- $0 \leq p(i, j) \leq 1$ for each i, j.
- $\sum_{i,j} p(i, j) = \sum_i \sum_j p(i, j) = 1$.
- $F(x, y) = \sum_{i \leq x} \sum_{j \leq y} p(i, j)$.

The marginal mass functions are easily calculated from the joint mass function:

$$p_X(x) = P(X = x) = \sum_j p(i, j), \quad p_Y(y) = P(Y = y) = \sum_i p(i, j). \quad \text{(A.36)}$$

Example A.15 Suppose a coin is tossed three times consecutively. Let X be the total number of heads in the first two tosses, and Y the total number of heads in the last two tosses. Assuming that all 8 outcomes are equally likely, that is,

$$\begin{aligned} P(\{HHH\}) &= P(\{HHT\}) = P(\{HTH\}) = P(\{THH\}) \\ &= P(\{HTT\}) = P(\{THT\}) = P(\{TTH\}) = P(\{TTT\}) = \frac{1}{8}, \end{aligned}$$

the values assigned by X and Y to these outcomes are

$$\begin{aligned} X(HHH) &= 2 & Y(HHH) &= 2 \\ X(HHT) &= 2 & Y(HHT) &= 1 \\ X(HTH) &= 1 & Y(HTH) &= 1 \\ X(THH) &= 1 & Y(THH) &= 2 \\ X(HTT) &= 1 & Y(HTT) &= 0 \\ X(THT) &= 1 & Y(THT) &= 1 \\ X(TTH) &= 0 & Y(TTH) &= 1 \\ X(TTT) &= 0 & Y(TTT) &= 0 \end{aligned}$$

This gives the joint mass function to be

$$p(0,0) = P(X = 0 \text{ and } Y = 0) = P(\{TTT\}) = \frac{1}{8}$$

$$p(0,1) = P(X = 0 \text{ and } Y = 1) = P(\{TTH\}) = \frac{1}{8}$$

$$p(0,2) = P(X = 0 \text{ and } Y = 2) = P(\emptyset) = 0$$

$$p(1,0) = P(X = 1 \text{ and } Y = 0) = P(\{HTT\}) = \frac{1}{8}$$

$$p(1,1) = P(X = 1 \text{ and } Y = 1) = P(\{HTH\} \cup \{THT\}) = \frac{1}{4}$$

$$p(1,2) = P(X = 1 \text{ and } Y = 2) = P(\{THH\}) = \frac{1}{8}$$

$$p(2,0) = P(X = 2 \text{ and } Y = 0) = P(\emptyset) = \frac{1}{8}$$

$$p(2,1) = P(X = 2 \text{ and } Y = 1) = P(\{HHT\}) = \frac{1}{8}$$

$$p(2,2) = P(X = 2 \text{ and } Y = 2) = P(\{HHH\}) = \frac{1}{8}$$

and the marginal mass functions by

$$p_X(0) = P(\{TTH\} \cup \{TTT\}) = \frac{1}{4}$$

$$p_X(1) = P(\{HTH\} \cup \{THH\} \cup \{HTT\} \cup \{THT\}) = \frac{1}{2}$$

$$p_X(2) = P(\{HTH\} \cup \{THT\}) = \frac{1}{4}$$

$$p_Y(0) = P(\{HTT\} \cup \{TTT\}) = \frac{1}{4}$$

$$p_Y(1) = P(\{HHT\} \cup \{HTH\} \cup \{THT\} \cup \{TTH\}) = \frac{1}{2}$$

$$p_Y(2) = P(\{HTH\} \cup \{THT\}) = \frac{1}{4}$$

When random variables X and Y are both continuous, we define the joint density function by

$$f(x,y) = \frac{\partial^2}{\partial x \partial y} F(x,y). \tag{A.37}$$

The joint density function has the following properties:

- $f(x,y) \geq 0$ for all x, y.
- $\int_{-\infty}^{\infty} \int_{-\infty}^{\infty} f(s,t)dtds = 1$.
- $F(x,y) = \int_{-\infty}^{x} \int_{-\infty}^{y} f(s,t)dtds$.

The marginal density functions are easily calculated from the joint density function:

$$f_X(x) = F(x, \infty) = \int_{-\infty}^{x} \int_{-\infty}^{\infty} f(s, t)dtds,$$
$$f_Y(y) = F(\infty, y) = \int_{-\infty}^{\infty} \int_{-\infty}^{y} f(s, t)dtds.$$

Example A.16 Let X and Y be continuous random variables with ranges $(0, \infty)$ and $(0, \infty)$, respectively, and joint density function

$$f(x, y) = \begin{cases} xe^{-x(y+1)}, & 0 \le x < \infty, 0 \le y < \infty \\ 0 & \text{otherwise.} \end{cases}$$

The marginal density functions are given by

$$f_X(x) = \int_0^{\infty} xe^{-x(y+1)}dy = xe^{-x} \int_0^{\infty} e^{-xy}dy = e^{-x}, \quad 0 \le x < \infty$$

and

$$f_Y(y) = \int_0^{\infty} xe^{-x(y+1)}dx = \frac{1}{(y+1)^2}, \quad 0 \le y < \infty.$$

A.4.5 Conditional Distributions

For two random variables X and Y with joint distribution function $F(x, y)$, and marginal distribution functions $F_X(x)$ and $F_Y(y)$, respectively, we define the conditional distribution function of X given Y as

$$G_{X|Y}(x|y) = \frac{F(x, y)}{F_Y(y)} \tag{A.38}$$

provided $F_Y(y) > 0$. Whenever $F_Y(y) = 0$, $G_{X|Y}(x|y)$ is not defined. Similarly, we define the conditional distribution function of Y given X as

$$G_{Y|X}(y|x) = \frac{F(x, y)}{F_X(x)} \tag{A.39}$$

provided $F_X(x) > 0$. Whenever $F_X(x) = 0$, $G_{Y|X}(y|x)$ is not defined. In terms of conditional probability, $G_{X|Y}(x|y)$ and $G_{Y|X}(y|x)$ are, respectively, $P(X \le x|Y \le y)$ and $P(Y \le y|X \le x)$.

If X and Y are both discrete random variables, we can define the conditional mass function of X, given that $Y = j$ as

$$p_{X|Y}(i|j) = P(X = i|Y = j) = \frac{P(X = i \text{ and } Y = j)}{P(Y = j)} = \frac{p(i, j)}{p_Y(j)}, \quad p_Y(j) > 0. \tag{A.40}$$

The condition mass function of Y, given that $X = i$, $p_{Y|X}(j|i)$ is defined similarly.

Example A.17 Suppose we perform the following experiment. First, we roll a fair die and observe the number of spots on the face pointing up. Call this number x. Then, a fair coin is tossed x times, and the number of resulting heads is recorded. We can think of this experiment as defining two random variables X and N, where X is the first number selected and N is the number of heads observed.

The marginal mass function of X is given by

$$p_X(x) = \begin{cases} \frac{1}{6} & x = 1, 2, \ldots, 6; \\ 0 & \text{otherwise.} \end{cases}$$

The conditional mass function of N given X is

$$p_{N|X}(n|x) = P(N = n|X = x) = \binom{x}{n}(\frac{1}{2})^x, \qquad n = 0, 1, \ldots, x.$$

Thus the joint mass function of X and N is given by

$$p(x, n) = p(n|x)p_X(x) = \binom{x}{n}\left(\frac{1}{2}\right)^n \cdot \frac{1}{6}, \qquad x = 1, 2, \ldots, 6, \ n = 0, 1, \ldots, x$$

and the marginal mass function of N is given by

$$p_N(n) = \sum_{x=1}^{6} \binom{x}{n}\left(\frac{1}{2}\right)^n \cdot \frac{1}{6} \qquad n = 0, 1, 2, \ldots, x.$$

In the case that X and Y are both continuous random variables, we define conditional density functions of X, given that $Y = y$, and of Y, given that $X = x$ analogously:

$$f_{X|Y}(x|y) = \frac{f(x, y)}{f_Y(y)} \qquad \text{and} \qquad f_{Y|X}(y|x) = \frac{f(x, y)}{f_X(x)}$$

provided, respectively, that $f_Y(y) > 0$ and $f_X(x) > 0$.

Example A.18 Consider the joint density function of Example A.16. For this case,

$$f_{X|Y}(x|y) = \frac{f(x,y)}{f_Y(y)} = \frac{xe^{-x(y+1)}}{1/(y+1)^2} = x(y+1)^2e^{-x(y+1)}, \qquad 0 \le x < \infty,\ 0 \le y < \infty \tag{A.41}$$

and

$$f_{Y|X}(y|x) = \frac{f(x,y)}{f_X(x)} = \frac{xe^{-x(y+1)}}{e^{-x}} = xe^{-xy}, \qquad 0 \le x < \infty,\ 0 \le y < \infty. \tag{A.42}$$

When the random variables are clear from the context, we will drop the subscripts of the conditional distribution, mass, and density functions.

A.4.6 A Mixed Case from Queueing Theory

There are many cases of interest that involve the joint distribution of a discrete and a continuous random variable. All of our results will carry over to this *mixed* case. In this section, we will work through an example from queueing theory that illustrates the use of a mixed density function.

Suppose that individual jobs arrive at random to a single machine for processing. We will call the sequence of arriving jobs the *arrival stream*. Jobs are served one-at-a-time in the order of arrival. When processing is complete, the jobs depart for finished goods inventory. Those jobs that arrive while the machine is processing another job wait in a queue until the machine becomes available and all previously arrived jobs are completed. Let us define A_t as the random number of jobs that arrive to the machine in the time interval $[0, t]$, where t is a fixed time. Note that A_t is a discrete random variable that can take on values $0, 1, 2, \ldots$. Suppose we model the probability distribution of A_t as a *Poisson* distribution; i.e., we assume the mass function of A_t is given by

$$p(a) = P(A_t = a) = \frac{e^{-\lambda t}(\lambda t)^a}{a!}, \qquad a = 0, 1, 2, \ldots, \tag{A.43}$$

where λ is a given positive constant (we will justify this particular choice of mass function later).

Another random variable of interest to us is the length of time it takes for a particular job to be processed on the machine. Note that here we are measuring the time from start to completion of processing of the job; we are not including the time that the job may wait in queue before processing begins. We will assume that all the jobs are *statistically* identical and independent of each other; that is, the processing time of each job is selected independently from a common distribution function. We define T as the time it takes to process a particular job, and we assume that T is a continuous random variable that follows an *exponential* distribution; i.e., we assume that T has density function

$$f(t) = \begin{cases} \gamma e^{-\gamma t}, & 0 < t < \infty \\ 0, & otherwise, \end{cases}$$

where γ is another given positive constant.

With these definitions, let us attempt to find the distribution function for a third random variable N, which is the number of jobs arriving during the service time of a particular job. We begin by considering the pair (N, T), where N is a discrete random variable and T is a continuous random variable. Note that if the actual value of T were known (say, t), then N would have the same mass function as A_t. Thus, the *conditional* mass function of N, given that $T = t$ is

$$f(n|t) = P(N = n|T = t) = \frac{e^{-\lambda t}(\lambda t)^n}{n!}, \qquad n = 0, 1, 2, \ldots,$$

The *joint* density function of (N, T) is then obtained by multiplying this conditional mass function by the marginal density function of T; i.e.,

$$f(n, t) = f(n|t) f(t) = \frac{e^{-\lambda t}(\lambda t)^n}{n!} \cdot \gamma e^{-\gamma t} = \frac{\gamma e^{-(\lambda+\gamma)t}(\lambda t)^n}{n!}, \quad n = 0, 1, \ldots, \; t > 0.$$

To find the marginal mass function of N, we integrate the joint density function over all t:

$$\begin{aligned} p_N(n) = P(N = n) &= \int_0^\infty f(n, t)dt \\ &= \int_0^\infty \frac{\gamma e^{-(\lambda+\gamma)t}(\lambda t)^n}{n!} dt \\ &= \frac{\gamma \lambda^n}{n!} \int_0^\infty e^{-(\lambda+\gamma)t} t^n dt \\ &= \frac{\gamma \lambda^n}{n!(\lambda + \gamma)} \int_0^\infty t^n (\lambda + \gamma) e^{-(\lambda+\gamma)t} dt \end{aligned}$$

Note that the integral of the right-hand side is the nth moment of an exponential random variable with parameter $\lambda + \gamma$; hence

$$p_N(n) = \frac{\gamma \lambda^n}{n!(\lambda + \gamma)} \left[\frac{n!}{(\lambda + \gamma)^n} \right] = \left(\frac{\lambda}{\lambda + \gamma} \right)^n \left(\frac{\gamma}{\lambda + \gamma} \right), \qquad n = 0, 1, 2, \ldots.$$

All these manipulations carry through in spite of the fact that N is discrete and T is continuous. Notice that N follows a *geometric* distribution. Can you provide any intuitive justification for this result?

A.4.7 Independence

We have seen that the probability of any event generated jointly by random variables X and Y can be computed via the joint distribution function. That is, the joint distribution function encapsulates not only the probability structure of each random variable separately, but also of their relationship. In general, it is not possible to deduce the probability of an event generated by both X and Y if we only know the marginal distributions of X and Y. This section considers a particular kind of relationship (namely, independence) between random variables that does allow us to deduce the joint distribution from marginal distributions. We first define the idea of *independent events*.

Definition 61 Two events F_1 and F_2 (defined on the same probability space) are said to be independent if

$$P(F_1 \cap F_2) = P(F_1)P(F_2). \tag{A.44}$$

Written in terms of conditional probability, the definition yields the following: *Two events* F_1 and F_2 *are independent if and only if*

$$P(F_1|F_2) = P(F_1)P(F_2)/P(F_2) = P(F_1) \tag{A.45}$$

and

$$P(F_2|F_1) = P(F_1)P(F_2)/P(F_1) = P(F_2). \tag{A.46}$$

The definition of independent events leads to an analogous definition of independent random variables.

Definition 62 Two random variables X and Y are independent if the probability of any event generated jointly by the random variables equals the product of the probabilities of the marginal events generated by each random variable; i.e., for any subsets R_1 of the range of X and R_2 of the range of Y,

$$P(X \in R_1, Y \in R_2) = P(X \in R_1)P(Y \in R_2). \tag{A.47}$$

Since the joint distribution function yields the probability of any event generated by X and Y, and the marginal distributions yield the probability of any event generated by X and Y separately, the above definition is equivalent to the following statement. *Random variables X and Y are independent if and only if*

$$F(x, y) = F_X(x)F_Y(y) \quad \text{for any } -\infty < x < \infty, -\infty < y < \infty. \tag{A.48}$$

In terms of the mass or density functions, the above statement is equivalent to the following statements.

Discrete random variables X and Y are independent if and only if

$$p(x, y) = p_X(x)p_Y(y) \qquad \text{for any } x, y. \tag{A.49}$$

Continuous random variables X and Y are independent if and only if

$$f(x, y) = f_X(x)f_Y(y) \qquad \text{for any } x, y. \tag{A.50}$$

Determining whether X and Y are independent involves verifying any of the above conditions.

Example A.19 Suppose the joint density function of X and Y is given by

$$f(x, y) = \begin{cases} 2e^{-x-y}, & 0 \le x \le y, 0 \le y \le \infty \\ 0, & \text{otherwise.} \end{cases}$$

Notice that $f(x, y)$ can be written as $f(x)f(y) = (2e^{-x})(e^{-y})$. But

$$\begin{aligned} f_X(x) &= 2\int_x^{\infty} e^{-x-y}dy \\ &= 2e^{-x}\int_x^{\infty} e^{-y}dy \\ &= 2e^{-2x} \\ f_Y(y) &= 2\int_0^{y} e^{-x-y}dx \\ &= 2e^{-y}[1 - e^{-y}]. \end{aligned}$$

Clearly $f(x, y) \neq f_X(x)f_Y(y)$, and hence X and Y are not independent.

A.5 Bayesian Analysis

A.5.1 Bayes' Theorem

Bayes' theorem is a particularly useful statement regarding conditional probabilities. Let the events $B_1, B_2, \ldots$ make up a *partition* of the sample space Ω. Now suppose we are able to observe from an experiment that the event A has occurred, but we do not know which of the events $\{B_j\}$ has occurred (because the B_j's form a partition, one and only one of them occurs). *Bayes' theorem*, which is a simple restatement of the definition of conditional probability (Eq. A.4) and the law of total probability

(Eq. A.7), allows us to refine our guess at the probabilities of occurrence of each of the B_j's:

$$P(B_j|A) = \frac{P(A|B_j)P(B_j)}{\sum_i^n P(A|B_i)P(B_i)} \tag{A.51}$$

Bayes theorem is of particular importance in modeling experiments where new information (in terms of the occurrence of an event or empirical evidence in the form of data) may lead us to update the likelihood of other events. Speaking somewhat informally, suppose we are interested in estimating some property of a probabilistic mechanism that we will term a *system state*, and suppose we have available to us some empirical output of that probabilistic mechanism that we will term a *sample*. Then Bayes' theorem can be used to help refine our estimate of the system state as follows:

$$P(state|sample) = \frac{P(sample|state)P(state)}{\sum_{\text{all states}} P(sample|state)P(state)} \tag{A.52}$$

Beyond the formal use of Bayes' theorem in Eq. A.51, this interpretation allows us to use the result to refine our model of the probabilistic mechanism based on observed output from the mechanism. Clearly, this expression may have important applications when modeling damage accumulation. The following section provides further details.

A.5.2 Bayesian Inference and Bayesian Updating

Bayesian analysis refers to a collection of procedures in which Bayes' theorem is used to refine estimates of event likelihoods as new evidence become available. Bayesian analysis includes Bayesian inference, Bayesian updating, Bayesian regression, and many other techniques. This approach has found wide application in many fields and is often contrasted with frequentist reasoning, which assumes that observations (data) are the product of a statistical mechanism (distribution) whose design is known a priori and remains constant as data are accumulated. Bayesian analysis, on the other hand, asserts that the statistical mechanisms producing observed data are themselves probabilistic in nature, so that, in particular, their parameters are random and can be estimated and updated as observations are revealed. In simple terms, the frequentist approach holds that data are realizations from a mechanism whose parameters are fixed (and thus data are potentially infinitely repeatable), while the Bayesian approach holds that available data from a particular study are fixed realizations from an unknown (random) mechanism, and thus as additional data are revealed, our understanding of the random nature of the mechanism changes. Bayesian analysis provides a means of updating the estimates of the statistical properties of the mechanism (i.e., its parameters).

Suppose a probabilistic mechanism produces a random variable X, and suppose the distribution of X involves a parameter Θ that can take on only discrete values $\{\theta_1, \theta_2, \ldots\}$). In Bayesian analysis, the parameter Θ is taken to be a random variable, and we begin with a *prior* distribution, which conveys the probability law of the parameter prior to observing any data. If the parameter takes on discrete values, the prior distribution can be described by a probability mass function p, i.e., $\{p(\theta_i) = P(\Theta = \theta_i), i = 1, 2, \ldots\}$. The choice of the *prior* distribution may be based on any already available information, such as previous studies or other data sources, expertise or intuition, or simply convenience. In practice, it is common to assume a uniform distribution for the prior distribution, which is commonly referred to as *diffuse prior* [1].

Consider now that new information e becomes available as a realization of the probabilistic mechanism. Then, conditioned on the new information, the updated *pmf* of Θ, denoted by p', where $p'(\theta_i) = P(\Theta = \theta_i|e), i = 1, 2, \ldots$ can be obtained from Bayes' theorem as [1]

$$p'(\theta_i) = \frac{P(e|\Theta = \theta_i)p(\theta_i)}{\sum_j P(e|\Theta = \theta_j)p(\theta_j)}, \quad i = 1, 2, \ldots, \tag{A.53}$$

where $P(e|\Theta = \theta_i)$ is the conditional probability of the information given that the parameter takes on the value θ_i. The pmf p' is known as the *posterior* probability mass function; i.e., the new *pmf* for Θ given the observations.

The expected value of Θ, computed using the posterior distribution, is known as the Bayesian (updated) estimator of the parameter Θ, and is computed as

$$\hat{\theta}' = \mathbb{E}[\Theta|e] = \sum_i \theta_i p'(\theta_i) \tag{A.54}$$

The new information e leads to a change in the *pmf* of Θ, and this change should be reflected in the evaluation of the probability of the random variable X. Based on the theorem of total probability (Eq. A.7) and using the *posterior pmf* from Eq. A.53, we obtain the distribution function of X as follows:

$$P(X \leq x) = \sum_i P(X \leq x|\theta_i)p'(\theta_i) \tag{A.55}$$

Similarly, for the continuous case, we can define $f(\theta)$, $-\infty \leq \theta \leq \infty$ as the *prior* density function for Θ. Then, when additional information e becomes available, the posterior probability density function f' can be computed as follows [1]:

$$f'(\theta) = \frac{P(e|\Theta = \theta)f'(\theta)}{\int_{-\infty}^{\infty} P(e|\Theta = \theta)f'(\theta)d\theta} \tag{A.56}$$

where $P(e|\Theta = \theta)$ is the conditional probability of the information (data) given $\Theta = \theta, -\infty \leq \theta \leq \infty$. This is commonly referred to as the *likelihood function* of Θ and it is denoted by $L(\theta)$. Then, the updated estimator of the parameter is

$$\hat{\theta}' = \mathbb{E}[\Theta|e] = \int_{-\infty}^{\infty} \theta f'(\theta) d\theta \tag{A.57}$$

and similar to Eq. A.55

$$P(X \leq x) = \int_{-\infty}^{\infty} P(X \leq x|\theta) f'(\theta) d\theta \tag{A.58}$$

The posterior distribution can be used to develop Bayesian inferential statistics, such as Bayesian confidence intervals. As an aside, one of the primary differences between the frequentist and Bayesian approaches is how confidence intervals are interpreted. In the frequentist case, confidence intervals are interpreted in terms of *coverage*; an α-level confidence interval means that in a large number of repeated trials with the same number of observations, approximately $\alpha \cdot 100$-percent of the computed confidence intervals contain the true parameter. In the Bayesian case, we interpret the confidence interval in terms of *probability*; an α-level confidence interval means that, based on the information provided, the parameter is in the computed confidence interval with probability α.

Reference

1. A.H-S. Ang, W.H. Tang, *Probability Concepts in Engineering: Emphasis on Applications to Civil and Environmental Engineering* (Wiley, New York, 2007)

Index

M. Sánchez-Silva and G.-A. Klutke, *Reliability and Life-Cycle Analysis of Deteriorating Systems*, Springer Series in Reliability Engineering,
DOI 10.1007/978-3-319-20946-3

MIX
Papier aus verantwortungsvollen Quellen
Paper from responsible sources
FSC® C105338

If you have any concerns about our products,
you can contact us on
ProductSafety@springernature.com

In case Publisher is established outside the EU,
the EU authorized representative is:
Springer Nature Customer Service Center GmbH
Europaplatz 3, 69115 Heidelberg, Germany

Printed by Libri Plureos GmbH
in Hamburg, Germany